Praise for *Organic Mushroom Farming and Mycoremediation*

"This is a reference book for the next generation of DIY mycologists. It is a great practical guide to mushroom cultivation, starting with basic concepts and building from there to mycoremediation and experimental strain development. Tradd Cotter is a man with a mission, who has done and thought about all this a lot; he has learned to explain it with great clarity and in a simple and well-organized manner."

—**SANDOR ELLIX KATZ**, fermentation revivalist and author of
The Art of Fermentation and *Wild Fermentation*

"Tradd Cotter has written a clear, comprehensive guide that is a gift to amateur as well as professional mushroom growers. The pages are enlivened by Cotter's enthusiasm for the many possibilities that fungi offer, and his obvious familiarity with growing these marvelous creatures—not just theoretical knowledge—makes the book particularly valuable. This book opens the doors wide to a diverse and fascinating fungal world."

—**TOBY HEMENWAY**, author of *Gaia's Garden:
A Guide to Home-Scale Permaculture*

"Finally, an accessible resource covering a wide variety of mushroom-cultivation approaches. Tradd Cotter's book fills an enormous need—I've been wishing for a resource like this for a long time. This is the kind of book I'll keep nearby and will turn to often over the years. Any farmer or gardener who wishes to garner food or medicine value from wood needs to understand and cultivate mushrooms. And this is the best all-around manual I've seen."

—**BEN FALK**, author of *The Resilient Farm and Homestead*

"Wow! Tradd Cotter is a genius of organic mushroom production. His step-by-step instructions and beautiful photography make this a must-have book."

—**ROBERT ROGERS**, author of *The Fungal Pharmacy:
The Complete Guide to Medicinal Mushrooms
and Lichens of North America*

"Mushroom cultivation should be playing a much bigger role in our gardens and farms. Tradd Cotter's *Organic Mushroom Farming and Mycoremediation* provides low-cost, easily accessible techniques for growing mushrooms indoors and outdoors, from home to commercial scale."

—**ERIC TOENSMEIER**, author of *Paradise Lot* and *Perennial Vegetables*

"Tradd Cotter has done a wonderful job sharing his practical experience in a well-organized way with illustrations that clearly underline the topics. *Organic Mushroom Farming and Mycoremediation* is an invaluable resource for teaching students about mushroom cultivation."

—**PETER OEI**, author of *Mushroom Cultivation* and director of horticulture innovation at InnovatieNetwerk, Dutch Ministry of Economic Affairs, and founder of MeattheMushroom.nl and spore.nl

"*Organic Mushroom Farming and Mycoremediation* is a guide and inspiration for new and experienced mushroom cultivators alike. Tradd Cotter has done a great job of combining the complexity of mushroom cultivation with the intuitive simplicity of 'small steps.' Highly recommended for fungophiles as a great read and reference!"

—**JIM GIBSON**, past president, Colorado Mycological Society

Organic Mushroom Farming and Mycoremediation

Simple to Advanced and Experimental Techniques for Indoor and Outdoor Cultivation

Organic Mushroom Farming and Mycoremediation

Simple to Advanced and Experimental Techniques for Indoor and Outdoor Cultivation

TRADD COTTER

Chelsea Green Publishing
White River Junction, Vermont

Developmental Editor: Brianne Goodspeed
Project Manager: Patricia Stone
Copy Editor: Nancy Ringer
Proofreader: Laura Jorstad
Indexer: Shana Milkie
Designer: Melissa Jacobson

Printed in the United States of America.
First printing August, 2014.
10 9 8 7 6 5 4 3 2 1   14 15 16 17 18

Our Commitment to Green Publishing

Chelsea Green sees publishing as a tool for cultural change and ecological stewardship. We strive to align our book manufacturing practices with our editorial mission and to reduce the impact of our business enterprise in the environment. We print our books and catalogs on chlorine-free recycled paper, using vegetable-based inks whenever possible. This book may cost slightly more because it was printed on paper that contains recycled fiber, and we hope you'll agree that it's worth it. Chelsea Green is a member of the Green Press Initiative (www.greenpressinitiative.org), a nonprofit coalition of publishers, manufacturers, and authors working to protect the world's endangered forests and conserve natural resources. *Organic Mushroom Farming and Mycoremediation* was printed on paper supplied by RR Donnelly that contains at least 10% postconsumer recycled fiber.

Library of Congress Cataloging-in-Publication Data
Cotter, Tradd, 1973–
  Organic mushroom farming and mycoremediation : simple to advanced and experimental techniques for indoor and outdoor cultivation / Tradd Cotter.
     pages cm
  Includes bibliographical references and index.
  ISBN 978-1-60358-455-5 (pbk.) — ISBN 978-1-60358-456-2 (ebook)
1. Mushroom culture. 2. Mushrooms—Organic farming. 3. Fungal remediation. I. Title.

SB353.C83 2014
635'.8—dc23
                           2014015959

Chelsea Green Publishing
85 North Main Street, Suite 120
White River Junction, VT 05001
(802) 295-6300
www.chelseagreen.com

Contents

PART IV
Meet the Cultivated Mushrooms

For a more complete list of common names, see individual profiles for each genus

Introduction

When someone asks me if I grow magic mushrooms, I always reply by asking, "Aren't all mushrooms magical?" I have been growing, culturing, researching, hunting, and learning everything I can about mushrooms for the last twenty years. I work with all kinds of mushrooms, and I am fascinated by every single one. The more you learn, the more your belief in their magic will grow.

My journey with mushrooms did not start out auspiciously. Although I loved the outdoors as a kid, I was far more interested in walking down through the woods to my grandparents' lake in North Carolina to go fishing than in paying attention to the mushrooms growing around me. By twenty, I was living with my parents—trying to balance college classes and singing in a band—and one day my mother suggested that I stop by a nearby mushroom farm for a tour. She knew I was interested in biology, it seemed like something different and cool to do, and maybe it was her subliminal way of telling me to get out and find a job.

I knew nothing about mushrooms. Zero. Not even basic varieties at the supermarket, which in 1994 were white buttons and, newly, portabellas. But I called the farm anyway. I remember it sounding very noisy and active and the owner sounded out of breath. "Sure," he said, "come on by later this afternoon and I can show you around quickly if you want to see how mushrooms grow." When I arrived, the building seemed very plain, with cinder-block walls and a metal roof, and not very exciting. I wasn't at all impressed with the looks of things.

The owner greeted me and quickly led me around, showing me the entire place, from the sterilizer unit cooking the growing media to colonization rooms and, finally, the place where the magic hit: the fruiting room. I will never forget the moment when I walked into that strange, foggy space, like something out of a dream, and down aisle after aisle of fruiting shiitake mushrooms growing on sawdust blocks. This was intriguing, and overwhelming, and I had so many questions running through my mind. My mouth opened all on its own and started peppering the owner with questions, so many that in retrospect I realize that my incredible inquisitiveness must have been annoying. But I was in shock. Amazed.

Ten minutes later the tour was over. The owner thanked me for stopping by and gave me a pound of shiitake mushrooms. I felt like I had won a miniature lottery. I walked out the farm door with my brain buzzing. It was love at first sight . . . with mushrooms. I sadly returned to my car and climbed in, cranked up the engine, and started pulling away when a sudden loud bang hit the rear of my vehicle. What I thought was a tire blowout was the owner of the mushroom farm, who had chased my car down the driveway and was trying to get my attention. Did I leave something there? No. I rolled down the window and asked what was going on. The owner, now panting, asked, "Would you like to work here?"

I often think back to that moment when the owner ran after me. If he hadn't caught up with my car, he would have had no way to contact me. I was probably just seconds away from missing out on the future I would go on to explore with mushrooms—starting with a job at the mushroom farm. That entire tour had lasted ten minutes, but it triggered events that have lasted me a lifetime.

Over twenty years later, I now frequent food and sustainability conferences, lecturing and teaching the values of environmentally responsible, low-tech and no-tech mushroom cultivation projects that anyone can apply to their home or small farm. I am against the use of all chemical pesticides and synthetic fertilizers; I think nature knows best. Using mycorrhizae and composted mushroom substrate filled with worm castings, my gardens thrive and are highly productive without compromising the soil and water quality. I believe in creating perpetual and circular food systems, using sustainable techniques such as water harvesting and no-till cultivation, and using passive energy or seasonal production to minimize the energy use on my farm.

My work has also evolved into research, such as creating mushroom rescue modules (discussed in chapter 12) for use in regions struggling with poverty or devastated by natural disaster, where shattered habitats and cultures struggle to recover. On a recent trip to Haiti, for example, I introduced mushrooms to a group of children I met in the village of Cange. They were intrigued when I told them that some of mushrooms I had with me tasted like chicken and that I could teach them how to cultivate these wonderful mushrooms on paper or cardboard that we collected in the street. The idea that they would fruit in just a few weeks seemed magical to them. That trip was one of the most memorable I've ever taken. Here, I felt, mushrooms could make a real difference as a potential food source. I later returned to help set up a commercial production facility and spawn production lab to keep the process perpetuating.

Through such experiences, I've found that sharing what I know about mushrooms has become an imperative. Mushrooms are an excellent source of protein, and they have a wide range of medicinal properties. With an estimated 1.1 million fungi on the planet and only 150,000 collected and described (never mind being screened for their potential), you can easily see the implications for food production and medicine. In these ways fungi have the potential to shape our future for millennia to come.

This book is a compilation of knowledge I've gained from my experiences, from when I cultivated my first mushrooms up to the present, in which I now conduct mushroom research and own my own mushroom business. As you are reading, I must warn you that you are embarking on a path that may change the way you see yourself fitting into this life. In choosing mushrooms, you have decided to cultivate a wonderful food using what most would consider waste or by-products of many industries. I hope this book serves you well in giving you the skills necessary to explore mushroom cultivation and empowering you to dream up experiments and ideas on your own. Part skill, part art, part intuition, mushroom cultivation will give you a lifelong relationship with this incredible kingdom of life.

How to Use This Book

This book is designed to help you build skill and confidence, starting in part 1 with a thorough foundation in both indoor and outdoor mushroom cultivation using purchased spawn (a form of mycelium that is physically "plantable," being packaged in sawdust, grain, or a wooden dowel). Although my eventual goal is to help you become more self-sufficient by culturing and cultivating your own spawn (as well as to teach you how to grow mushrooms perpetually on just about anything), using purchased spawn will help you develop your skills and gain experience with a variety of cultivation techniques before you make larger investments of time and money. While the information in part 1 can be considered more foundational than the material in the latter parts of the book, my hope is that even more experienced growers will find value in these chapters. Rather than simply focusing on yield as an end goal, I've strived for a more holistic approach, one that pays careful attention to the mushroom life cycle, to ecology, to fungi's relationships with the other kingdoms of life, and to developing the kind of intuition that will teach you more about cultivating mushrooms than a book or a workshop ever can. The focus of this book is primarily edible mushrooms, but you will find a great deal of information on medicinal, industrial, and mycoremediation applications as well. Once you develop solid cultivation skills, you can apply them to grow whatever kind of mushrooms suit your goals, or fancy.

The chapters in part 2 are designed to help you apply that foundational information to incorporate mushroom cultivation into your life and landscape, in whatever way reflects your goals. It includes information for both urban and off-the-grid growers, on making value-added products from mushrooms (including mushroom-infused beer, wine, and spirits), and the incredible and largely untapped potential of mushrooms to provide high-quality protein for people in poverty- and disaster-stricken regions of the world. Part 2 is full of useful and largely low-tech ideas for bringing mushrooms even further into your life. The goal here, as in part 1, is not only to help you cultivate mushrooms successfully (though that's a big part of it), but to help you develop an understanding of the complex relationships mushrooms have with bacteria, plants, and animals (including humans). I believe that the more we develop that understanding, the more potential we have for successful cultivation, and the more we focus on linear goals of maximum production, the more we risk failure—in more senses than one.

The chapters in part 3 cover more advanced and experimental techniques such as basic lab construction, sterile culturing, and techniques for mushrooms that are extremely difficult to cultivate, like morels. While the material in part 1 and part 2 is mostly low-tech, requiring relatively small investments in infrastructure and equipment, the material in part 3 requires more refined skill and bigger decisions. Some of it is research-in-progress, which I've included in the hope that as you build on your experience as a mushroom cultivator, you will contribute your own experiments and experiences to the body of knowledge about mushrooms. There is still so much to learn, and the more we understand about mushrooms and the fungi kingdom, the more we can use that understanding to grow high-quality food and medicinals, remediate polluted land and water, and replace plastics and other industrial materials with fungus-based textiles, building materials, and other consumer goods. Although many researchers and cultivators are protective of their knowledge, the only way we can really build a collective body of knowledge is through collaboration.

Don't be too quick to rush to an advanced chapter or scale up your operation based on the information in this book. The only way to improve and succeed at your goals is to learn the specific and subtle needs of each mushroom you grow. Treat each one as an individual, like someone you know (and want to know better), understanding its individual needs and differences. This takes time and patience and, inevitably, some failure. Give yourself the opportunity to experiment before the stakes are too high. Seek hands-on workshops, attend mushroom walks, and join mushroom hunting clubs to meet like-minded people and share knowledge and experiences.

Many people have commercial aspirations for mushroom cultivation, and I have tried to include as much information as possible that can be applied to small-scale and environmentally responsible commercial operations (including a chapter on marketing your product). Again, weigh this decision carefully. Only you can decide when, if, and at what scale it's right for you. But my hope is that you'll have all the tools you need to scale up if you choose to.

As you proceed through the book, you'll notice a focus on shiitakes and oyster mushrooms. This isn't because those are the only mushrooms worth growing! I use oysters and shiitakes frequently to illustrate specific phenomena or techniques because they are two of the easiest and most satisfying mushrooms to grow, and many people are familiar with them. If you are a beginner, they are great mushrooms to start with. But if you flip to part 4, "Meet the Cultivated Mushrooms," you will also find profiles of nearly thirty mushrooms, with growing parameters and suggestions for each. While most of these mushrooms are primarily edibles, some have wonderful applications for use as medicinals, in mycoremediation, or potentially in industrial capacities. Spend some time looking through the profiles and familiarize yourself with the possibilities. Although you may want to start with oysters or shiitakes, you can then apply much of what you learn to the other mushrooms, factoring in each individual mushroom's needs.

To me, this book is much more than a cultivation guide. It is about healing the people and the planet, one mushroom and one cultivator at a time, reversing destructive cycles into creative forces. If we think with an opportunistic yet minimalistic approach, much like a mushroom, taking what it needs to survive and then returning resources to its ecosystem so they can be used by others, the future looks like somewhere I want to be. Spend as much time as possible cultivating, collecting, and observing the natural cycles of mushrooms, no matter how small they are. From old-growth forests to mulched urban sidewalks to fruiting growths on debris floating out at sea, fungi are everywhere, and there's much to be learned from them.

PART I

The Fundamentals of Mushroom Cultivation

CHAPTER 1

The Ecology and Life Cycle of Cultivated Mushrooms

Although the ancient Egyptians are credited with pioneering the use of yeasts to create beer, wine, and bread, and historical records indicate that cultivation of many edible and medicinal mushroom species dates back over four thousand years in Japan and China, humans are not the only—or even the first—fungal cultivators on the planet. Recent discoveries have estimated that South American leaf-cutting ants have been actively culturing fungi for forty-five to fifty-five million years (Currie, 2011). In their colonies, specialized worker ants harvest and shred leaves to make a fungal growing medium. The larvae feed on the fungi; the mycelium is rich in protein and provides the ants with a natural antibiotic that helps them combat a dangerous pathogen. When a new queen rises to start a fresh colony, she carries with her a pellet of mycelium, much like a starter culture, stored in an infrabuccal pouch (a cavity in her mouth). Like this new ant queen, guardian of the mushroom spores and thus of the capacity for perpetual food production, humans too have been bestowed with the gift of mushrooms. We just need to learn to use it.

Although fungi are often somewhat neglected as a kingdom—perhaps in part because they tend to be less visible than plants and animals—they have critical ecological roles, and they interact with their environments in compelling and sometimes surprising ways. Mycorrhizal fungal relationships, for example, are obligate partnerships between plants and fungi at the root interface underground. Some examples of mycorrhizal mushrooms include truffles (*Tuber* spp.), chanterelles (*Cantharellus* spp.), and porcini (*Boletus edulis*). This specialized relationship allows fungi to thread into and around cell walls in the root tips, increasing the surface area within the cell and in the surrounding soil, where nutrients are absorbed and transported to the plant roots. I call this the original carbon trading scheme, where the mycelium collects a resource that the plant has a difficult time procuring and trades it for sugar, which the plant produces as a product of photosynthesis. Within these soil interfaces are countless layers of interkingdom interactions that connect bacteria, fungi, plants, and animals to maintain a dynamic microcosm of constant nutrient exchange and balance.

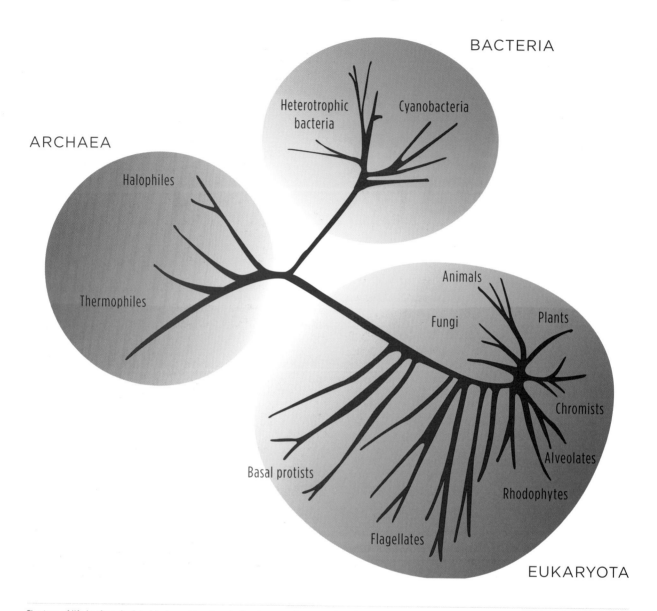

BACTERIA

Heterotrophic bacteria

Cyanobacteria

ARCHAEA

Halophiles

Thermophiles

Animals

Fungi

Plants

Chromists

Alveolates

Rhodophytes

Basal protists

Flagellates

EUKARYOTA

The tree of life is often depicted by branching, terminal ends separating organisms into distinct groups based on biological functions and, more recently, molecular data. Although many of these terminal ends seem distant from each other, a three-dimensional model would be more accurate in depicting complex relationships among organisms.

Not all fungal relationships are mutually beneficial, however. Some types of fungi are capable of attacking other living organisms, such as *Cordyceps* spp., which attack and mummify many kinds of insects and then fruit out of the insects' body. Some molds, including some *Trichoderma* spp., are considered mycoparasites, or fungal pathogens. They possess the enzyme chitinase, which breaks down chitin, a compound found in the cell walls of fungi and various insects and soil organisms. In short, they are designed to attack and digest their fellow fungi. Many strains of *Trichoderma* in my collection are now being used in trials at a local vineyard investigating their ability to help grapevines combat leaf, trunk, and root pathogens. The growers inoculate the vines with *Trichoderma*, and the mold imparts an immune response to the plants, which then synthesize their own compounds designed for targeting a wide spectrum of fungal pathogens.

Eumycota, or "true fungi"—as opposed to slime and water molds, which are not technically part of the fungi kingdom—share many common characteristics, including reproduction by means of spores, a lack of chlorophyll, and the presence of chitin—a hard natural substance that provides structure and protection—in their cell walls. Fungi produce and secrete many different kinds of extracellular enzymes, such as lignin peroxidase, manganese peroxidase, laccases, amylases, and cellulases. Think of these enzymes as "chemical scissors" or "molecular keys" that cut or unlock the bonds of large molecules, such as lignin, embedded in woody plant tissue. As the mushroom's enzymes break down its growing substrate, the smaller, essential chemical units in the substrate, such as carbon, organic nitrogen, minerals, and other trace elements, are released and are transported through the fungi's cell walls for use as energy sources and for metabolic function.

In this process, fungal cells first stream outward into the environment—the "infantry cells," as I like to think of them—and communicate back to the body of the mycelium, instructing it to produce enzymes specific to the kind of food available and what will be needed to break down those particular compounds. Fungi sweat these "cell-free" enzymes, meaning that the enzymes are able to saturate and move freely into the environment to degrade the organic substances before the actual fungal filaments reach them. Bacteria, by contrast, must contact their food source directly, making use of surface receptors that disassemble and transport their food source directly across their membranes. Fungi's amazing chemical consciousness and profound ability to adapt and react quickly to the environment are part of what makes them so captivating.

Mushrooms feed in different ways on organic compounds, varying from species to species, depending on the mode of nutrition and their genetics, strains, or ecotypes. Most of what we'll cover in this book are classified as true fungal saprophytes, or mushrooms that decompose dead organic material, such as dried plant waste. Edible but parasitic fungi, which often attack their hosts and can spread rampantly

Cordyceps species have evolved to attack and mummify many kinds of insects, including this unfortunate wasp. Here, a fruiting body, or teleomorph, has emerged from between the wasp's head and the thorax.

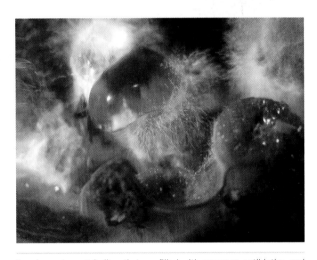

Fungi secrete metabolites that are filled with enzymes, antibiotics, and other waste by-products. Here, shiitake mycelium is sweating out metabolites (the amber-colored substance), likely as a defense mechanism in response to an environmental contaminant.

through a forest ecosystem, such as honey mushrooms (*Armillaria* spp.), are not covered in this text, and I discourage propagating them since they can easily outcompete most other fungi.

Oyster mushrooms (*Pleurotus* spp.) are notoriously aggressive saprophytes due to their rapid reaction mechanisms for detecting different compounds and their ability to manufacture a much wider spectrum of enzymes, not only for their own metabolic needs but also for making nutrients available for other organisms such as bacteria and plants. Other fungi are more limited in their ability to produce enzymes, reducing their capability of colonizing and fruiting from different food sources. Maitakes (*Grifola frondosa*), for example, can break down only very particular types of wood—mostly oak, and, in fact, certain species of oak and even certain wood densities—which limits their colonization and, by extension, your cultivation if those resources are not locally available.

Most mushrooms can be classified as either brown rot or white rot fungi based on their mode of feeding on wood-based substrates. Brown rot fungi tend to concentrate their feeding on the cellulose rather than the structural lignin, leaving behind the visually browner components of the plant tissue, sometimes cracking it into the fissure-like patterns you may see on fallen logs in the forest. White rot fungi operate in the reverse, instead focusing the bulk of their enzyme degradation on the lignin first, bleaching the woody tissue white, before feeding on the bulk of the cellulose available in the woody tissue. (White rot fungi make excellent candidates for biobleaching and for breaking down very complex man-made molecules in a process called mycoremediation.)

That's the broad view. Of course, mushrooms feed on more materials than just cellulose and lignin. The good news—or bad news, depending on your goals—is that every strain of mushroom varies in its ability to adapt to or break down specific substances into food. Mating new strains in a lab and isolating strains from the wild offer us an infinite supply of mushrooms with differing appetites.

In fact, mushrooms prefer to vary their diets slightly. If you get to the point of expanding spawn (see chapter 18), you'll want to adjust the fruiting formulas slightly every now and then. Otherwise your fungi may experience what's called strain senescence—weakening of the strain due to the fact that the fungi consistently overproduce and overuse a particular combination of enzymes. Like I ask my workshop attendees, "How would you like it if I fed you nothing but oatmeal for a few years?" Your body

would respond by manufacturing only those gut bacteria needed for that diet, reducing your microflora to a monoculture-type, rather than biodiverse, ecosystem—never a healthy model. As you develop and progress in your cultivation techniques, try to tune in to the fungal consciousness. Give your fungi what they need to survive the colonization process and for the life cycle to complete itself.

THE MUSHROOM LIFE CYCLE

Most mushrooms share the same basic stages of development, although there are a few variations and exceptions. Typically a mushroom matures and releases its spores; the spores germinate on a suitable growing media; they colonize the environment in an effort to capture as much territory as possible to build up a competitive biomass; and when the growing mycelium begins to feel confined, mushrooms fruit again. When the mushrooms are mature, they produce spores, and the cycle begins again. One of the reasons many cultivators use spawn as their mushroom starter is that it gives them a head start on the process, allowing them to grow mycelium directly, which gives them a better chance of outcompeting germinating spores from the wild.

Understanding all stages of development will help you become a more intuitive grower. You'll learn to recognize not just the individual needs of each stage, but the specialized needs of every single mushroom you grow, so that you can develop a unique strategy for each. I spend a great amount of time observing the natural cycles of these mushrooms in the wild in the hope of catching clues from the mushrooms in their natural habitat, and I encourage all growers to spend more time collecting fungi to build an understanding of the ecological connections that trigger them to fruit.

Sporulation

Fungi produce not seeds but spores, which are microscopic packets of fungal DNA, packaged in a hard chitin coating to protect them from heat, drought, and damage. They produce the spores on the surface of their gills or pores through the process of meiosis.

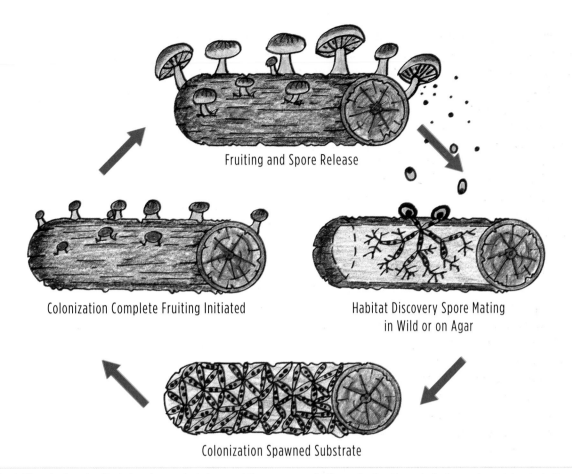

Fruiting and Spore Release

Colonization Complete Fruiting Initiated

**Habitat Discovery Spore Mating
in Wild or on Agar**

Colonization Spawned Substrate

The basic life cycle of most cultivated mushrooms includes sporulation; spore germination and cell mating; colonization; and complete colonization and primordia formation. When a mushroom matures, sporulation occurs again and the cycle repeats.

Meiosis is a form of sexual reproduction, meaning that instead of a cell simply dividing, with each new cell containing a complete package of DNA in its nucleus (asexual reproduction), it forms "parent" cells, wherein each nucleus contains just half the necessary DNA. When two parent cells combine, the new cell contains the complete DNA package (this is the same mechanism by which a human zygote is formed). So each mushroom spore carries only half of the genetic information needed for reproduction. When a mushroom is mature, it forcibly ejects its spores in a process known as sporulation.

Understanding spores and sporulation will be valuable to you as a cultivator. At this stage, you can collect spores from either cultivated or wild mushrooms for observation and germination by taking a spore print.

Spores can be harvested both for culturing and also to assist with identification of fungi. This spore print on aluminum foil was taken from a mature king stropharia mushroom.

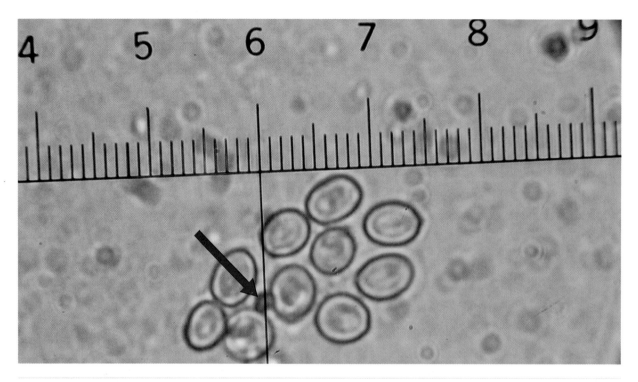

Spores germinating. Most cultivated mushrooms require two compatible spores to mate before forming a fruiting structure, or mushroom.

To take a spore print, place the stem of a mature mushroom through a slit of paper, sliding the paper up as close to the gills or pores as possible. Since you don't always know what color the spores will be and, hence, whether they will show up best on black or white paper, it can be helpful to take your spore print on paper that is half white and half black, ensuring that the spores will show up on one side or the other. Glass is another option, as is aluminum foil, which is my preference because the spores are easy to remove from it for culturing.

Spores do not simply fall out or drop from the gill surface; when a mushroom is mature, it will forcibly discharge them—with water pressure from inside the mushroom—and will leave a pile of spores on the paper within hours if left undisturbed. Locate the glass, paper, or aluminum foil in a draft-free area to ensure a nice thick spore print.

Spores can remain viable for several years; to save them, fold the print to enclose them, place it in a plastic bag, and store in the refrigerator. The cold temperature and lack of light will keep them viable for many years with a minimal loss of germination.

Spore Germination and Cell Mating

In the case of most mushrooms, when a spore germinates, it is monokaryotic (meaning "one nucleus per cell") and carries only half of the genetic information needed for sexual reproduction. After germination and sporulation, a spore's hyphae (mycelial strands) explore its immediate environment, looking for hyphae from another germinated spore suitable as a mating strain. When compatible hyphae meet, they fuse and combine genetic material. This fusion, called karyogamy, brings together two nuclei per cell, making the cell dikaryotic. Once fused, hyphae branch out and growth explodes.

Chapter 18 describes the manipulation of this process in a lab setting. When you streak spores onto an agar-laden petri plate, those spores are in an ideal environment to mate with adjoining spores and germinate.

Colonization

Also called "nutrient capture," colonization is the stage at which fungi secure their territory, competing for the nutrients in their environment. The infantry cells push their way through wood, compost, and soil until they encounter a competitor or barrier. Once that happens, the mycelium branches heavily to increase its surface area and begins aggressively digesting the substrate, recharging its biological battery in preparation for fruiting. I like to think of this process as paving a major highway; once the highway is done, the fungi begin paving all of the side roads, dirt roads, and pathways, until colonization is complete.

During colonization, the mycelium fans outward to capture nutrients.

Complete Colonization and Primordia Formation

The fruiting cycle begins—stimulated by environmental changes and nutrient availability—when colonization is complete and the fungus feels threatened enough to put its energy into a structure that will produce spores—mushrooms! Nutrients and water from the captured biomass stream to spots that the mycelium has "chosen" as good fruiting locations, based on favorable gas exchange gradients, consistent humidity or access to water, and the availability of light. Primordia, which resemble little knots, are the first observable sign that mushrooms are starting to fruit. The fungal life cycle comes full circle when a mushroom matures and releases its spores. At that point, the spores are free to explore the world in search of a mate and a new home to set up a colony.

Mushrooms form when colonization is complete. Here, king oyster (*Pleurotus eryngii*) primordia are fruiting on sawdust.

FUNGAL METABOLISM: BY-PRODUCTS

In the process of this life cycle, what do fungi require for food and what are the metabolic by-products? For food, fungi simply require a carbon source, or a feedstock—wood chips, sawdust, coffee grounds, agricultural wastes—which they break down using their "digestive" enzymes. Since they're essentially sweating these digestive enzymes into the environment, I like to think of fungi as "humans turned

The mushroom life cycle is complete and will begin again when a mushroom produces and forcibly ejects spores. Spores are adhesive when they are discharged and sometimes stick together to form long chains, as with these oyster mushroom spores.

Mushrooms need oxygen to develop correctly. Elevated carbon dioxide levels can influence the shape and length of the stems in most mushrooms. Here we see oyster mushroom primordia stretching in a fruiting room with poor ventilation and therefore excessive carbon dioxide.

Chicken of the woods (*Laetiporus sulphureus*) mycelium grazing on methicillin-resistant *Staphylococcus aureus* (MRSA).

inside out" (even though this isn't even close to being biologically correct). Instead of eating food, digesting, and then excreting waste, fungi excrete enzymes into the environment, which digest their food using oxygen, and then the fungi absorb the resulting dissolved nutrients directly through their cell walls. The fungi then excrete and essentially swim through their own waste as it accumulates. The by-products, just like our waste, are toxic to the fungi and accumulate in their immediate surroundings, so the mycelium must also secrete other secondary enzymes and by-products that attract and encourage bacteria that recycle the waste into substances less toxic to the fungi.

Like animals, fungi produce heat, carbon dioxide, and water, in the form of metabolites. For this reason, they are the perfect biological companions for the plant kingdom. In fact, integrating mushroom cultivation with enclosed plant production systems such as greenhouses can improve growing conditions for both sets of organisms.

Heat

Fungal enzymes produce heat when combined with available oxygen. The enzymes cleave molecules into smaller and smaller compounds, resulting in the bioavailability of smaller chains of cellulose and other energy-rich compunds as food sources. Much like in our stomachs, temperatures increase due to the spike in digestion or breaking down of the food source into usable material. This temperature increase, though slight, is relevant because you'll need to account for it when you're planning your growing space; packing too many mushrooms together can overheat and "cook" the mycelium during colonization, creating large dead pockets and anaerobic zones where molds and other thermophilic bacteria will thrive. These bacteria and molds will marble the growing medium with unused volumes of substrate, decreasing yields and increasing your risk of other containers nearby becoming infected. Monitor and moderate the heat levels and you will not have any problems.

Carbon Dioxide

As fungi consume the growing medium, they release enzymes and stimulate the combustion of oxygen, which creates and releases carbon dioxide. Much like humans, mushrooms need fresh air constantly, especially when they are forming actual fruitbodies. Excessive levels of carbon dioxide result in stem elongation, an undesirable result for growers because caps are typically more valuable than stems.

Since carbon dioxide is heavier than oxygen, when a fungus excretes it, the gas pools in undisturbed pockets of soil and depressions. These invisible pockets of carbon dioxide cause developing mushrooms to stretch, staying very narrow with a reduced cap size, which helps them navigate up through duff in the forest, or through straw or wood chips in your garden. If mushrooms formed large caps early in their development before having the chance to push through these barriers, they would not be able to release their spores effectively into the wind. Because spore dispersal is critical to mushroom survival, some mushrooms have developed other adaptive mechanisms to help them disperse spores. Truffles, for example, evolved to rely on insects and animals for spore dispersal.

Other By-Products

Every fungus is chemically unique, and the discovery of new compounds that fungi produce is rapidly increasing their use in many applications and industries. These compounds can be found either embedded in the mycelium or floating about in the "sea" of fungal metabolites. Fungi produce these different compounds in varying amounts in reaction to environmental changes and shifts in their available food sources. An observant cultivator can customize the growing parameters to optimize the levels of desirable by-products.

Fungal Growth: Requirements

Mushroom cultivation is not difficult if you can tune into the natural growing requirements and try to recreate what natures seems to do so effortlessly. Whether your mushroom cultivation projects are located indoors or outside, there are four key components that comprise the essential needs of a mushroom to produce prolifically; they are food, water, gas exchange, and light. If any one of these variables is missing or neglected, the mycelium and mushroom

biomass suffers greatly, so perfecting the subtle differences within each one of these factors for every mushroom species is your goal. This is a constant learning process until you perfect it, so be patient while mastering these requirements below.

Food

Fungi are heterotrophs, or consumers of food just like animals; however, mushrooms have adapted their enzymes to break down complex organic compounds found in their habitats such as logs, leaves, straw, manure, and much more. Many mushrooms prefer wood and plant debris as their main food source, while others consume manure-based substrates. I typically classify mushrooms into two categories when it comes to cultivation, the primary decomposers and the secondary decomposers. These classifications help a cultivator design a production system based on sequencing the growth of specific mushrooms first on one food source, then adding a second species of mushrooms when the food source has been altered or decomposed. Supplements can also be used in bulk-growing medium formulas; however, the more you use, the higher the risk of contamination competing with the mycelium for the prime real estate, so use supplements sparingly or not at all. Most supplemented media are used in sterilized substrates, which is easier to control for contamination since the media are autoclaved and spawn transfers are performed in a sterile environment.

Water

Mushrooms need high moisture and humidity levels (90 to 95 percent) to sustain their growth and supply the mycelium with enough water to spread into the environment. The "sweating" of enzymes lubricates the paths of the spreading mycelium, allowing it to "swim" through the growing medium, even when conditions are relatively dry. In this respect, fungi can breathe in humid air, condensing it into water internally in their cells, and redistributing the water wherever the body of mycelium needs it most, whether for nutrient capture or for mushroom production efforts. Once the colonized growing medium has charged its

Hydraulic pressure can force mushrooms up through concrete, asphalt, and heavy gravel.

mycelium full of nutrients, it is ready and waiting for a good soaking, where the water pressure generated by some fungi can actually break through concrete and asphalt, an amazing feat for a compressed bundle of mycelium!

Be aware that mushrooms can hyperaccumulate heavy metals in tissue, so be sure to test the water you are using for lead, mercury, and arsenic, at the minimum, to make sure the fruiting bodies are not going to be toxic. City water, although resplendent with chemicals and chlorines, is usually not a problem for mushrooms and is fine to use, but I still recommend testing any water source no matter how clean you think it may be, including pond, well, and roof-reclaimed water sources.

Gas Exchange

It must be understood that when mushrooms are colonizing, they are embedded in wood, under mulch, or streaming through compost piles, with very little oxygen needed until they are ready to fruit. The entire biomass uses the outer edges of the mycelium, in contact with fresh air, transporting oxygen deep within and channeling carbon dioxide out. If no oxygen is present, or if the medium is too wet or deep, the mycelium may not be able to supply enough oxygen to its entire body, creating dead zones. Mushrooms form where oxygen is plentiful, and will emerge from the fruiting medium wherever there is access to air, such as a tray surface or holes

One day later, these *Coprinus* mushrooms have pushed through an asphalt driveway in the completion of their life cycle.

in plastic bags and buckets. Keeping mushrooms too wet during fruiting can drown them, by coating them with a continuous film of water; their mycelium cannot breathe

The colonization room should have less oxygen than a fruiting room to trigger the natural stimuli that mushrooms need for prolific fruitings; most commercial farms keep colonization rooms at around 5000 parts per million (ppm). When colonizing mushrooms in bags on a small scale, it is possible to use the same room for fruiting, though this is not ideal since diseases can spread to uncolonized media.

Fruiting rooms by comparison should be ventilated at all times to carbon dioxide levels below 1000ppm; this promotes primordia formation and avoids stem elongation. Adding plants in growing rooms is a great idea to help offset carbon dioxide.

Light

Don't all mushrooms grow in the dark? Actually, only a small handful of species—including the *Agaricus* species (portabella, white button, et cetera)—can mature and develop true to form in the absence of light. All other mushrooms require diffuse natural or fluorescent light. For a cultivator, the wavelengths that are most important are in the blue-green spectrums. They are responsible for metabolic pathways that regulate most energy and growth requirements for mushrooms, as well as initiating mechanisms for creating higher levels of protein, vitamin D, and

Most mushrooms need light to prevent stem elongation and to develop cap cuticle color. These golden oyster (*Pleurotus citrinopileatus*) fruiting blocks were both initiated at the same time. The one on the left was left in darkness, and the one on the right in natural, diffused light. Fungi also use different wavelengths of light to synthesize vitamins and medicinal compounds.

medicinal properties. This allows low-wattage LED lights to be used in a fruiting room to save energy.

For most edible mushrooms, a lack of light creates the same effect as elevated carbon dioxide conditions: long stems and small caps. Mushrooms stretch from those dark places underneath mulch and in depressions, their caps small and narrow, enabling them to weave upward and into a well-lit environment. Mushrooms are also phototrophic (turning toward light), and they are capable of utilizing ionizing radiation from sunlight for the manufacture of many chemical compounds, including vitamin D and melanin, similar to the processes in human skin. This means that a mechanism exists that allows fungi to react and use radiation for their survival and biochemical triggers, a strange phenomenon since they lack the pigment chlorophyll, the pigment plants use for that purpose. This fungal production of chemical by-products has important implications for humans; by discovering new ways to cultivate mushrooms, we are able to take advantage of these "fungal factories" for producing new medicines, food, fuel, fiber, and much more.

MUSHROOMS IN SPACE?

In a NASA-funded experiment at Cornell University, mushrooms were exposed to varying wavelengths of light at different intervals so that researchers could observe the effects on and needs of cultivated mushrooms. I suspect these efforts were paving the way for a study on the feasibility of successfully cultivating mushrooms in space using low-energy LED lights powered by solar energy. The same technology can be used to make a growing operation here on earth less dependent on nonrenewable energy with the use of solar panels equipped with a battery and transformer. A simple setup like this could supply the minimal amount of light that mushrooms need in an enclosed or underground environment, provided oxygen is present, whether from fresh air intake, supplied by accompanying plants, or recycled by bubbling growing-room exhaust through algae tanks.

CHAPTER 2

The Seven Basic Stages
of Mushroom Cultivation

There are seven basic stages of mushroom cultivation: media preparation; inoculation and container filling; the spawn run; complete colonization; initiation and pinning; maturation and harvest; and rest. These stages are based on the stages of the mushroom life cycle, so if you understand the life cycle and you also become fluent in these stages of cultivation, you can apply that information to many different kinds of mushrooms on many different kinds of growing media, from easy-to-cultivate mushrooms like shiitakes and oysters to more difficult or experimental mushrooms.

Though there are countless methods, species, and materials you may choose from, what must stay constant is your adherence to these seven stages. Deviating from the essentials that mushroom mycelia need to develop edible mushrooms can result in extremely poor yields or even total crop failure. When you are cultivating a mushroom for the first time, or if you are trying a new method, try growing the mycelium on the smallest scale possible and then scale up when you're ready—it's like building a prototype car before you spend the money and resources on constructing an entire fleet. This gives you the time and flexibility to make corrections to your system before mass-producing mushrooms in a way that is less than optimal.

STAGE ONE:
MEDIA PREPARATION

All mushrooms require an energy source, or food, just like humans. This is called their growing medium, or substrate. (It can also be referred to as biomass.) It is also true that, like people, some mushroom species are omnivores and will feed (and fruit) on just about anything; for example, the phoenix oyster (*Pleurotus pulmonarius*) can be grown on wheat straw, coffee grounds, hardwood, conifers, and more. Other mushrooms are more finicky and have specific dietary requirements; for example, in my part of the country (upstate South Carolina), maitake prefers mostly oak. Some species can be "trained" to experiment with other food sources, but this takes months and reasonably sophisticated lab skills to accomplish.

For any mushroom you want to cultivate, you'll need to think about what growing medium to use and how to prepare it. Chapters 5 through 7 go into great detail on the many options, whether you want to grow mushrooms outdoors on logs and stumps, or whether you want to cultivate mushrooms indoors using a variety of methods on types of sawdust or agricultural by-products. And part 4 will tell you which

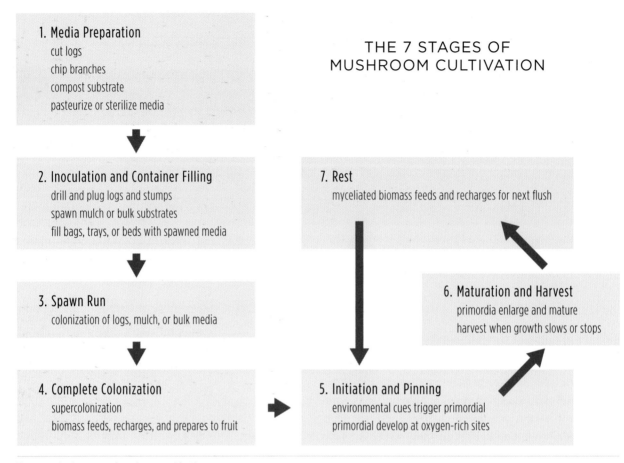

THE 7 STAGES OF
MUSHROOM CULTIVATION

1. Media Preparation
cut logs
chip branches
compost substrate
pasteurize or sterilize media

2. Inoculation and Container Filling
drill and plug logs and stumps
spawn mulch or bulk substrates
fill bags, trays, or beds with spawned media

3. Spawn Run
colonization of logs, mulch, or bulk media

4. Complete Colonization
supercolonization
biomass feeds, recharges, and prepares to fruit

5. Initiation and Pinning
environmental cues trigger primordial
primordial develop at oxygen-rich sites

6. Maturation and Harvest
primordia enlarge and mature
harvest when growth slows or stops

7. Rest
myceliated biomass feeds and recharges for next flush

The seven basic stages of mushroom cultivation.

growing media each of the mushrooms I've profiled prefers. While the options may seem limitless, there are some considerations you'll want to weigh before you choose and prepare your growing medium.

Choosing a Growing Medium

The most common substrates for anyone starting outdoors are typically hardwood chips, manure-based composts, and hardwood logs and stumps. For indoor cultivation, cultivators typically choose to use shredded and pasteurized cereal straw (wheat, oat, rye, or rice) or sterilized sawdusts for making artificial or "synthetic" logs. Supplementing bulk-growing media is like developing a recipe, or formula, for producing a specific mushroom on a mixed-food source, such as sawdust supplemented with wheat bran. The supplemented formulas generally are intended for

pasterurization or sterilization, since formulas that contain higher percentages of nutrients are prone to contamination and must be handled differently than natural log or wood chip cultivation. To start I would encourage you to follow the tried-and-true methods and substrate formulas listed with each mushroom, and then you can attempt to improve them based on the materials and methods you choose to employ.

Handling the Growing Medium

One concern for the growing medium is the possibility that it will be contaminated by "weed" fungal spores, bacteria, or other organisms that could potentially outcompete the mushrooms you are attempting to cultivate. If you plan to grow mushrooms outside on logs or chips—most likely on deciduous hardwood trees and the chipped debris—the wood should be fresh. If you're using wood chips, they should be used immediately after chipping and not allowed to sit around for more than two to three weeks; otherwise you risk growing everything but what you inoculated. If you're growing on logs, you can probably wait longer, depending on the wood type, temperature, and moisture conditions. A general rule is softer hardwoods (sweetgum and poplar) have a limited window and must be inoculated within a month or so, whereas more dense hardwoods, such as oak and maple, can be inoculated up to two months. Obviously any trees that are downed and already growing bracket fungi, like turkey tails, should not be used, nor firewood that has been split and aged.

If you are cultivating using bulk agricultural substrates—by which I mean dried vegetable wastes such as cereal straws (rye, oat, rice, and wheat), hulls of shelled beans, dried fruit peelings, and other by-products—if stored properly, they can be stored for many years. Try to keep them dry and out of the weather; otherwise they will become contaminated with molds and "weed" fungi before you even begin. If you have no way of keeping the medium dry, buy only what you need to use per batch. You can save space by shredding and bagging your medium if you don't plan to use it right away (see the discussion of shredding below). At our growing operation, we store all of our growing media in a small greenhouse in sealed drums or in bags on pallets, so ground moisture doesn't wick up into them.

Sawdust, fresh cut directly from a sawmill, is resilient to contamination and actually benefits from being rained on and aged for a few months. Fresh sawdust is hydrophobic, meaning that it repels water, and will not absorb the optimal moisture needed for mushroom cultivation. Soaking or aging the substrate can greatly improve the water-holding capacity of the wood particles, which translates directly into higher yields, since the mushroom mycelium draws a considerable amount of water from the growing substrate. Sawdust substrates are typically supplemented with nutritive ingredients and sterilized for mushroom cultivation, thereby destroying any molds and other potential competitors that have made a home for themselves.

Fresh hardwood chips from a local sawmill make a perfect growing substrate for outdoor cultivation of species such as king stropharia (*Stropharia rugoso-annulata*) and other terrestrial wood-inhabiting mushrooms. If you don't inoculate the chips with your chosen spawn within two or three weeks, though, the spores of wild mushrooms will land here and compete for the valuable real estate.

Fresh-cut hardwood logs work well as the substrate for many types of edible and medicinal mushrooms. This white oak (*Quercus alba*) is a highly desirable species for cultivating shiitakes (*Lentinula edodes*), among others.

Always store your growing media in a dry location to prevent mold and bacteria from getting a head start over your mycelium. I store these wheat straw bales on old pallets in a spare greenhouse.

AVOID PRESSURE-TREATED LUMBER

I do not recommend using sawdusts or by-products from treated lumber, as they may contain trace elements that can translocate into the fruiting bodies, such as heavy metals. If you happen to have treated lumber or sawdust as a by-product from building activity, you can collect it in separate bins and inoculate it with a species of mushroom that tolerates resins and volatiles. Pine and conifer decomposers, such as *Neolentinus lepideus*, can grow on railroad ties and promote the decay of its preservatives and soil generation, but again I would not advise the consumption of these mushrooms without first thoroughly testing them for hyperaccumulated or absorbed compounds.

A simple but effective home shredder can be used for small-scale commercial production. This dry wheat straw is being shredded to decrease the particle size and increase the density of the growing media, which will improve yields of many mushroom species. A large, vented bag or cage on the discharge will catch the debris, which can be used to prepare a mushroom growing medium.

Preparing the Growing Medium

Most mushrooms like their growing medium to be fine and somewhat dense, rather than fluffy and airy; this way the mycelium is not forced to expend all its energy in building surface area as it colonizes the substrate. Think of the expansion of the mushroom mycelium like the building of a network of highways, with bridges spanning the gaps between particles of substrate. The mycelium must devote considerable resources and energy to build and maintain these bridges, so the less space that needs to be spanned (read: the smaller the particles in your substrate), the less energy the mycelium spends on building its highways. Particles that are too small, however, will become compacted; in this case the mycelium will become choked and—because carbon dioxide can't escape—the growing environment will become anaerobic. It is your mission to find the sweet spot—a particle size that is small enough to allow the mycelium to easily span the spaces between, but not so small that it results in overcompaction.

Wheat straw perfectly shredded on the left, compared to the coarse straw straight from the bale. Achieving the right particle size in your growing medium can dramatically improve your yields.

The simplest and least expensive way to match your mushroom's preferred substrate density is to use a growing medium that has a relatively small particle size to begin with, such as peanut or cotton hulls. But if you do need to reduce the substrate's particle size, shredding is usually your best option. You can purchase or rent a shredder. How often you use it will determine your choice of machine. For our cereal straws, we use a chipper-shredder with a 10:1 reduction ratio; when you are purchasing or renting one, consider the labor and hassle of shredding your material twice with a 5:1 model, not to mention the gas and emissions of repeating the process. Designate the shredder for use only with vegetation intended for use as growing media; we use our shredder only for cereal straws (wheat, oat, and rye), which helps keep it clean and free from contaminants. Be sure that your substrate is dry; otherwise it will clog the shredder. To dry out your substrate before shredding, place it in a sunny area and let it dry for at least a week, turning it often. We use our spare greenhouse for drying everything from squash plants to water hyacinth to use later as a growing substrate.

Pasteurizing and Sterilizing the Media

If you decide that the growing medium must be treated to remove bacteria, molds, or other "weed" fungi, the type of treatment you choose—if any—will depend upon the mushroom species you intend to cultivate, the type of substrate you're working with, and most importantly your equipment, budget, and other resources. Fast-growing, aggressive species of mushrooms, such as oyster mushrooms, can be grown with minimal treatments since they are capable of outcompeting most competitor molds and mushrooms. Others, such as shiitake and maitake, are much slower growing or lack the ability to grow on pasteurized substrates, preferring a supplemented sawdust medium and a sterile colonization environment. Each of these is a limiting factor, but all must be carefully considered to produce mushrooms in any particular area of the world effectively.

Though we'll discuss them in detail in chapter 7, the five common methods I would recommend for preparing your growing medium are:

- Hot water immersion pasteurization—a two-hour process that also hydrates dried vegetation in preparation for cultivation
- Steam pasteurization—a twelve-hour process often used for finishing manure-based compost media or any presoaked fruiting formula
- Solar pasteurization—a six- to eight-hour process used by off-the-grid cultivation systems
- Chemical treatments—designed to target competitor fungi and bacteria while being tolerated by the mushroom mycelium
- Steam sterilization—a two-hour process favored by larger operations because it's fast and efficient, used to completely destroy all organisms in the media and when using a large amount of supplements (such as when growing on sawdust blocks indoors)

Sterilization (heating to 250°F/121°C) kills all of the organisms present in the growing media, while

PRECAUTIONS

Never use your shredder to shred spent growing media; molds and other contaminants will stick to the walls and blades of the shredder and infect the next stream of media you prepare! Keep the shredder covered and in a dry place so any small particles of straw do not hydrate and encourage mold colonies to multiply and perpetuate into your growing media before you have even started.

Always wear eye and ear protection and a dust mask when using a shredder. Shredding agricultural wastes and woody by-products will create a plume of particles that are easily inhaled. Read the manual, know your machine, and sharpen your blades. A dull blade can clog, reduce efficiency, and spit out media that are not consistent. To keep it sharp, all you need is a wrench, a hard file or grinding stone, and a willingness to look and learn.

SANITATION PROCEDURES

At our farm, to prevent contamination of the growing media and spawn cultures, we have set up a a a separate area for each stage in the production process. Workers move from room to room in a sequence, never going backward. This limits the opportunities for a worker to accidentally introduce contamination from the fruiting area back to the initial stages of development, which could be disastrous, for example if green molds take over the media.

We begin with our lab, where we produce spawn. We move our spawn from the lab into our substrate preparation and inoculation room (where spawn and pasteurized or sterilized media that have cooled from a heat treatment the day before are mixed and bagged in batches for different species). The inoculated media are then carried on a wheeled rack through two doors into a spawn run room. Once the mushrooms begin to form, the containers are brought further down the line to the fruiting rooms. Walking from the fruiting room (where fungi spores abound) back into the spawn run room or, even worse, the inoculation room is prohibited.

In our production line, if we need to do any work with unpasteurized media, such as shredding straw, we do it at the end of the day before we need it since it is a dirty task that creates volumes of spore-contaminated dust clouds, which should not be around in the morning when workers return clean to the farm. The following morning only involves the treatment process, inoculation, and container filling. After that the workers are freed up to pick mushrooms and clean. The final step of the day is to inspect the colonizing media that you have made days or weeks earlier and remove any contaminated containers carefully—and then never to touch the next day's work, such as the shredded media or anything in the inoculation room.

These small steps can dramatically improve an operation. I once consulted for a commercial start-up that had contamination problems. After the workers cleaned the fruiting room, they would walk back through the spawn run room with their bags of garbage—which were, of course, contaminated. When that practice ended, so did the contamination downstream.

This is an ideal setup; however, most growers will not be involved in the level of production or research our facility is involved in, and it is probably unnecessary for smaller operations. Just concentrate on keeping things simple and clean, while focusing on personnel as the worst contamination vectors, and remain observant of your process. Organize your operation to keep the movement forward, the first steps being the most clean and critical, the last few steps being the removal of compost, contaminated growing containers, and other waste. Mistakes at any stage can be costly, they can deflate your enthusiasm, and if you plan to run a commercial operation, they can put you quickly out of business.

pasteurization (heating to 160°F/71°C) leaves behind those microbes with some heat tolerance. Some mushrooms, such as shiitakes, require sterilized media. Others do best in pasteurized media. (Part 4 gives the preferences for all the mushrooms profiled.)

Everyone's schedule and situation are different, of course, but at our farm we pasteurize and sterilize our growing medium the day before we plan to inoculate (see stage two, below). This allows it to cool slowly overnight. However and whenever you do it, always keep the pasteurized or sanitized medium covered with a sheet of clean plastic or a sanitized tarp to prevent contamination. It may take a bit longer to cool, but by covering it you minimize the amount of contact it has with the unfiltered air around it. This is the stage at which the medium is most vulnerable to contamination, and a little extra effort will pay off quickly.

Stage Two:
Inoculation and
Container Filling

Although not physically demanding, inoculation is the most difficult step to master in the mushroom cultivation process, since sanitation measures are critical to avoid contamination of your prepared growing medium. You will be in direct contact with the (possibly pasteurized) growing medium and your pure culture spawn, so cleanliness here is essential to your success: clean tools, clean hands, clean floors, clean containers, clean everything. It only takes one set of dirty hands to ruin an entire day, or more, of work.

Outdoor cultivation typically only involves inoculation and not necessarily container filling, unless you are filling up raised beds with wood chips or compost. For bulk substrates indoors, inoculating the growing medium basically entails mixing spawn (mycelium) into the substrate, transferring the mixture to containers, and moving the containers to the spawn run area. A simple method for a basic inoculation of shredded straw with oyster mushrooms, for example, is to allow the media to cool after pasteurization on a clean, washed tarp, thoroughly mix the spawn into the media, and then fill your containers immediately after. When inoculating outdoors such as in fresh-cut wood chips, you would simply wet the chips and mix in the spawn thoroughly, relying on the mycelium's ability to outcompete all the other organisms. But streamlining the process is important, because the more the growing medium is moved or comes into contact with anything else, the greater the risk of contamination.

If you are growing mushrooms on untreated or pasteurized media, you can inoculate in a nonsterile environment, so long as the mixing surface is clean, such as a lightly bleached tarp or table. But if you're growing mushrooms on a sterilized substrate, you'll need to invest in a so-called clean room, with a HEPA (high efficiency particulate air) filtration system and laboratory space as described in chapter 17. Pasteurized substrates are easier to work with and less prone to contamination than sterilized substrates, and the cost

SETTING UP AN INOCULATION ROOM

Some growers have converted sheds into dedicated inoculation rooms for pasteurized media, giving them the control and cleanliness they need to keep growing media free from contamination, as well as providing a comfortable and sheltered place to inoculate. The floor should be sweepable, either concrete or professional landscaping fabric. Airflow should be minimized; any incoming air, if it's needed for cooling or heating in extreme climates, should be prefiltered with a particulate filter to reduce the number of mold spores being transferred from the outdoors into the cleaner indoor environment. Larger structures can be partitioned with walls or makeshift plastic curtains to reduce air movement and limit contamination between interior work areas.

Inoculation involves mixing and filling containers, so give some thought to ergonomics. Surfaces that are easy to clean and maintain are preferable. This could mean a clean, chlorinated tarp, a concrete floor, or a waterproof table designed to hold a lot of weight. Locate your inoculation area at or next to the pasteurization or sterilization room to minimize the distance and time during which the media are exposed to external factors.

and energy input are much lower. But of course not all mushrooms will thrive on pasteurized substrates.

Getting Ready

Organization skills here are a plus. To begin, you'll want a list of tasks you need to accomplish during this phase, such as cleaning all surfaces and tables that are going to be used to mix the spawn and prepared growing media, sanitizing the tools that will contact the media while you're mixing or filling, and stocking all the materials you will need at each stage of medium preparation up to the point of container filling (such as gloves, rubbing alcohol, and twist ties, for example). Inoculation for pasteurized media, either indoors or out, can be mixed with spawn in several ways; most small-scale growers simply lay the media out and use a clean rake that has been bleached. Keep a small tray of 50 percent bleach close by; dipping the rake in the solution is easier than spraying and more efficient. The same bleach solution can be placed in a spray bottle for misting hands that are mixing and filling bags, or you can use isopropyl alcohol (70 percent). Once the medium has cooled, you have bags or clean containers ready to fill, and the other items you need have been sanitized, you are ready to start inoculating your growing media.

Inoculation

When you inoculate your substrate with mushroom spawn, you can mix it into the media with your hands provided they are clean and have been sprayed lightly with isopropyl alcohol (70 percent). If you feel more comfortable, use latex or tight-fitting rubber gloves (the tight fit will give you more dexterity). I try to minimize the use of disposable products in my production process, so my preference is for reusable rubber gloves.

If you are working on a larger scale, you may prefer to use a clean tool, such as a rake that has been dunked in diluted bleach or wiped with isopropyl alcohol. (Have a rake solely designated for mushroom cultivation; do not use a rake that you also use for your yard or garden.) Spread the media out in an even layer about 3 to 4 inches in depth. Break your spawn into the smallest fragments it allows and broadcast it evenly across the surface of your media. Lightly rake it in. If you have more media or spawn, add another layer and repeat the process. If you keep the tools in the mixing area and they do not touch anything except the media and spawn, they do not need to be cleaned until after you are finished with the container filling. If the rake accidentally leaves the clean space or touches anything

Keep your hands clean when touching pasteurized substrate. Use 70 percent isopropyl alcohol or diluted bleach (half strength mixed with water) to clean your hands and all surfaces and tools that will come into contact with the substrate.

AUTOMATING INOCULATION

If you are going for pure volume production, it's relatively easy to automate the inoculation process, but it's not cheap. Many companies offer commercial production equipment such as ribbon or paddle mixers, substrate elevators, and conveyer belts (I have included their contact information in the references) that can save you labor, but be aware that automation can also complicate your process, not only by magnifying your production volume, but also by magnifying your problems, such as low-yielding formulas, strain degeneration, and possible contamination if you have not streamlined and perfected every step of the operation. Expanding mistakes can be costly, so analyze each step of the production process and eliminate any potential for failure while optimizing improvements before committing to such a huge investment.

If you have succeeded in small-scale production and are serious about committing to a larger-volume operation, then my suggestion is to seek out a professional consultant who can make some recommendations before you commit to a large purchase. And visit a farm equipment show, where you can see some of the existing products. But keep in mind that you may not need a lot of capital tied up in expensive and specialized machinery. With a little innovation and creativity, sometimes you can modify used agricultural machines, retrofitting them to perform the same function as an expensive new machine for a fraction of the price and energy. Used paddle or ribbon mixers, for example, can be fitted with steam inputs to help pasteurize or sterilize bulk media. I have even seen a used concrete mixing truck modified with injection ports to pasteurize large volumes of media! There are also the methods I describe in the chapter 12, which may seem primitive, but they meet the basic needs of mushroom cultivation and can be adapted to almost any production system.

other than the medium, sanitize it before returning it to the medium area. When you're done mixing, use your rake to push the inoculated media into a pile to make them easier to bag or place into trays.

At my farm, we keep all of our tools, bags, hole pokers (for aerating the growing containers), and bottles of isopropyl alcohol near the inoculation area and never remove them for any other purpose. We wash them with fresh water and sanitize them with alcohol after each inoculation to avoid contamination or mixing of mushroom species.

Container Filling

Once you've inoculated your substrate, you're ready to fill your clean, sanitized containers. Chapter 8 goes into fruiting and cropping containers in detail, but for the purpose of understanding the stages of cultivation, keep in mind that if you are cultivating indoors your containers will most commonly be bags, pots, or trays.

I have seen many variations depending on what is locally available; in general, as long as the ratio of volume to exposed surface area is respected (see chapter 8), you can find suitable containers anywhere.

If your containers are reusable, be sure you've cleaned them a day ahead and have given them time to dry if you are using a diluted bleach solution. If you are using opaque bags or containers (which I recommend; again, see chapter 8), you might consider complementing them with a few clear bags or containers, so that you can monitor each batch's progress and check for contamination as the spawn run approaches complete colonization; we call these "spy bags."

Fill the containers with the inoculated substrate, packing them densely to avoid air pockets, where some unfortunate mushrooms may decide to form, but end up trapped and unable to mature. Aborted mushrooms present a problem, since they rot and can attract flies, harbor molds, and create bacterial blooms. If you're using plastic bags, fill them until they stretch, to the point just before they rip or break. As the mycelium consumes the media, the contents will diminish and air pockets may form; having a tight fit at the onset ensures that air pockets are minimized. Slamming the bags or containers onto the ground or using a tamping tool can help pack down the media. If you are stacking bags or pots you can press the top ones down to compress the lower ones.

When you're done, label your containers with the mushroom species and the date of inoculation. You'll also need to poke some holes in your containers; see chapter 8 for details.

Filling containers is one of the most laborious parts of the mushroom cultivation process—especially if you are have a larger operation—so if you see a way to automate any of your steps, I recommend it.

STAGE THREE:
THE SPAWN RUN

From the moment you inoculate logs and wood chips outdoors or mix your spawn and prepared growing medium together, the spawn will take time to fully colonize the medium; this period is called the spawn run or colonization period. This is the time when the separated bits of mycelium are exploring the medium, securing territory, and building the biomass they need to support fruiting. Remember that the fungus is collectively a single organism; by breaking up the spawn during the inoculation of bulk substrates, you make thousands of individual colonies that are genetically identical. The same thing applies for inoculated logs or stumps: Fill holes with plugs, sawdust, or inoculated cardboard; every little colony starts as a small island radiating outward to join and form a large, single colony. Once they start to spread, they can sense each other from great distances and attempt to reconnect into a solid, continuous mass of cells.

As you'll recall from the discussion of the mushroom life cycle, during colonization the infantry cells stream outward at a rapid pace. The mycelium organizes itself and communicates throughout the biomass by sending signals (chemical cues) from the advancing cell tips, or hyphae, back to the rest of the colony. These cells consume only enough of the media to thrust themselves deeper into the substrate. As they progress, they

Once it's mixed in with the growing medium, the spawn begins to explore and "leap off" into the substrate. Here oyster mushroom mycelium is spreading through spent coffee grounds and pencil shavings.

release enzymes that break down the molecules of the growing medium into smaller, digestible pieces, and thereby building heat in the medium (thermogenesis).

Heat and humidity are important factors during the spawn run, but can be different priorities for indoor and outdoor growers. For outdoor cultivation, such as on logs and chips, an occasional misting or weekly watering is typically enough to maintain optimum moisture during spawn run if it hasn't rained in a week or two. The rest of the information described here is for indoor cultivation on treated growing media, where humidity should ideally be in the range of 60 to 70 percent, moist enough to keep the substrate from drying out but not so high that it initiates primordia formation. The temperature should ideally be kept in the range of 70 to 80°F (21–27°C). Many commercial growers have dedicated spawn run or colonization rooms, where they can more easily control heat and humidity levels. In smaller operations using pasteurized or sterilized growing media, you could simply have a single space dedicated to both colonization and fruiting (stage five); just move the containers of inoculated media directly into a fruiting room kept at around 70°F (22°C) and allow the natural heat generation of colonization to incubate the mycelium. Then when they cool down they are already in position to fruit, thus reducing costly labor and moving about. This setup has some risk associated with exposing uncolonized media to potential contaminants, but if you maintain a clean operation and keep insects and molds to a minimum, it can be a great way to get started without needing to invest in building so many different rooms.

During the spawn run you should monitor the thermogenesis, the heat created by the mycelium as it feeds and spreads. A digital probe thermometer works well. If the temperature is too low, it may slow the spawn run slightly and allow competitor fungi to proliferate; still, this is a better option than overheating during a spawn run. If the temperature is too high, the growing medium will cook itself and create an anaerobic core or allow thermophilic molds to dominate. Adjusting the temperature of the spawn run space is critical. Using a small space heater with a thermostat or adding an air-conditoning unit to the room may be necessary

THE IMPORTANCE OF MONITORING

Monitoring all stages of your mushroom cultivation operation is an important daily task. To begin, spend a few bucks on a few cheap digital thermometers; you can use them in multiple areas of your operation. Put one near the pasteurization or sterilization area and another in the spawn run and fruiting room, and keep a spare handy in your pocket as you inspect your operations daily. Sanitize the tip every time you insert the probe into a different culture to prevent contamination.

Identifying contamination early in its development is critical. You need to be tuned into the different stages of competing molds, yeasts, and bacteria to prevent them from outcompeting the mushroom you are growing. Some mold will simply "marble" or occupy space inside your mushroom blocks or bags, while others (some *Trichoderma* spp.) are mycoparasites, meaning they will outcompete your spawn and wreak havoc in your growing operation if you do not catch them early. You must be able to recognize the differences between your mushroom's mycelium and other contaminants in order to make quick decisions. Spend time observing even the minor details of your operation.

during some times of the year. Growing media that have been supplemented greatly will generate much more heat than, say, plain sawdust or straw, so be aware that you will need to adjust the air temperature of the spawn run room to level the core temperature of the bags or trays at about 80 to 85°F (27–29°C). As the fungus completes colonization of the media, its enzymatic digestion begins to produce less heat, signaling to the mycelium that it's time to shift gears and concentrate on feeding rather than colonizing in order to support mushroom formation and production.

During this phase, it's important to be vigilant. Make a point of observing your fruiting room(s) daily so you can respond to any problems quickly and effectively. Mold outbreaks and fly infestations don't happen overnight; they gradually manifest and amplify. Capturing these outbreaks in their infancy will keep your operation running smoothly. If you do have a mold outbreak or fly infestation, dumping a batch of mushrooms may be your best and only option; if you decide to gamble by keeping containers that are partially colonized by these pests, you risk that problem becoming worse. Remove problematic containers at the end of the day, to minimize the risk that you'll carry the contamination on your clothing or shoes to other areas of your operation. If the problem can't wait, seal up the contaminated containers before walking them through your fruiting and colonization rooms.

STAGE FOUR: COMPLETE COLONIZATION

Finally! You have achieved the goal of creating a fully colonized fungal biomass, capable of fruiting mushrooms. Your spawn run is complete, and now you only

have to wait a few days to a few weeks before mushrooms will appear. Complete colonization is similar to the "resting" phase (stage seven) in that there is a metabolic shift and preparation. During this time the fungus concentrates on feeding rather than spreading; it is recharging its battery, so to speak, to give it enough energy to fruit as many mushrooms as possible.

You should continue to carefully examine your growing medium for competitor molds or fungi, insects, and other unwanted contaminants, which can quickly magnify in numbers and snowball into a huge problem. As always, remove suspect fruiting containers carefully and dispose of them. By inspecting every day and catching the problems in the very beginning stages, you will have an advantage; the longer you leave them in the growing space, the more the balance of power shifts, and not in your favor.

STAGE FIVE: INITIATION AND PINNING

Once your growing medium is fully colonized, it will take a few days, or weeks depending on species, to produce mushrooms. Initiation is the point at which

Colonization is complete when the spawn has completely run through the entire substrate and hits a physical barrier, limiting its growth and signaling the mycelium to concentrate on fruiting.

Mushrooms need oxygen to form and mature properly. These crowding oyster mushroom (*Pleurotus ostreatus*) primordia are trying to escape their bag through an airhole. This is a good sign, though a cultivator must make sure that there is adequate fresh air coming into the fruiting room.

either Mother Nature or the cultivator gives the colonized mycelium the green light to move forward and produce mushrooms. Now mushrooms begin to form on the exposed surface area of your colonized mass, a process termed "pinning," where primordia enlarge and rupture from the bark on your logs, pushing through the holes in the bags, and scattering along the surface of pots or trays.

Initiation can be prompted by changes in temperature, humidity, and light—all factors you can control in an indoor environment to keep batches of mushrooms flushing consistently. Outdoor cultivators are limited by Mother Nature's cycles, but they can exert some degree of control by covering a portion of the beds or logs to limit rainfall and thereby offset flushes, or rotating watering to stimulate fruiting over a percentage of the operation. Indoor operations that are environmentally controlled have two options: either keep the conditions as steady as possible in one large fruiting room, or create smaller rooms to fluctuate gases, temperatures, and watering to vary the timing of flushing cycles.

Temperature

Following the spawn run, a mushroom strain will want to fruit when it reaches the ideal temperature window, whether on logs outdoors or in bags indoors. For outdoor cultivation all you can do is wait for the season to come around, but for indoor growing you can lower the temperature in your fruiting room just a few degrees once colonization is complete and the mycelium is fully charged up with the nutrients; this simulates seasonal temperature fluctuation and can quickly stimulate mushroom formation. Some cool weather species (lion's mane, maitake, and enoki, for example) grown indoors in controlled environments require a cold shock as low as 40 to 50°F (4.5–10°C) for one to several days, depending on the strain or species; then the temperature can be brought back up to much warmer conditions to speed mushroom formation. You will need to find the lowest number of chill hours needed to trigger the fruiting. Some species, such as blewits and morels, require near-freezing to freezing periods to complete their temperature

requirements in order to fruit, so I strongly advise picking mushroom species to cultivate throughout the year and planning the inoculations to vibe as harmoniously as possible with your climate by varying species and growing with the seasons.

Humidity

High humidity levels (90 percent or more) are essential to initiate and sustain primordia formation. Hand watering or automated misting that is timed to coincide with the projected appearance of pins is critical. Micronized droplets of water, such as fog and dew, on the primordia can signal to the mycelium that there is enough moisture in the fruiting environment for the mushrooms to fully develop into maturity. Evaporation at the exposed surface of the substrate (such as the tray surface or the airholes in bags or buckets) has a cooling effect, further stimulating initiation. If the fruiting room is too dry, is receiving direct sunlight, or experiences too much air circulation, the surface areas can dry out and the mycelium will not produce mushrooms. You may wish to place a greenhouse humidity sensor in your fruiting room, especially if you are using automatic watering systems that provide intermittent misting to prevent drying. Outdoor fruiting beds would also appreciate an occasional watering and misting, timed to coincide with the fungi's flush cycles.

Light

Most cultivated mushrooms require light to initiate and fruit, but it does not have to be much at all. For outdoor cultivation shaded, indirect light is ideal. For indoor cultivation fluorescent or LED lights are ideal; they supply the color spectrums needed and offer energy saving options. Although light is not required in the spawn run stages, an increase in light is essential to healthy mushroom development after the pins have emerged from the growing media. You cannot overexpose mushrooms to artificial light if the temperature is cool and there is ample humidity in the fruiting room. Direct sunlight or hot lamps can damage primordia if they dry out. Some mushrooms are more photosensitive than others, so even the fruiting container could become a factor in initiation and pinning. I tend to use opaque containers (so that the fungi feel at home in the dark between fruiting cycles) with some holes for localized light exposure; this gives the fungi incentive to fruit and mature in a well-lit fruiting room. Keep in mind that mycelia in nature hide under bark, mulch, and leaves until the fruiting stage; don't stray too far from these natural inclinations, or you will be disappointed with low yields or no mushrooms at all. It is okay to use clear bags or containers so that you may watch the action of colonization and observe potential contamination, but many species prefer the seclusion of dark containers because light stimulates the entire growing medium to produce mushrooms that cannot find a way to escape their container.

Nurturing Primordia

All pins, or baby mushroom primordia, look different depending on species, so pay close attention after colonization is complete. Primordia can emerge in just a few hours after a soaking or increased misting, when the swelled biomass has absorbed enough water to devote itself to a fruiting cycle. It is a great idea to photograph these baby mushrooms, sometimes twice a day (try a time lapse for incredible footage), so you can familiarize yourself with the rapidity and beauty of this process.

After a few days of nurturing, the pins will have become big enough that they can afford to dry out a little between watering, which is important because it limits the risk of bacteria and rot for mushrooms that have been picked and placed in refrigeration awaiting a buyer.

Pinning is a critical period for the cultivator, and daily inspections are necessary to ensure that the primordia are well taken care of. If for any reason these baby mushrooms should dry out, the cycle will abort and you will have no crop. This can be disappointing at best, and costly for commercial mushroom producers who rely on consistent production. There is nothing worse than discovering a huge flush of dried-out baby mushrooms. The good news is that this is avoidable if you inspect your operation daily and provide the humidity and misting your baby mushrooms need at this critical point.

STAGE SIX: MATURATION AND HARVEST

One of the most frequent questions I answer is "How will I know when my mushrooms are ready to pick?" The answer depends on your ability to judge the stage at which the mushrooms are at maximum weight, but not beyond the point when they stop growing and begin to decompose.

The trigger for me is the maturation of the margin (outer edge) of the cap. Most gilled mushrooms form with their caps in a convex shape, to protect the gills from rain, flies, slugs, and other predators. The cap's edge will change shape as it approaches maturity, and it is this feature that you should focus on. When I start to see the margin turn upward slightly or begin to ruffle, it is time to pick—and maybe even slightly past time if I see a lot of ruffling of the cap margin of most gilled mushrooms. Nongilled mushrooms, such as polypores like reishi (*Ganoderma lucidum*), will have a lighter, actively growing margin, and their caps will elongate as they develop, with the color disappearing when they are ready for harvest.

In general, each mushroom species has a unique appearance at maturity that can help you time the harvest. Oyster mushrooms, for instance, will mature rapidly from pinning to harvest, taking as little as three days. Once oysters approach full expansion, their cap's margin begins to stretch and ruffle in an undulating pattern and their color fades.

Mushrooms that are picked late tend to attract hungry flies, promote heavy spore loads in the fruiting room, and have a short shelf life. Picking mushrooms before they're ready, or while they're pinning, will reduce overall yields. It's a fine balance. Simply put, in addition to observation and judgment, the art of cultivation includes developing an intuition for when a mushroom is ready to harvest.

Some strains fruit singly and can be picked individually, but most commercial strains are clustering cultivars chosen for their yield and ease of harvest. Solitary fruiting strains are more labor-intensive, but sometimes the flavor and quality are superior. In

A rainbow of oyster mushrooms! Warm blue, golden, and pink oysters all fruiting in unison. These mushrooms double in size every day and flush every two to three weeks for eleven weeks, and sometimes longer.

clustering strains, all the fruiting bodies fuse together at the base as they emerge from the substrate. Consider this like the maturation of clustering fruits. Would you pick one grape at a time as each matures, or would you rather pick the whole cluster? The cluster, of course! Harvest your clusters when the largest mushroom of the group begins to flatten out, or when the cap edge starts to turn upward, and when the group stops doubling in size over a period of twenty-four hours.

To harvest, twist the mushrooms, tightly pinching or grasping the cluster at the base. Using a sharp knife, or even better a pair of clean scissors, trim off the substrate-covered bases into a small bucket next to your containers to keep the mushrooms from dirtying

themselves in the harvest tray. Don't remove too much, however, since stems are also valuable products that can be used for powdering and flavoring. Set the mushrooms in a harvesting tray, all oriented in the same direction to keep the dirty bases from coming into contact with the caps.

If you're harvesting for commercial production, it's wise to place paper slips into each harvesting tray identifying the batches you've picked. With this identification, you can weigh and record your harvests to determine yields from specific dates and formulas, and this, in turn, helps you analyze the efficiency of your fruiting substrates, especially if you are using supplements to enhance yields. If you collect harvesting data for each batch of mushrooms you produce throughout the fruiting cycles, you can better understand the timing of the flush cycles and calculate the cost of the production versus the actual yields you are generating.

When packing mushrooms for sale, arrange them carefully in perforated boxes or containers. There is nothing more unappealing than a box of mushrooms that were simply dumped in lazily, with no respect to quality. Take the time to neatly align the mushrooms in the same direction or to organize them by size so consumers and chefs can locate the mushrooms they want quickly, without having to dig through a pile, which further damages the product.

Never pick or pack mushrooms while they're wet. Wet mushrooms will rot in days, sometimes hours, before you can eat or sell them. If you use an automatic watering system, be sure to turn it off in the morning a few hours before harvesting to allow the cap surface time to dry. If the mushrooms still happen to be wet when you harvest, set the harvest trays out in dry spot and allow the excess moisture to evaporate before boxing them up. Once picked and packed, mushrooms should be refrigerated.

STAGE SEVEN: REST

After their hard work, your mycelia are in need of a well-deserved period of rest. They've performed well, so give them a vacation before the next flush. Once the mushrooms have been harvested, the fungi will benefit from warmer, drier conditions and less light than what was required for fruiting. Reduce the watering, increase the temperature slightly, and darken the room if possible; this simulates the natural cycle that occurs in the wild.

Forcing mushrooms to fruit before the mycelium is fully charged can reduce yields, mushroom size, and your profits. Patience is the key here; if you are in need of more mushrooms per week, spawn more growing substrate per week! Never force your fungi to fruit unless they have had adequate time to recharge and prepare for the next flush. What if someone only allowed you to sleep for two hours a day and worked you straight for the other twenty-two?

Naturally you would be tired, and weaker and weaker as the days went on.

The length of this rest period depends on the species you're growing and the temperatures you're working with. The cooler the temperature, generally the longer the wait. The rest period is critical for successful successive fruiting cycles—and you might get two to five flushes, depending on species if you give the mycelium time to feed and charge itself for the next explosion of mushroom production.

Depending on species and temperature, you can calculate a period of rest, with concomitant reductions in temperature, watering, and light conditions. Then resume watering and begin the cycle again, duplicating this sequence for every flush afterward. Record keeping is essential for timing these initiation sequences and optimizing yields, especially in commercial operations. Subsequent flushes will generally yield, in weight, half of the previous flush. Each species also has an approximate total number of flushes you can expect before you'll want to swap in fresh cultures, replacing the older cultures with new, higher-yielding flushes. (See part 4 for the particulars on each species.)

AFTER THE HARVEST: STORAGE OPTIONS

Do you want your mushrooms to be fresh or dried? If you are growing for your own personal use, it simply depends on how soon you plan to eat them! If you're growing commercially, it depends on your market and production cycles, but hopefully you are selling most of the product fresh, since most fleshy fungi are composed of water, and drying them reduces much of their weight and, as a result, their value. (Of course, you'll charge considerably more, per pound, for dried product.) Also many mushrooms lose their appeal when they're dried, though some, when rehydrated or reconstituted, are almost identical to their fresh form. Dried shiitake mushrooms (*Lentinula edodes*), for example, are a bit leathery even when rehydrated, while reconstituted maitakes (*Grifola frondosa*) retain their

These harvested oyster mushrooms were lightly trimmed. Later they'll be stored in ventilated cardboard boxes to preserve moisture and allow for gas exchange, which prevents spoilage.

crunchy, meaty texture even after years in storage. In general, I only recommend drying your mushrooms if you grow too many, so that you can at least recover the cost of production and offer a companion product to your fresh mushrooms that can be used for flour, making teas and extracts, and other products. Chapter 15 discusses some of the ways you can market, package, and display your items if you are going to sell them; you might be surprised how many different ways mushrooms can be sold. Be creative and dream up something new to make your operation different!

Refrigeration and Shelf Life

If you're storing fresh mushrooms at 38 to 42°F (3–6°C)—the ideal temperature for long-term storage of fresh mushrooms—store them in cardboard boxes that have a few holes, similar to produce boxes. Since these boxes aren't cheap, consider asking local restaurants or produce distribution companies for their used boxes and then use them over and over again. You'll get many uses out of the boxes; they only become unusable when they get wet, so store them in a clean, dry place.

It's not uncommon for larger commercial growers—just like some vegetable producers—to add chemicals to their mushrooms to prolong their shelf life. These can be hormone-related compounds, such as 10-oxo-trans-8-decenoic acid (ODA) (Mau and Beelman, 1996), that have not fully been tested for consumer safety. If your goals include mass production and large-scale distribution, you may want to consider alternative preservation methods such as a light application of aqueous citric acid prior to storage or storing mushrooms in sealed, nitrogen-filled containers. Anything sealed, however, poses a higher risk of contamination by food-borne pathogens such as *Clostridium botulinum,* the bacterium responsible for botulism. Of course, one advantage to producing for your local market is that you don't have to deal with that; selling your beautiful mushrooms just a day or two after harvest requires nothing more than refrigeration and packing the mushrooms into slightly aerated cardboard boxes to keep them from sweating or drying out.

Drying Mushrooms

If you're drying your harvest, you need to provide air circulation and low heat to dehydrate the mushrooms evenly and without burning or damaging the cellular structure and embedded proteins, vitamins, and medicinal compounds. A simple home dehydrator or commercial unit will work, but you could also take advantage of that beautiful, glowing disk in the sky. Solar-drying mushrooms has proven to increase the vitamin D content in many gilled mushroom species, a bonus compared to conventional electric dehydrators (Kalaras and Beelman). Setting the mushrooms on wire racks on a wooden frame, or old screens from windows, in the sun for a few hours during the day is usually sufficient. Keep the mushrooms whole and trim minimally; this keeps the mushrooms alive and metabolically active. Place the mushrooms with the gills facing the sun to allow them to synthesize the vitamin D. After a few hours the vitamin D is locked in and the mushrooms, if large, can be sliced to speed the drying process. If you leave them overnight in humid climates, be aware that they may just rehydrate and revive, so bring the racks inside or stow them in a dehumidified space if they need another day's drying. Once your mushrooms are dehydrated, bag them immediately in airtight bags or containers. These dried mushrooms will store indefinitely. I have noticed many beetle species that can chew through plastic bags, enter, breed, and reduce the mushrooms to powder, so I recommend storage in glass containers.

What About Freezing Mushrooms?

Freezing can also be a good long-term storage method, but you have to blanch them first, or else they'll become mushy when defrosted. To blanch, bring water to a boil, drop in your mushrooms, boil for a few minutes until they soften, and then plunge into cold water to stop the cooking. Then freeze, in portions equivalent to what you plan to use or sell on a regular basis.

Sometimes frozen mushrooms develop freezer burn and a chewy texture. You can improve the texture by adding some water to your freezer containers and freezing the mushrooms suspended in this solution. Mushrooms frozen in icy blocks tend to keep much of their original color, flavor, and texture. Be sure to experiment with blanching and freezing on a small scale before you commit your entire harvest.

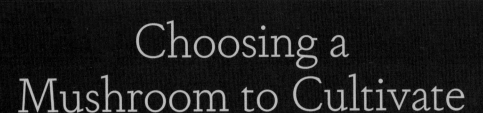

CHAPTER 3

Choosing a Mushroom to Cultivate

Now that you understand the mushroom life cycle and stages of cultivation, it's time to choose a mushroom to grow. The overall challenge of cultivating any mushroom ultimately depends on your skill level and the infrastructure you possess. Although you'll likely be excited and ambitious as you launch into mushroom cultivation, it's important to also be realistic, especially before you invest a lot of money. Be patient—a series of projects that offers an incremental increase in difficulty will soon increase your skill and confidence and give you greater flexibility in what you choose to cultivate. The beauty is that there is no finish line in mushroom cultivation; there is always a way to improve your process or yields. But as you begin, the most important factor is ease of cultivation. So let's see what makes some of these mushrooms easier, or more difficult, than others, and then we'll examine the other factors that will guide your choice of mushrooms.

Ease of Cultivation

My recommendation for anyone starting out in mushroom cultivation is to begin with the easiest species and the most readily available substrates. When I give workshops, I encourage people to start with three particular species on three basic substrates: shiitakes on logs; king stropharia on wood chips; and oyster mushrooms on agricultural waste. Practicing and perfecting your techniques and skill with these more simple cultivation combinations is a good strategy for initial success, which will build your confidence, capabilities, and intuition. As you gain experience and possibly even earn some income from your mushrooms, it's then easier to justify spending time and capital on larger and more ambitious projects.

Though the three species just cited are relatively easy to cultivate, some mushroom species require a good deal of orchestration of complex environmental qualities to coax them into predictable fruiting patterns. Maitake (*Grifola frondosa*), for example, is a desirable edible and medicinal mushroom, but it has several stages that make it difficult to cultivate. To begin, it requires supplemented sterilized sawdust as a growing medium. Once it has fully colonized this substrate, it prefers a mild cold shock (a drop in temperature of 10 to 20°F, or 5.6–11°C, for several days), followed by a period of initiation under high carbon dioxide and humidity levels, and finally a high-oxygen phase when the fronds of the mushroom flatten and develop. Morels are another extremely difficult species, with additional factors such as flooding, freezing, and even a sclerotia stage, when it forms a compact mycelium, like a tuber, an anatomical phenomenon most mushrooms lack.

Each mushroom has its own unique nuances, or "pickiness" factors—though not many are as extreme as the examples just cited—and I've tried to reflect this fact in the difficulty rankings I've assigned to the mushrooms profiled in part 4.

EASE OF IDENTIFICATION

The ease with which you can identify a mushroom can be an important factor, especially for mushrooms that you grow outdoors on logs, wood chip piles, and composts, because they can share their growing habitat with other wild mushrooms that compete for the same food. You have to be able to distinguish "weed" mushrooms—unwanted species popping up where you don't want them—from your crop mushrooms. Weed mushrooms will not be as prevalent when you're cultivating indoors because, if you pasteurize or sterilize the growing medium, you eliminate any other species in the medium, and because your spawn has a competitive advantage over sporulating mushrooms since it is already in the colonization state by the time you inoculate the medium with it. Likewise, when you're cultivating more than one type of mushroom in the same fruiting room, there is little risk that species will cross-contaminate, since the spores will not have time to land, mate, and colonize before the spawn has already run through and taken over the substrate, preventing any unwanted colonies from forming.

Although most of the cultivated mushrooms I've included in part 4 are easy to identify, I've included warnings for species that can pose difficulties for beginners. *All* mushrooms should be positively identified before they are harvested and consumed. If you're at all unsure, ask someone skilled in identification to assist you.

BROAD VS. SPECIFIC SUBSTRATES

As you'll remember from the discussion of metabolites and enzymes in chapter 1, many mushrooms are gifted at breaking down a variety of substrates to produce food. Others, however, are more limited in their ability to manufacture these specialized enzymes, or "chemical keys," needed to break down chemical compounds and convert them into the basic nutrients they need for energy and reproduction. In addition, mushroom species often have many strains and ecotypes, or strains that have evolved to grow in particular ecological niches, and these strains may exhibit a specific affiliation for a particular substrate. Researching which mushroom species, strains, and even ecotypes are particular to your region and the substrates you can obtain will be one of your first steps in determining which mushrooms are best for you to start with.

The most limiting factor in your choice of which mushrooms to cultivate might be the raw materials that are available to you for growing media. As part of your preliminary decision making, identify the materials you could use for mushroom production. See what might be available as waste from local industry and agriculture. In particular, look for anything that could be a compatible substrate for the species that are known to fruit in your area; culturing and cultivating your native ecotypes on locally available materials has proven to be the best strategy for success, since the mushrooms are familiar with the fluctuations and extremes of your particular climate and are most likely compatible with native substrates.

TEMPERATURE RANGE AND SENSITIVITY

Mushroom strains vary in their ability to initiate fruiting or even survive at specific temperature thresholds. Ideally, you should seek local ecotypes, which are already acclimated to local conditions and cycles. For those cultivating in tropical climates I suggest pairing this book with two others, *Mushroom Cultivation* (Oei, 2006) and *Technical Guidelines for Mushroom Growing in the Tropics* (Quimio, Chang, and Royse, 1990), which cover low-tech methods for growing mushrooms that thrive in extremes. I have included many heat-loving species in this manual, since I live in and conduct most of my research in the southeastern

United States, and because I also often travel abroad to tropical locations such as Haiti to establish mushroom cultivation systems.

Attempting indoor cultivation of a mushroom that does not normally thrive in the temperatures of your particular climate can prove difficult, since you will have to maintain the temperature artificially using air-conditioning or heat year-round, adding great expense to your operation. Indoor cultivation is used primarily to create ideal conditions—either to customize gas exchange for different stages of development or to maintain a high oxygen level—and provide optimal humidity while also protecting your crop from insect damage, not to control temperature. So heating or cooling incoming air just to make it a suitable temperature for cultivating a mushroom that probably does not thrive in your area could be a losing battle.

Choosing Substrates and Species: Temperature Key

Cold: 32–59°F (0–15°C)

- Logs and stumps: birch polypore, brick top, cauliflower, chicken of the woods, enoki, lion's mane, maitake, nameko, oyster (cold-tolerant ecotypes), shiitake (cold-tolerant ecotypes), tiger sawgill
- Wood mulch or chips: blewit, brick top, nameko
- Composts: blewit, shaggy mane
- Agricultural waste, straw, plant debris: oyster (cold-tolerant ecotypes)
- Sawdust: enoki, lion's mane, maitake, nameko, oyster (cold-tolerant ecotypes), shiitake (cold-tolerant ecotypes), tiger sawgill

Temperate: 60–85°F (16–29°C)

- Logs and stumps: beefsteak, birch polypore, black poplar, cauliflower, chicken of the woods, hairy panus, oyster (temperate ecotypes), reishi, scaly lentinus, shiitake (wide range of ecotypes), tiger sawgill, turkey tail, wood ear
- Wood mulch or chips: brick top, king stropharia, parasol
- Composts: almond portabella, king stropharia, parasol, shaggy mane

GROWING MUSHROOMS IN HAITI

The tropics have a wide spectrum of biomass suitable for mushroom production that is virtually untapped, such as waste by-products from the forest and agriculture. In March 2013, I traveled to Haiti as a consultant with Clemson Engineers for Developing Countries (CEDC) to explore applications for mushroom cultivation and mycoremediation in Haitian communities. We discovered that a peanut operation was shelling and composting the peanut hulls, which could have been used to produce edible oyster mushrooms. Upon returning to South Carolina, we found peanut hulls, pasteurized them, and cultivated oyster mushrooms in under two weeks, which is extremely fast. Although we have not been able to sustain the mushroom operation yet, this clearly demonstrates the ability to fruit mushrooms on agricultural waste by-products anywhere in the world if spawn is available and the techniques are distributed freely.

Working in areas where infrastructure is minimal, forcing us to use extremely low-tech methods of cultivation, has improved my understanding of and appreciation for the absolute minimal requirements for growing mushrooms productively and efficiently. In these "survival" situations, I'm challenged to push the limits of what is possible to produce edible protein from garbage, paper, and other waste in just a few weeks. And it *is* possible. Given the knowledge and skills, just about any region in the world could build a thriving food production system using little or no infrastructure, while improving local soil fertility by composting the spent mushroom media.

- Agricultural waste, straw, plant debris: elm oyster, king stropharia, parasol, oyster (temperature ecotypes), shimeji

- Sawdust: black poplar, beefsteak, elm oyster, hairy panus, king stropharia, maitake, nameko, oysters (temperate ecotypes), reishi, scaly lentinus, shimeji, shiitake (wide range of ecotypes), tiger sawgill, turkey tail, wood ear

Tropical: 86–100+°F (30–38°C)

- Logs and stumps: hairy panus, oysters (tropical ecotypes), reishi, scaly lentinus, shiitake (tropical ecotypes), turkey tail, wood ear
- Wood mulch or chips: king stropharia
- Composts: giant milky, paddy straw
- Agricultural waste, straw, plant debris: giant macrocybe, oysters (tropical ecotypes), paddy straw
- Sawdust: giant macrocybe, hairy panus, oysters (tropical ecotypes), reishi, scaly lentinus, shiitake (tropical ecotypes), turkey tail, wood ear

TIME TO MATURATION AND YIELD

Some mushrooms fruit rapidly and produce a significant yield in a short time. Others take considerably longer and may only produce a fraction of what other mushrooms can produce. Of course, mushrooms that take longer and yield more poorly generally demand more value at market, since it costs more to grow them, by simple virtue of the fact that, for an equivalent harvest, they take up more space in your growing operation, and for a longer period of time. (If you're pricing your harvest for sale, be sure to take into account not just the difficulty of growing a particular mushroom but also the time and space you must dedicate to it.)

INFRASTRUCTURE

I address method and infrastructure together because one often influences the other, and vice versa. Each method of cultivation has its basic needs and limitations, with environmental factors ranging from completely natural outdoor cultivation (which you might use for shiitakes) to tightly controlled indoor environments with extremely precise gas exchange requirements, such as cultivating maitakes indoors on sawdust. The best strategy is to start with cultivating mushrooms using whatever basic equipment and infrastructure you have already available and a method that matches your skills and goals.

Begin your cultivation efforts by surveying the machinery, vessels, fuel sources, water sources, and growing substrates available to you. If logs and tree debris are all you have, you may be limited to species that can be grown outdoors, locating the fruiting area in a shady spot and planning for just minimal inputs to control any aspect of the fruiting cycles. If you have steel barrels and dried, shredded plant debris, then cultivating oyster mushrooms could be your best option.

If you plan to cultivate mushrooms indoors on a small scale, you might simply use an extra room, closet, or bathroom as both your colonization room and your fruiting room. You can promote selective fruiting by misting only the mushroom containers that are ready to fruit by hand and placing a humidity tent over them. This can be as simple as a large, clear plastic bag poked with holes and draped loosely over the forming mushrooms. Some growers create humidity-controlled "fruiting chambers" by wrapping modular shelving in clear plastic sheeting, although these must be ventilated with holes both near the top and at the base to allow the heavier carbon dioxide the mushrooms produce to slowly seep out, pulling fresh oxygen into the top.

For larger operations, a more uniform and controlled environment is ideal, where the grower can fine-tune the gas exchange and humidity requirements for each stage of development. Commercial farms that devote separate rooms to every batch of mushrooms maintain a higher carbon dioxide level up to pinning, introduce fresh air throughout the mushroom maturation process while maintaining high humidity, and then cut back on watering altogether during the resting stage, making the process highly efficient and extremely predictable, which is important when you're counting on calculated yields on specific schedules for buyers.

Whatever your infrastructure, you'll want to stage your operation in a way that minimizes the number of steps—steps of the process and actual physical steps. How many steps you need to take in order to run and maintain your operation can add up and become a surprisingly important factor in how easy, or difficult, it is to accomplish your goals. If you use equipment such as shredders, conveyors, or mixers, consider positioning them in a sequence that resembles an assembly line. Organizing equipment and rooms in sequence, in the order in which you'll use them, will optimize work flow and reduce unnecessary steps and costs, while maximizing the use of space. Remember, automation magnifies your problems as well as your strengths, so eliminate any unnecessary work in your system before setting it in motion.

If you are, or have dreams of becoming, a commercial operation, cost will likely factor into whether you modify an existing structure or construct something new. Many commercial growers locate and then modify existing spaces that have some of the characteristics they need. Old barns, chicken houses, indoor hog farms, warehouses, and textile mills can be surprisingly well suited for mushroom growing with some minimal modifications. Even if all that remains of the building is the frame and a concrete floor, you may find that recovering it can be cheaper than constructing something new.

The structure you use for a fruiting room must be able to withstand high humidity, so a wood-framed structure will not be immediately suitable (although it would be fine for a lab or office). To retrofit a wood structure, the fruiting space should be sealed with a waterproof covering. Plastic wall panels and waterproofing paint with low VOCs are commonly used; they not only protect the wood structure and any insulation against water damage but also make periodic wash-downs and cleaning much easier.

The fruiting room walls and ceiling should be white in color, so that they will be highly reflective and help suffuse the room with light. As discussed in chapter 2, the lighting itself should be natural, fluorescent, or LED in the blue-green wavelength. It does not have to be bright, but it should bathe the entire fruiting room with light. As a rule of thumb, if you can comfortably read a newspaper or book at arm's length, there is enough light for healthy mushroom maturation.

In an indoor fruiting room, you'll also need to control the gas exchange—venting carbon dioxide (produced by the mushrooms) and bringing in fresh air—to maintain high oxygen levels. Position incoming air ducts and vents high in the rooms so that fresh air is evenly distributed throughout the fruiting space. All air that comes into the fruiting room needs to be humidified if you are conditioning it (cooling or heating it). Exhaust fans should be much lower to the ground, to target the heavier, carbon dioxide–rich air that needs to be expelled, the fans should be waterproof. You can also place fans at intervals throughout your fruiting room in order to evenly distribute fresh air. This keeps the gases from layering, which can cause mushrooms growing at lower levels (on the lowest shelves of racks, for example) to suffer from lack of oxygen, which will be quite evident from their stem length, compared to the stems of mushrooms fruiting on the top shelves.

If you want to get really sophisticated, consider pretreating the incoming air by channeling it through a plant greenhouse for the added oxygen, or percolating the exhaust back into an algal tank, which will create oxygen while producing a valuable and oil-rich biomass that can be used for biofuel.

If you are cultivating multiple mushroom species in a large space, it can be beneficial to partition the space into smaller, individual fruiting rooms with separate air inputs to customize the gas exchange, air temperature, humidity, and light if the mushrooms have different requirements. You can, of course, combine multiple species in the same growing space if they require similar growing conditions. But a larger space, even if you are growing just one species in it, can be problematic because diseases spread more easily through larger open spaces. Converting these spaces to smaller, modular rooms with solid or movable partitions can reduce the movement of molds, insects, and bacteria. (See chapter 9 for more details.)

For cultivation in tropical climates, heat is the enemy, especially during the colonization process when there is a heat spike from the mycelium itself. In these regions I encourage earthen walls or coverings, which offer splendid low-tech insulation against heat transfer (for example, you could invest in a shipping container and bury it to make a mushroom "bunker"). Reflective foil shade cloth and roof coverings can also reduce the heat transfer. In colder latitudes growers should be thinking the opposite, by covering their structure in black fabric or other heat-accumulating materials.

Your Strategy

As you start out in mushroom cultivation, your strategy should be to identify a mushroom strain that is compatible with your climate, is able to grow on substrates available in your region, has a rapid production or flushing cycle, and requires minimal cost and infrastructure to sustain harvests consistently. Even if you have access to higher-tech equipment, I still advise beginners to tackle the basic cultivation methods first so that they can learn the subtle needs of every species before automating a process. Ultimately you will develop a sense of intuition that will guide you in predicting outcomes or making calculated risk-versus-reward decisions about cultivating a new species of mushroom or using new materials or equipment. So remember: Start small, gain skill, then go for it.

CHAPTER 4

Choosing, Handling, and Storing Spawn

Spawn is simply a form of mushroom mycelium that is ready to plant into a growing medium, such as straw, sawdust, or compost. The spawn is broken up into pieces and scattered in the growing medium. Each piece begins to produce hyphae, thread-like strands that radiate outward until they meet an oncoming hypha from another spawn piece. At this point the colliding mycelia recognize each other as copies of themselves and fuse together, increasing the size of the spreading biomass. The process continues until the substrate has been fully colonized.

This phenomenon is strain-specific. Even if you use spawn from the same species of mushroom, the mycelia will only fuse if they are genetically identical. Two non-identical strains will remain independent of each other, leaving a small uncolonized zone between them.

Spawn can be either collected wild or purchased, in the form of dowels, grain, and sawdust. Different cultivation techniques and growing media work best with different spawn types; part 4 provides specific details for all the mushrooms profiled there.

Although most spawn is generated in commercial or university laboratories, some individuals have successfully invested in small-scale equipment for creating their own spawn. For more information on experimenting with laboratory methods and producing cultures from mushrooms to make your own spawn, refer to chapter 18.

Given the cost of spawn—approximately $2 to $4 per pound—many growers opt to buy their spawn in small quantities and then expand it, or multiply the mycelium to make more spawn, as efficiently as possible to obtain more yield for the cost of spawn. Using spawn to produce more spawn can be tricky, since the quality and purity need to be maintained, and it's easy to expand contaminants as well as desirable mycelium unless you're able to purify the culture in a laboratory. It can be done (again, see chapter 18), but you're just starting out, I recommend using pure cultured spawn to increase your chances of success.

Choosing Spawn

As you know by now, cultivators need to make a number of decisions based on their given situation and their goals. Just as you need to decide which species of mushroom to grow and what kind of substrate to use, you also need to make decisions about spawn. There are many kinds of spawn, so it's important to understand what type of spawn you should use in a given situation and, more importantly, why.

Plug Spawn

The most common form of spawn for beginning cultivators growing outdoors on logs and stumps is plug spawn. The plugs are usually made of hardwood, most commonly birch, and are, universally, 5/16 inch in diameter and about 1 inch in length. Typically when you buy them, they have been boiled and sterilized in an autoclave for several hours, allowed to cool, and spawned with a pure culture in a laboratory under sterile conditions. The plugs typically have grooves that run the length of the dowel or, preferably, a spiral groove that wraps and descends around the plugs, which harbors fluffy mycelium in the minute trench that quickly colonizes the drilled hole after being hammered in.

Many growers make their own plugs by purchasing bulk grain or sawdust spawn and transferring it to wood plugs. To make your own plugs, purchase wooden dowel pins that are approximately 1 inch in length, soak them in hot water to hydrate the dowels, sterilize them, and then add a grain spawn culture. Some growers have had success expanding commercial spawn plugs into a larger volume of their own

If you buy plug spawn, look for mycelium growing in the grooves as a sign of life. This is a good sign. Your plugs may lose their "fuzziness" after being shipped but will recover in a few days at room temperature.

sterilized plugs, but be aware that contamination can magnify every time you expand spawn. Just remember to experiment on a small scale before committing to a process. Most commercial spawn suppliers make plug spawn in labs, but you can attain some success by using fresh spawn cultures and proper plug treatments.

Sawdust Spawn

Sawdust spawn is the choice of many commercial growers because it is typically the least expensive and most practical form of bulk spawn for a wide range of substrates. It also resists contamination and does not attract molds and insects the way that grain spawn does, since it is less nutritive. It also stores well and takes a long time to over-incubate and become unusable, which makes it ideal for long-distance shipping.

Sawdust spawn is usually not just a bag of colonized sawdust, but rather a blend of hardwood sawdust and various nutritive supplements, such as wheat bran, rye grain, powdered calcium, and other proprietary ingredients. The formula is designed to have many more colonizing particles than grain spawn, while providing additional nutrients from the supplements.

If you are using the sawdust spawn to cultivate mushrooms outdoors on logs and stumps, sawdust spawn inoculators (or plungers) are handy and can be had from spawn producers for typically about $30 $50, depending on the model and source. A plunger is basically a small-diameter, spring-loaded pipe that pushes and packs sawdust into the holes you drill into your logs or stumps. In twenty years, I have bought two of these inoculators, but I prefer to use an inexpensive funnel and dowel to inoculate wood—not only because I don't want to spend the money on an inoculator, but also because I find it simpler.

If you go this route, you'll want a small funnel that fits into the holes you drill; these holes should be slightly larger than standard plug spawn holes (typically 5/16 inch), to accommodate the slightly bigger tip of the plunger or funnel (usually 3/8 to 1/2 inch). And you'll want a dowel with a diameter about 1/4 inch smaller than that of the holes, for tamping in the sawdust. (If you're going to be filling a lot of holes, you

Sawdust inoculators can range from a fancy tool to a funnel and a wooden dowel. I prefer the funnel for cost and ease of use (with practice).

may want to affix a comfortable handle to the dowel.) Position the funnel over a hole and fill it with what looks like enough sawdust spawn to fill three or more holes—you'll be packing the spawn in tightly. Use the dowel to tamp down the sawdust spawn, sliding it in and out of the funnel at all angles.

Grain Spawn

Grain spawn is exactly what its name suggests: mycelium that has been carefully grown in a laboratory on sterilized cereal grains such as wheat, millet, milo, or rye. It is sold in jars or bags fitted with a filter that allows it to breathe while keeping contamination out. It has a rapid "leap off" rate, drawing on the rich bank of nutrients in the grains to propel the mycelium outward in search of other colonizing grains of spawn that you have distributed throughout a growing medium. Grain spawn can be more expensive than sawdust spawn for inoculating bulk substrates, but it provides important nutritional supplements to the growing medium. The grain size is important; the smaller the grains, the greater the volume of growing medium the spawn can colonize, and the faster the colonization process will be.

Grain spawn spoils easily, usually from a bacterial contaminant that makes it smell like apple cider or musty grapes, and is best kept either cool (for tropical species) or under refrigeration for most temperate mushroom species during storage or shipment. Grain spawn is susceptible to insect and mold contamination, so it is best suitable for indoor cultivation in controlled, clean environments on pasteurized or sterilized media,

Grain spawn is usually more expensive than sawdust and plug spawn, but it supplements the medium with nutrients from the grain.

Transplanting wild stem bases and mycelia can provide beneficial microbes that some mushrooms need to trigger a fruiting response.

and should never be used in outdoor beds that will be in contact with soil or for wood log or wood chip cultivation, where insects can find and eat it before it has a chance to spread forth mycelium. One exception to this rule is outdoor beds of manure susbtrates that have been steam-treated or composted to a high heat. These can be inoculated with grain spawn effectively, along with paddy straw cultivation using low-tech techniques.

Naturalized or Wild Spawn

Spawn that has been transplanted or grown from the bases of wild mushrooms is called naturalized or wild spawn. Although using natural spawn can often work in a pinch when laboratory spawn is unavailable, you wouldn't want to use it if you were running a commercial operation. It has benefits, but also potentially disastrous disadvantages for a large operation. The downside is that naturalized spawn comes along with millions of microbes. If you've foraged mycelium from the wild, how do you know you are expanding the right type of mycelium, when most mycelium is white and indistinguishable? You could be propagating an unwanted or even poisonous species.

One advantage, however, is that the mycelium most likely contains microbes that stimulate fungal metabolic function and contribute to greater fruiting. Mushrooms such as king stropharia, blewit, almond portabella, parasol, and a few others have a greater fruiting response when there are specific microbial communities present that help trigger mushroom production. And some species of mushrooms, such as king stropharia, have a distinct strand rhizomorphic mycelium that is easily identified in wood chips. Harvesting the myceliated substrate with the parent mushrooms attached can help verify the identity of the colonized biomass you intend to use as spawn, and can also lead to some great discoveries that help promote fruiting. You can make a slurry out of the stem bases by pulverizing them in a blender and adding this mixture later to your outdoor patch. (See chapter 20 for more information on microbial slurries with wild-harvested spawn.)

HANDLING SPAWN

Spawning methods vary for each type of mushroom and for each method of cultivation. The following chapters provide more specific details for using spawn when you're cultivating on logs, stumps, or wood chips (chapter 5), on compost or livestock waste (chapter 6), or on pasteurized or sterilized media (chapter 7). And of course the mushroom profiles in part 4 describe the idiosyncrasies of particular species and strains. But regardless of what kind of spawn and cultivation method you choose, it's helpful to know how to handle spawn prior to inoculation.

When you're working with nonsterile substrates (that is, untreated or pasteurized substrates), the only sanitary requirements are that you use clean equipment. That includes your hands; you should

A handful of king stropharia (*Stropharia rugoso-annulata*) spawn ready to sprinkle onto a newly constructed wood chip patch layered with cardboard.

wash them and them rinse them twice with rubbing alcohol (70 percent isopropyl alcohol). You can wear rubber gloves if you like. Sterilized substrates should never touch anything unsanitized; you'll pour the spawn directly from their bag into the sterilized growing medium or mixing container. The reason for this is that sterilizing a growing substrate removes all the organisms, both those that are aggressive competitors and those that are considered beneficial, which means that any contaminants that come into contact with the substrate, whether from the air or from unsanitized tools and containers, have a wide-open field, theirs for the taking, which could result in a contaminated growing medium. This is why pasteurized substrates are more forgiving than sterilized substrates; nonaggressive species that survive the pasteurization keep the substrate from being taken over by competitor microbes, allowing the mycelium a window of opportunity to begin colonization with no obstacles.

Sometimes, during shipment and storage, spawn gels into a solid block. Plugs that have stuck together are relatively easy to separate; just pull them apart with your fingers when you need to use them. You'll find it easier to break up a solid block of sawdust spawn while it is still sealed up in its packaging, rather than after you've opened it. Just lay it on its side on a solid surface and strike it a few times with your fist or a mallet to break it into chunks. Then use your fingers to squeeze and rub the chunks apart, until they are well separated. Your aim is to get the spawn broken down to the smallest particle size possible. You can treat grain spawn the same way, though you want to be a bit careful not to smash the grains into mush.

Even after I think I've thoroughly broken down the spawn, I tend to find small chunks that I missed after I open the bag, so I try to continue separating the spawn even during inoculation, as I spread the spawn into the growing medium. The more that you break it apart, the greater the number of colonizing particles, or individual colonies, you will spread through the growing medium. I sometimes think of these particles as miniature lost islands looking for other islands. They will eventually find each other and colonize all the territory in between, joining into one big landmass.

If you receive your spawn and it is broken apart, that is fine. Spawn sometimes gets shaken up and separated in shipment. If you are suspicious of the spawn's quality, leave it out for a few days at room temperature to see if the spawn recovers and begins to recolonize and bind the particles back together with the color of mycelium typical of the mushroom, which can vary from white and thread-like for most mushrooms, to orange and powdery (chicken of the woods), to brown and furry (morels). If it does, simply store it as described below (and depending on the species). If it doesn't, throw it out and ask for a refund.

It's important not to use too much or too little spawn during inoculation. Although using too much will speed colonization and limit competitors and contamination, it's not cost-effective. Use too little and the spawn run is slower, which opens a larger window of opportunity for problem species to move in. Inoculation rates vary on a case-by-case basis, and you will need to experiment to find the right amount based on your method of cultivation and chosen mushroom species. The colonization speed and resiliency of the species you choose will of course play a major role in determining your inoculation rate. The nutritive value of the spawn is also an important factor; grain spawn, for example, is much more nutritive than the same amount of sawdust spawn. What the sawdust has, in contrast, is possibly more surface area and points of inoculation, making the mycelia more likely to find each other quickly.

Never leave your spawn in full sun. Spawn is generally sealed in plastic bags, and direct sun can increase the temperature inside the bag to well above 110°F (43°C). At this temperature, it doesn't take long to "cook" your spawn, destroying the cells and rendering it useless. When it's not in cold storage, take the precaution of keeping your spawn in a cool or at least shady area. When I'm inoculating logs or spawning beds of wood chips outdoors, for example, I keep my spawn under a table, out of direct sunlight.

Shipping and Storing Spawn

Mushrooms vary not only in the temperature at which they prefer to fruit, but also in the temperature at which their spawn needs to be stored. Part 4 gives spawn storage and fruiting temperatures for each mushroom. Most species are temperate; everything that fruits at less than 80°F (27°C) can presumably be stored in a regular refrigerator or walk-in cooler for many months at temperatures ranging from 38 to 45°F (3–7°C). But since tropical mushrooms have no experience with colder weather, they are vulnerable to refrigeration and should be stored at milder, or even room, temperatures; the paddy straw mushroom, for example, fruits at 90°F+ (32°C+) and is ideally stored at 60°F (16°C). If there is no power available, storing your spawn in a root cellar or even placing the spawn outside during cooler or cold months is a reasonable alternative.

Many spawn companies provide storage information when you place an order. Try not to order tropical species if you are shipping to a freeze-prone region; their sensitive spawn might have to sit for days in frozen trucks and unheated warehouses. If you have any concerns, contact your spawn supplier. With time and experience, you will remember which times of the year are best for shipping a few of the temperamental species.

I don't recommend freezing your spawn, although many mushrooms can colonize, survive, and grow in freezing temperatures. But in general, frozen spawn does not recover well and dies easily due to the crystallization of water in the cells, which makes them burst. I have tried to freeze spawn experimentally using several species, but they usually only last a day or so. Freezing laboratory stock cultures, such as isolates that have known purity and are for the sole purpose of expansion, is a different story, and parameters are discussed in chapter 19, but even in that case you'd need to use a cryoprotectant, such as glycerol, which replaces water within the cellular matrix, to protect the cells from bursting as they come in and out of crystallization.

Keeping Records of Your Spawn Usage

You will find it helpful to keep a production journal for mushroom cultivation projects, where you can record the details of each batch. This journal should include your dates of inoculation, growing media (composition or formulas), methods used, container type, hole spacing, weights, spawn type, spawn origination codes (if available), amount of spawn used, and any other information that could vary from batch to batch for your operation. If you plan to run experimental trials, you will definitely need to keep records in order to compare yields and improve production. This is very typical of both small and larger commercial operations.

Plug spawn typically does not have laboratory tracking codes since it is used primarily by beginner and small commercial growing operations, but sawdust and bulk spawn cultures usually do. The lab tracking codes are assigned by your spawn supplier in order for you to be able to trace the age and origin of the cultures, or spawn generation batch numbers, that you have received. These codes can be helpful if you are having either poor yields or bumper crops, since they allow you to match the strains and eliminate or identify spawn quality as a variable. If you are having good luck with a particular strain, your spawn company can check its records and see which strains you purchased to help you formulate your growing operation. Likewise, if you have an issue with contamination or lower yields, the codes can be used to trace contaminants and determine whether the supplier is accountable. Good communication with your spawn producer will improve your growing skills and help you reach your goals.

CHAPTER 5

Cultivating Mushrooms Outdoors on Logs, Stumps, and Wood Chips

Cultivating mushrooms outdoors is one of the simplest and most natural forms of mushroom cultivation. It dates back thousands of years and can be applied just about anywhere in the world, using local trees and wood debris, so it is an excellent place to start. However, since you are at the mercy of a fluctuating climate, including temperature and rainfall, you will likely have somewhat varied yields rather than the precise, calculated harvests that come with indoor cultivation, where you can control growing conditions and better streamline your processes. In comparison to the rapid flushing cycles of indoor fruiting, outdoor patches are slower to produce, but they can eventually fruit several times a year for many years, and they require minimal labor and cost (compared to the infrastructure of indoor cultivation), making them well worth the investment.

Log and Stump Cultivation

People have been growing mushrooms on logs and stumps for hundreds, possibly thousands, of years. It is one of the simplest ways to cultivate edible and medicinal forest mushrooms such as shiitakes, maitakes, and oysters. In fact some mushrooms, such as maitakes, are easier to grow outdoors under natural conditions than indoors.

Outdoor log and stump cultivation is generally low-tech, but when practiced on a large scale it requires a lot of initial labor in getting the logs or stumps set up. For this reason the log-grown mushroom industry has developed more quickly in countries where labor is plentiful and inexpensive and the technologies required for indoor cultivation are limited. Overall, I don't recommend outdoor log and stump cultivation for commercial enterprises. It is better suited for home-scale and hobby growing, or as supplemental income for a small commercial enterprise.

Cultivating mushrooms on stumps is similar to cultivating them on logs, although if you are a beginner, I recommend starting with logs, since stumps can take many years to fruit. (Once they start, they can fruit for a decade or more.) You use the same inoculation techniques, such as drilling and plugging with dowels or sawdust spawn. The main difference is

THE IMPORTANCE OF IDENTIFICATION

How do you know the mushrooms you harvest from your outdoor beds are the species you spawned and thus safe to eat? When you're cultivating outdoors, chances are good that spores from a few competitive species will land and colonize small areas of your growing substrate. As an outdoor mushroom grower you will need to pair your cultivation skills with a few basic identification practices, such as taking spore prints and learning the basic anatomy of mushrooms, so that you can differentiate the good from the bad. After you inoculate your mushroom gardens, it will take several months for them to fruit, giving you plenty of time to redirect your energies toward identification skills. The good news is that cultivated mushrooms are relatively easy to identify, especially with the help from your spawn supplier and field guides specific to your area.

Never identify mushrooms from an online search. Many of the images that are posted on websites and in online photo galleries are mislabeled or incorrectly identified. My advice is to photograph your primordia, or baby mushrooms, as your mushroom gardens begin to fruit and send the photos to your spawn supplier for verification. Take the time to watch the mushrooms as they develop and note the progression of color change and other features as the mushrooms fully develop. Once you verify the identity of the mushroom and have watched it grow throughout all of the stages of development, you will have no problem identifying future harvests of that same mushroom. I also encourage you to join any of the vast network of mushroom clubs and foraging enthusiasts who can help with all aspects of mushrooms for food, fuel, and fiber. For a complete list of mushroom clubs in North America, visit the North American Mycological Association website at www.namyco.org.

that you plug as much of the visible barked wood that remains aboveground as possible and you also spawn the outer ring of the top, cut face of the stump. Many mushrooms, such as maitakes, chicken of the woods, reishis, enoki, oysters, and beefsteaks, benefit from stump rather than log cultivation. (Part 4 details the particulars for each species.)

What Kind of Wood Can I Use?

Generally, hardwood species just about anywhere in the world are suitable for mushroom cultivation. In temperate climates, deciduous trees are a good choice. Although cultivators often overlook invasive and non-native tree species, there's no reason not to capitalize on their availability, especially since using them can help you manage their presence on your land or in your community. Different strains of mushrooms vary in their ability to break down or fruit on specific wood types, so if you have a question, ask your spawn supplier if a type of wood you have available is appropriate.

Evergreens in temperate climates are typically conifers, many of which do not support mushroom production, with the exception of a few species: cauliflower (*Sparassis crispa*), tufted conifer lover (*Hypholoma capnoides*), and a few strains of chicken of the woods (*Laetiporus conifericola*). Some mushrooms grown on conifers can have a sap-like flavor and may cause gastric upset, compared to those strains cultivated on hardwoods, so I tend not to eat many mushrooms found on conifers, cauliflower mushrooms being the exception.

Be sure your logs or stumps have the bark intact all the way around, and try to avoid tree species that drop their bark when cut, such as elms. Mushrooms form between the inner and outer layers of bark. The mushrooms use this small space as a nursery where the primordia can safely develop a defined cap and stem base before they're exposed to the elements. As the primordia grow, they swell with water and break through the outer bark. Although this seems like an

If you are using plug spawn, match the depth and diameter of your holes to your plugs; for most plug spawn on the market, you'll be drilling a hole 1 to 1½ inches deep with a ⁵⁄₁₆-inch bit. If you are using sawdust spawn, you may need a slightly larger diameter depending on the tool you use to insert the sawdust.

Tap the plug spawn in gently. Watch those fingers!

Cover the plug with wax to preserve the moisture and keep contamination and insects out. This gives the mycelium enough time to drill its way into the cavity and begin colonization.

A beautiful, meaty shiitake emerging form the bark surface eight months after inoculation. These logs will fruit for five to eight years!

A basic log and stump drilling kit should include a drill bit, wax, and a small brush for applying the wax.

unbelievable feat for a seemingly small, tender organism, some mushrooms can lift or split materials with a pressure greater than 80 pounds per square inch.

Split logs, such as firewood, can support mushrooms on their bark-covered sides, but they tend to dry out more readily than fully barked logs. If all you have is split wood, it may be best to bury the split side or at least rest it on the ground. This forces the split side to act as a water wick. It also makes for a compact fruiting area; since mushrooms will not fruit from the undermost side of a buried log, all the fruiting will take place on just the upper surface, which will explode with large numbers of mushrooms.

The best logs have thick bark with a lot of texture, which helps trap moisture and encourages the hydration of the mushrooms developing just below the bark. These small crevices and valleys are the perfect microclimates for superior mushroom formation.

How Fresh Does the Wood Need to Be?

Your logs and stumps should have been cut no more than a month ago, although the density of the wood can be a factor in the amount of time you have

before the wood becomes contaminated with wild fungi. Harder woods typically have a longer window before they become infected by wild fungi, whereas the softer woods have considerably less. Considering the variability of wood types and contamination risk windows, I prefer to simply use the freshest wood available so I don't have to wait and wonder if my money and effort were worth it.

That said, some tree experts say that after wood is cut, it remains "alive" for about a week, with its natural antifungal defense system still actively rejecting the growth and formation of fungi. Delaying inoculation for that period of time allows the inhibitory compounds to dissipate. After about a week, the mycelium will encounter no competitors or obstacles as it leaves the confines of the plug or sawdust spawn and drills its way into the wood.

When Is the Best Time of Year to Cut and Inoculate Logs?

During the spring and summer trees put most of their energy into photosynthesis, so in temperate climates wood from deciduous hardwoods will have more sugars available to the mycelium if you cut and inoculate them in the winter months, after their leaves have fallen. In tropical climates, where trees tend to be evergreen to semideciduous, it's more important to match up the kind of wood with the species and strain of mushroom you want to grow than to cut at a particular time of year. In these warmer regions there is generally little variation in the movement of tree sugars from a seasonal standpoint, and year-round inoculations are possible. Palm trees and their fibers do not make good cultivation material compared to hardwoods and are generally resistant to rot; however, some growers are using palm coir to replace peat moss in casing soils for indoor cultivation. (See chapter 20 for more information on casing soils.)

Because winter is the best time to inoculate logs and stumps in temperate climates, temperature can be a concern. Cold tolerance varies among species and ecotypes, but in general, extremely cold temperatures and prolonged periods of temperatures less than 18°F (−8°C) can negatively affect mushroom growth. I still think it makes sense to inoculate this wood rather than leave it, even though the yields may be less, since the wood is fresh and available and it's better than no yield at all. If you're worried about cold temperatures, consider bundling the inoculated logs and covering them with leaves or other natural insulative material until warmer weather returns.

Log and Stump Cultivation Step-by-Step

Step 1. Cut your logs into lengths that are easy to handle and carry. The most common lengths are between 2 and 4 feet, depending on how much weight you, or the people who are helping you, can carry. And aim for a diameter for 6 to 10 inches. Since you may be picking these logs up often, choose a size that best suits your strength and abilities.

Step 2. Next you will need to drill holes into the log at regular intervals at a depth of about 1 to 1½ inches. Most plug spawn on the market call for a ⁵⁄₁₆-inch drill bit, although this can vary; just match the drill bit to the plugs you're using. If you're using sawdust spawn, match the bit to the size of the packing tool you're using. A high-speed drill bit attached to an angle grinder with an adapter can be well worth the investment if you are inoculating a large number of logs. The higher rpm and speed makes drilling logs effortless compared to a regular electric power drill.

Drill your holes in a diamond pattern, all the way around the bark-covered surface of the wood, spacing the holes about 4 to 6 inches apart. Drilling and spawning more holes will not produce more mushrooms, but it will speed up the colonization process for earlier fruiting. I would not advise spacing your holes farther than 6 inches apart or you increase the risk of attracting spores from competing fungi. This doesn't "hurt" the mycelium you spawned, but it will reduce your yield by taking valuable real estate away from your culture.

Step 3. Insert spawn into each hole, tapping with a hammer if you are using plugs or tamping with a dowel if you are using sawdust. Fill the hole until the

Use a diamond drilling pattern all the way around the barked surface of the log. Closer spacing will speed up colonization, but I suggest spacing the holes 4 to 6 inches apart.

Another method of inoculation, which doesn't require a drill, is to make deep gashes in the wood with a chain saw or ax and fill the gaps with sawdust spawn or a cardboard culture (described in chapter 12).

Stack your logs in a shady spot for colonization. I keep them stacked tightly together to preserve moisture while they are colonizing and then restack them in this open-air log cabin style once they show signs of fruiting.

spawn is flush with the bark. Brush a small amount of melted wax over the spawned hole to seal in the spawn. Wax helps maintain moisture in the hole, supporting the mycelium as it drills its way outward from the sawdust or plug into the walls of the freshly drilled holes.

Some growers also wax the ends of their logs to lock in moisture, but that can also lock it out. Another option is to wax one end and not the other, allowing moisture to wick in the unwaxed side during rainfall, during an overnight soak, or when it comes into contact with the moist ground.

Step 4. Situate the logs in a shady area with dappled forest light, or on the shady side of a building if you're in an urban environment. Locate them near a water source for easy watering and soaking when necessary. The colonization phase can take six to ten months, depending on the type of wood, temperatures, and spawn spacing. Visit the logs periodically—at least once a month during the first six months. The wax you applied may disappear or crack and fall off, which is fine since the mycelium really only needs just a few weeks of protection after spawning. Dry conditions can slow colonization, so lightly water the stack once a week or so to maintain moisture if it has not rained. Pick up a few of the logs to feel the weight; if they are extremely dry and seem much lighter than they were at spawning, you may want to soak them overnight with water—but for no longer than twenty-four hours, or the mushroom mycelium can drown. Submerge the logs in water in a trash can, large plastic tote, or cheap baby pool. Use other logs or bricks to weigh down the logs, keeping them completely submerged. Another alternative is to mist them gently over a period of a week or more, to gradually allow the logs to wick in moisture. This option can waste water, however.

Step 5. About six months after inoculation, start inspecting the log ends once a week for signs of mycelium, such as white streaks on the open cut face, which signals that the mycelium is approaching full colonization and will soon be capable of fruiting. Optimum moisture is now more important than ever to ensure that the mycelium will be able to supply the growing mushrooms throughout the pinning and fruiting stages.

At this point, as the mycelia in the logs approach complete colonization, the stacked logs may begin to fuse together. Soak them overnight in water and then spread them out in a "log cabin" formation, or lean them against the trunks of living trees in a circular formation, with one end on the ground and the other resting against the tree.

Step 6. At long last, your mushrooms will begin fruiting. Logs will fruit two or three times a year, typically for as many years as they are inches in diameter, resting a few weeks in between each flush. Mushrooms may fruit near the spots where you set the spawn on the first flush, because it is an easy escape, but they can also fruit from just about anywhere around the entire bark surface. Once you notice baby shiitake mushrooms, which will split the bark and appear as dark brown to black buttons, mist them as much as you can; several times a day is ideal. A misting system is preferred for commercial log production, or simply locate your logs near your home to mist them in the morning and evening as you pass them by. If your logs have ground contact, check that area often, since they might direct their fruiting downward in the search for water. When the mushrooms have matured, it's time to harvest!

Keep records of the flushing patterns and establish a schedule for watering that coincides with those patterns; this will not only increase yields but improve your understanding of the mushrooms you are cultivating. Once mushrooms are colonized and ready to fruit, they will be waiting for adequate water to support fruiting. Time the watering for when the mushrooms have had enough rest and are fully charged to take advantage of the watering; this will give you the optimal fruiting with each flush.

Trenched Log and Log Raft Cultivation

Many mushroom species that thrive on wood fruit better on logs that are partially buried rather than stacked aboveground. The access to extra ground moisture, reduced fruiting surface area, and microclimate created by covering the logs with leaves or straw trigger and support prolific fruitings preferred by (but not limited to) reishi (*Ganoderma* spp.), nameko (*Pholiota nameko*), black poplar/pioppino (*Agrocybe aegerita*), brick top (*Hypholoma sublateritium*), and maitake (*Grifola frondosa*). There are basically two ways to do this: either you trench the logs, or you

When you see mushroom mycelium on the ends of the logs, the mycelium has fully colonized the logs and should soon begin fruiting.

A healthy flush of shiitake mushrooms.

Trenched log cultivation.

Log raft cultivation.

build a "raft" of them and cover it; they both produce the same results.

When you cultivate mushrooms on partially buried wood, the flushes typically only last for two to three years, so be prepared to build additional beds every two years to sustain a continuous harvest; that said, certain mushrooms can fruit heavily in that time span. I'll describe two methods; your choice will depend on your access to fresh wood chips. Whichever method you choose, first drill holes in your logs, insert plug or sawdust spawn into the holes, and allow them to colonize for three to four months aboveground; water these logs weekly for the first two months. Once the beds are prepared, you only need to water the logs and saturate the beds around the time that the mushrooms should fruit.

Trenching your logs means you'll need to excavate a space half the depth of your logs, long and wide enough to fit all the logs in tightly. After you place your logs in the trench, shovel the soil you removed back in between the seams and gaps and around the edges. Then water the logs so that the soil settles in firmly, leaving the clean upper bark surface exposed. (You could bury them entirely, but finding them again can be difficult, and you don't want to step on any mushrooms that may be hidden and working their way up through the leaves and mulch.) Place a stake or sign near your bed so you can mark where you have planted, and label it with the date and strain information for your records.

Log raft cultivation is similar to trenched log cultivation, only it takes place aboveground. Lay down a thin layer of wood chips, mulch, sawdust, or soil and then wiggle the logs into place so that they lie tightly together in a row. Fill in the seams and gaps, build up

Using the raft or the trenched log method, you can produce an overwhelming number of fruitbodies in a single flush. All the mushrooms on this entire colonized raft are considered one organism, since the mycelium fuses and connects.

A wood chip bed.

a layer of substrate all around the edges of the bed, and then water heavily. Wood chips are especially helpful for log raft cultivation; they give the bed an extra food source and help retain water for the developing mushrooms when they fruit.

Once the logs have been trenched or set up in rafts, water them once a month. When you begin to see fruit, mist them daily.

Wood Chip Cultivation

There are many popular and reliable edible mushrooms that people cultivate outdoors on wood chips, many of which incorporate beautifully into fruit and vegetable gardens. For example, king stropharia (*Stropharia rugoso-annulata*) is suitable to most climates and loves fresh-cut hardwood chips, preferably the heartwood or sapwood, rather than the outer bark layers commonly sold as mulch. Some species of terrestrial wood-inhabiting mushrooms need a thin (½ inch) layer of soil or compost on top of a wood chip bed to supply symbiotic bacteria or a different habitat interface to promote fruiting. (For all the details on particular species, see part 4.)

Wood chip beds can be spawned year-round, except in extremely cold climates, where you would want to spawn your bed in the spring to let the mycelium get well established and then mulch it in the fall to provide an additional layer of insulation and food for the following year. Regardless of when you spawn your bed, when winter approaches, cover any summer-fruiting strains with leaves and straw to insulate them from the freezing weather. Remove the covering before the mushroom's fruiting season arrives, but don't rake it all away; you need a thin layer to preserve moisture.

Wood chip beds generally fruit for one to two years. If you add more wood chips at the end of the year, you can extend the life span of the beds for several years, but I have found that worms often infiltrate the beds and consume the mycelium in the second year, so I encourage scooping out the beds and refurbishing them completely if no visible mycelium is detected with a little digging around. You can also simply add more cardboard to the existing bed and start the process again on top of the existing bed, which will build your soil and add inches per year of incredible compost to your gardening areas. Or, to perpetuate your operation, you can harvest the best-looking mycelium from select regions of the bed and move it into fresh chips. Whatever you choose to do, never add more chips to a bed when your mushrooms are about to fruit or it will delay fruiting, since the mycelium will become more interested in colonizing the fresh substrate.

Wood Chip Cultivation Step-by-Step

Step 1. For a tidy wood chip bed, build a frame out of hardwood logs that are about 6 to 8 inches in diameter. (You can inoculate the logs themselves with mushroom strains that prefer log cultivation, like reishi.) If you don't have logs available, you can use untreated lumber. Or just clear an area; you do not necessarily need edging or formal borders. The mushrooms will need some shade, so locate the site in a shady area or between rows of vegetable plants.

Step 2. Cover the bed with cardboard from flattened boxes. Water the cardboard until it is saturated. Then sprinkle sawdust spawn sparingly onto the cardboard, casting it in small islands, so the mycelia pieces can branch out and find each other to join forces.

Step 3. Add 2 to 3 inches of fresh hardwood chips. Scatter sawdust spawn over the wood chips. Rake the chips to even out the surface of the bed, and pack them down to get rid of air pockets. Sprinkle the bed lightly with water, enough to moisten the chips.

Step 4. Cover the wood chips with another layer of cardboard. For this layer, you'll want smaller pieces of cardboard to allow water to pass through to the wood chips. A single layer of newspaper will also work. Sprinkle another layer of spawn "islands" onto the cardboard or newspaper. Sprinkle lightly with water.

Step 5. Repeat steps 3 and 4 until the beds are at least 6 to 8 inches deep. Cover the bed with 1 to 2 inches of straw or leaves to preserve moisture and shade the chips.

Step 6. Water your bed every day for the first week. Water it every other day for weeks two through four, and once a month thereafter (unless it receives sufficient rain). After four to eight months the mycelium will have spread throughout the chips and penetrated the surrounding soil. Check back on your patch often! Mushrooms grow extremely quickly once they start fruiting. Your patches may flush several times a year, during the temperature window for the mushroom you are cultivating.

Sprinkle sawdust spawn sparingly onto a layer of wet cardboard.

Add fresh wood chips to a depth of about 2 to 3 inches and scatter spawn on that as well.

Using a rake, even out the chips. This will help integrate spawn into the wood chips.

In a few weeks, you can poke around and sneak a peek at the developing mycelium streaming outward from your spawn, as it enjoys its new home and food source.

King stropharia fruiting on a wood chip bed just three months after inoculation.

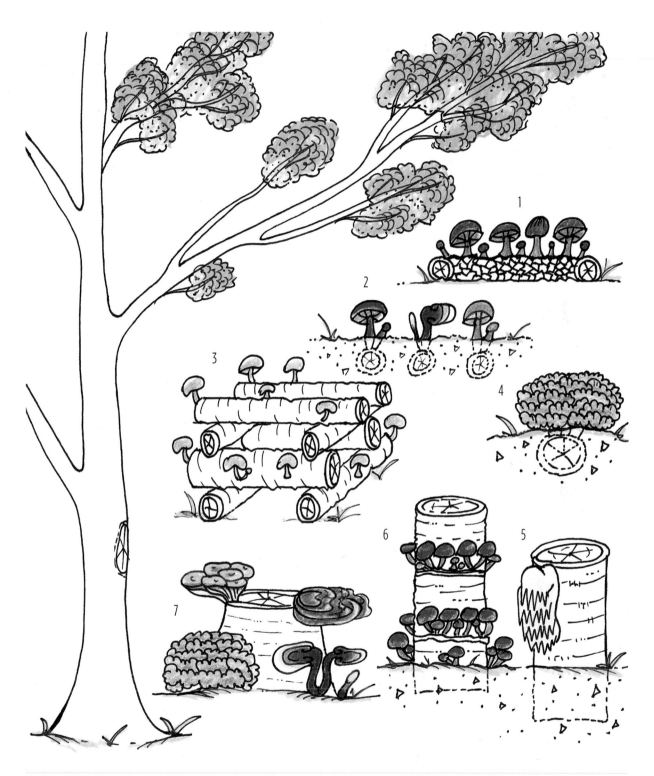

Using the whole tree from leaf to stump in an outdoor mushroom cultivation operation. 1. Wood chip bed: blewit, king stropharia. 2. Log trench cultivation: reishi, nameko. 3. Log cabin: shiitake. 4. Buried rounds: maitake. 5. Planted totem: lion's mane. 6. Cardboard stack: nameko, black poplar, oyster. 7. Stump: maitake (*bottom left*), reishi (*bottom right*), oyster (*top left*), chicken of the woods (*top right*).

CHAPTER 6

Cultivating Mushrooms on Compost and Livestock Waste

You can compost many types of organic waste, including manure, to produce a substrate that can support mushroom production either indoors or outdoors. (For certain of these substrates, such as manure, I advise only bringing them inside for cultivation if they are pasteurized or sterilized.) Different species of mushrooms possess different enzymes and methods of accessing the nutrients, so the yield can vary greatly depending on what materials, and in what amounts, you use. White button and portabella (*Agaricus* spp.), shaggy mane (*Coprinus comatus*), and blewit (*Clitocybe nuda*), for example, love to grow on manure-enriched compost. Although the availability of manure can be limited if you don't live on a farm, a determined homeowner can usually find a source to mix in with grass clippings and organic garden matter, which is ultimately what the mushrooms usually thrive on anyway: a complex but digestible matrix of organic debris. The best time to compost is in warmer months, so start in the spring if you live in temperate climates; in the tropics you can begin any time of year.

There are two methods for growing mushrooms on composts that I'll cover here. The first is for small-scale backyard cultivation that will produce a compost of grass clippings, garden debris, and other common yard waste (Harris, 1989). The second method will focus on commercial-grade compost involving ingredients (a more complex and carefully calibrated compost) and processes (pasteurization) that may be more involved than most home cultivators are looking for. Regardless of which method best fits your situation, I encourage you to read through both descriptions, because you may find a way to hybridize these techniques in a way that suits your particular circumstances.

METHOD ONE: A SIMPLE COMPOST PILE USING AVAILABLE MATERIALS

Designate a space you can use for three to four weeks and big enough for a compost pile 5 feet square and 3 to 4 feet high. Build a standard compost pile, alternating layers of green organic matter, such as garden and lawn clippings, and brown organic matter, such as dried leaves and chopped straw. (Do not use any organic matter that has been treated with pesticides.) Add layers of fresh or partially composted livestock manure if it's available. Water the pile as you layer—and if you are into marking your mushroom territory, add urine; the nutrients and urea are beneficial in the first two weeks. Once or twice a week turn the pile, mixing and shifting

Agaricus subrufescens fruiting on manure-enriched compost at Sharondale Farm in Virginia. *Photograph by Mark Jones, Sharondale Farm.*

the layers from bottom to top, middle to bottom, and top to middle if you can. This is laborious but well worth it; as the mixture becomes more homogeneous, the nutrients and composition become more consistent.

As the materials in the compost decompose, the pile will generate a large amount of heat (up to 140°F/60°C), depending on its dimensions, how well you've chopped up its components, and the aeration of the pile. You can use a digital thermometer to monitor the compost temperature and thus its progress, if you like (if you do not have a long probe, you can simply dig down into the center, quickly take a reading, and then push the material back). There are many variables that can affect the temperature of compost (and if you need more details, there are many good gardening guides with excellent instructions on managing a compost pile). But in general, when the compost temperature spikes for a few days and then drops below 80°F (27°C), it is about ready for colonization by fungi. Left unattended, the compost will play host to a variety of wild fungi that assist in

the decomposition of its materials. But if you're ready with your spawn, you can now insert your chosen species into the process.

By this point, the pile should have shrunk considerably, to about 60 to 70 percent of its original size, and it should covered with whitish flecks of beneficial bacteria called actinomycetes, signaling that it's time to add spawn. Mix your spawn into the compost and incubate at the recommended temperatures for each species (see part 4), following colonization with a casing soil if needed. (See chapter 20 for more information on casing soils and beneficial microbes.) This spawned compost can be used to fill woodland bed gardens outdoors or pots and trays indoors.

METHOD TWO: COMMERCIAL COMPOST CULTIVATION

Method two produces a more complex and pasteurized compost that should result in higher yields than

USING MUNICIPAL COMPOST

Municipal compost can be suitable as a mushroom growing medium. Recent studies have confirmed that it contains the necessary microflora for healthy plant and fungi cultivation, although studies have also confirmed the presence of human pathogens, meaning that municipal compost should be heat-treated in order to render it safe for mushroom cultivation (Bonito, Isikhuemhen, and Vilgalys, 2010). Municipal compost can be a wonderful resource if you don't have the space or material for composting. Just bear in mind that you don't always know and can't control what's in it, so your yield may be inconsistent and you probably wouldn't want to rely on it for a commercial formula.

the cruder method one. If you plan to grow mushrooms on compost commercially, you should nevertheless start with method one or with method two on a small scale, not at the volumes you will need to meet full-fledged production expectations. As with any cultivation method, start small to get experience and to explore your particular production parameters; then you can scale up. I recommend first running parallel trials and keeping records on the substrate formulas,

spawn strains, temperature fluctuations, and environmental conditions you use to evaluate which methods have the most success. I also recommend that you read chapter 7 on pasteurized or sterilized media, as well as chapter 20 on advanced techniques.

Making the Compost

As with method one, the first goal in method two is to prepare organic material for the mushrooms to use as a food source. In this case, however, you'll be much more deliberate about the "ingredients," or the component organic materials, to optimize levels of nitrogen, which the fungi use to produce protein in the form of edible mushrooms. Gypsum is commonly used as a calcium source and also to help coat the particles to ease in aeration and separation of the growing media, but will not alter the pH.

The nitrogen content of your raw materials should fall between 1.5 and 1.7 percent of the total dry weight. You can calculate this percentage by totaling up the dry weight of each of your raw materials and dividing that figure by the total weight of the nitrogen, calculated as a percentage of each ingredient. See, for example, the figures in table 6.1. The total nitrogen weight (0.94 pound) divided by the total dry weight of all ingredients (59.75 pounds) = 0.015745, which rounds off to 1.57 percent.

You can find out the nitrogen content of many common raw materials in your area using agricultural databases publishing nutrient breakdowns for livestock feed for ruminants (many of these databases are available online). Since vegetation varies from region

TABLE 6.1. Sample Formula for a Manure-Based Compost with a Variety of Nitrogen and Carbon Sources

Ingredients	Wet Weight (pounds)	Dry Weight (pounds)	Nitrogen Percentage	Nitrogen Weight (pounds)
Horse manure	80	50	1.20	0.6
Poultry manure	7.5	6	4	0.24
Brewers grains	2.5	2.5	4	0.1
Gypsum	1.25	1.25	0	0
Totals		59.75		0.94

Source: Adapted from Dr. David M. Beyer, Penn State Extension, "Substrate Preparation for White Button Mushrooms," 1997, http://extension.psu.edu/plants/vegetable-fruit/mushrooms/mushroom-substrate/substrate-preparation-for-white-button-mushrooms.

to region, locating the analyses for the local materials you will be working with is important and will save time and minimize blind experimentation. Whatever raw materials you decide to use, your goal is 1.5 to 1.7 percent nitrogen, so adjust your ratios accordingly. To confirm the nitrogen content, you can periodically take a sample, dry it, and test it for nitrogen using a soil test kit, or have it analyzed by a local university that specializes in testing composts.

With these parameters in mind, the compost-making process proceeds as follows.

Step 1. Determine your compost "ingredients." You can use a wide variety of locally available wastes to make a compost as long as the starting nitrogen content is in the correct range. For example, your compost could be:

49% shredded wheat straw
48% chicken manure
3% gypsum

Or

96% horse manure
2.5% gypsum
1.5% cottonseed meal

Step 2. For the first five to seven days, wet and turn your compost materials every day. Wet your compost by watering it thoroughly, until it is saturated, but not to the point of runoff. (If you do have runoff, you can capture this nutrient-rich waste and channel it into a mycoremediation area.) You can turn and flip the compost by hand, but even with a small pile this work can be laborious. Commercial composting operations form long piles of compost on concrete and use giant turning machines that wet and mix the compost at once as the machine drives down the length of each pile. The turners are an expensive investment since they are custom-made. Many small-scale farmers rely on tractors to flip and blend compost piles. However you turn your compost—whether by tractor, by turning

machine, or by hand—do the best you can to break down the pile, blending the layers so the outer material goes in, the inner moves out, and all of the biomass at some point spends time in the middle of the pile.

After the initial wetting and turning period, scale back to turning the pile just every other day. (By this point, you'll see the temptation to invest in expensive equipment!) The compost should be extremely moist, yet not dripping wet when you squeeze a handful. Water sparingly to keep moist when needed at turnings.

Step 3. Every day, monitor the heat of the compost using a compost thermometer with a long probe that reaches the center of the pile. Check the temperature in a few different locations. Your goal is to achieve an internally consistent temperature of around 150°F (66°C), which is the point at which water evaporates from the carbohydrates and the carbon becomes concentrated. If the compost is too hot or cold, adjust the size and depth of the pile; spreading it out more will cool it, whereas piling it more will heat it. Since the temperature will vary according to seasonal conditions, keep records of your work, so that in following seasons you won't have as many trial-and-error adjustments to make to your compost pile.

Step 4. After the compost has heated to 150°F (66°C), it should move into the "caramelization" phase, when the mixture is uniformly brown and gives off a strong smell of ammonia. The "tear and squeeze test" is commonly used here, meaning that if you can shred the straw or hay fibers with your bare hands and squeeze water from a handful of the compost at this state, it is ready to pack into trays or other containers for pasteurization. Wet, short compost straw or smaller particles should be packed loosely, while drier, longer straws or larger particles can be packed in more tightly to achieve the optimum density. The compost should finish dense, and not fluffy; otherwise it will not be able to achieve a high-density dry weight and may yield poorly. The amount of compression will depend on the particle size; you will need to perform trials to see how your particle size and compression rates affect yields.

Pasteurizing the Compost

Pasteurizing your compost eliminates insects, microbes, and other pathogens. For commercial mushroom production, the stakes are higher, beds are densely placed, and contamination can easily spread. A heat treatment is highly recommended and is an industry practice. The process also converts ammonia into protein and other nitrogenous compounds that the mushrooms use as a food source. Commercial growers use special insulated rooms, much like trailers or shipping containers, for pasteurizing, keeping their compost at a temperature between 130 and 140°F (54–60°C).

Step 1. To low-heat pasteurize compost, you must maintain it at a temperature between 130 and 140°F (57–60°C) for three to six hours. Do not exceed this temperature window; temperatures over 140°F (60°C) can inactivate or kill thermophilic fungi, creating opportunities for other competitor species once the compost cools down. Be sure to supply fresh air to the compost while maintaining the necessary temperature. A flame test can tell you if there is enough oxygen in the space; if a match will not light, increase the air exchange. (But don't be hanging out in this area!) Commercial growers use oxygen detection instruments to help them regulate the air exchange in their pasteurization chambers.

Step 2. After pasteurization, lower the temperature gradually (about 2°F [1°C] per day) until the core substrate temperature reaches 122°F (50°C). Once the substrate reaches 122°F, hold it at that temperature until ammonia falls to an undetectable level, which signals that elemental nitrogen has been created from the ammonification process orchestrated by the bacteria. Although there are expensive ammonia detection kits and sensors available, the ammonia detection strips available from horticulture and composting companies also do the job and are especially great tools for use on a smaller scale.

Using the Compost

Once the compost is finished, it can be spawned with your species of choice. Many mushrooms benefit from an application of casing soil that has been wet and pasteurized to stimulate fruiting (see chapter 20 for details).

CHAPTER 7

Cultivating Mushrooms on Pasteurized or Sterilized Media

When cultivating mushrooms on agricultural by-products such as sawdusts, cereal straws, and other forms of dried plant debris and vegetation, most growers choose to either pasteurize or sterilize the medium before adding the spawn. There are two reasons for doing this. If the medium is dry, pasteurizing or sterilizing with a hot water bath or injecting live, wet steam can help hydrate it, softening up the substrate to make it more digestible for the mushroom mycelium. And the heat or chemicals used to pasteurize or sterilize can decrease or eliminate populations of potential competitors that the spawn may encounter as it is colonizing the medium. In a sense, it buys time, enough time for the mycelium to capture as much of the substrate as possible.

Pasteurizing has some benefits over sterilization. The process of heating growing media to 160°F/71°C (pasteurization) rather than 250°F/121°C (sterilization) uses less energy, and the equipment used to pasteurize is less expensive. Heat-treating media to the point of sterilization kills all of the organisms present in the media; pasteurization does not, leaving the thermophilic (heat-tolerant) microbes. These microbes can be beneficial, since they are not competitive with mushroom mycelium and instead stand guard over the substrate so that other contaminants do not have the opportunity to take over.

The type of mushroom that you want to grow may dictate which process to use, but the kind of heat treatment you have available may also influence what kind of mushroom you cultivate. In this chapter, I'll cover pasteurization of agricultural waste such as cereal straws, which can be used for many species (the most well-known being oyster mushrooms), and sterilization of sawdust, which involves pressure-cooking or autoclaving sawdust in heat-tolerant bags (as is common for shiitakes cultivated on a large scale indoors). You can scale these concepts up or down depending on your production goals, so don't hesitate to start small. Whichever method you choose, I am sure you will find the process easier than it sounds. All it really involves is heating water for pasteurization or cooking media inside of a pressure cooker for sterilization.

Hot Water Immersion Pasteurization

Economical and very efficient, hot water immersion is the primary method we use to pasteurize straw

The growing medium is suspended in welded baskets, which are lowered and raised into the 55-gallon drums with winches attached to an engine hoist. The two yellow controllers move the baskets up and down, or left and right, to easily fill them with substrate materials and immerse them in hot water.

Electric or manual hoists can be used for lifting and lowering heavy baskets full of growing medium into and out of barrels.

THE TWO-DUNK RULE

Each barrel of water should be limited to two pasteurization batches (two dunks). You can pull out the first batch and reload the dark, soupy water that it leaves behind with a second batch, but not a third one. The water becomes laced with micronutrients that inhibit and reduce yields significantly. I like to make the first dunk my nitrogen source (cotton hulls and wheat bran) all by itself, which allows the dissolving of beneficial nutrients into the water. The second dunk gets the carbon source, in my case pure shredded wheat straw, which wicks up and absorbs the nutritive fluid and contributes to a higher-yielding substrate. I cool them separately and then mix the nitrogen source with the carbon before spawning.

and other medium types before spawning. To achieve pasteurization, you submerge your substrate materials (shredded, if needed) in water kept at about 165 to 175°F (74–80°C) for one to two hours. I recommend first heating the water up to about 180°F (82°C), so when you add the cooler substrate materials, the temperature falls right into the 165°F (74°C) range. Once the substrate has soaked for an hour or two, drain it and let it cool. Then bag it up for later use, or proceed directly to adding spawn.

For home-scale cultivation, pasteurization can be done in a large pot on a stovetop. With a small-scale commercial operation, for a setup cost ranging from $200 to $1,500, you can pasteurize enough growing media to grow hundreds of pounds of mushrooms a week using this method. Fifty-five-gallon drums are the norm for pasteurizing media at this scale; make sure they are food-grade and have a tight-fitting lid. I pick my barrels up from a local supplier in the fruit juice concentrate business. They only cost me $7 to $10 each and last for years. (I also use

Draining and cooling pasteurized straw on a table that was wiped clean with rubbing alcohol.

the drums for storing dried supplements and grains; their locking lid bands keep out bugs, rodents, and other critters.)

You can use an inexpensive propane burner or even a wood fire to heat the barrel. Set the barrel up on a level platform of bricks, and fill it two-thirds full with clean water. Start a fire beneath the barrel and heat the water up to at least 175°F (79°C). This usually takes a few hours, so keep an eye on the operation, making sure the ground around the setup is wet or moist to prevent any accidental fires.

Once the water has reached at least 175°F (79°C), you can add your growing medium. Dry material will want to float, so you may need to weight it down, but even then, your biggest challenge won't be getting the medium into the drums; it will be getting it out once it is pasteurized, hydrated, and extremely heavy. I used to weigh it down with a screen and bricks, fishing it out later with a manure fork, but holding sopping wet straw on the end of a fork while it drains is too exhausting for large volumes of media.

At my farm, we eventually developed custom-made aluminum baskets that can be packed with growing media, hoisted with a winch, and moved along a steel beam. This system cost around $400 to build. The beam we use is an engine hoist I bought from a local boat mechanic. (Counterweighted pivoting levers with removable weights could do a similar job if you are off the grid, and there are numerous ways to pulley baskets out of a vessel.) To make it easier to remove the packed, pasteurized straw after it drains, our baskets are built of two parts. The rigid outer frame is the shell, and the smaller inner wire frame forms a removable cylinder; when we hoist the basket out of the 55-gallon drum and turn the basket upside down, the cylinder slides out easily. Then we can unhook the ends of the cylinder and unfold it to release the medium.

When you pasteurize via hot water immersion, you'll need to monitor the temperature of the water periodically and maintain it in the range between 165 and 175°F (74–79°C) for one to two hours. I use cheap digital probe thermometers from the grocery store, with the probe poked through a hole in a piece of wood or insulation foam so it floats, and I don't have to hold them over the burning steam. Gloves are cheap; new skin is not. So keep a few thick pairs of gloves handy around the barrels. At our farm we organize and store all of our gloves, thermometers, and other tools in a waterproof bin and keep it close and handy to the pasteurization site. They should always be there when you need them, so make a point to put them back so you don't get tempted to try to pull off the steaming hot lid with your bare hands and get burned. (Yes, I have done this.) The lid may not seem hot, but as soon as you crack it, steam will jet outward and sear everything within a few inches, including the hand that is holding it. So be careful.

Once the growing medium has been immersed in the pasteurization-temperature water for one to two hours, it's done. Pull it out and let it drain. Sanitize a clean table or tarp by wiping it down with rubbing alcohol, and then spread the pasteurized growing medium out on it to cool before spawning.

STEAM PASTEURIZATION

In steam pasteurization, you hydrate the growing medium by submerging it in water or wetting it with overhead irrigation until it is completely saturated and then drain it briefly to allow excess water to run off. You load the saturated medium into a sealed barrel, container, or room that has been modified to receive live steam injection. This unit is generally not under pressure, so steam will have to be constantly injected. Some large commercial growers use an insulated shipping container with a porous floor; the steam is injected from below and works its way up and through the growing media.

A simple way to perform steam pasteurization on a small scale is to modify your 55-gallon drum. If you use a locking metal lid, you will need to bore a ½-inch hole in the top face of the lid to allow steam to escape. You can place a brick over the hole to allow pressure to build within without risk of exploding the barrel. This will conserve water and it

will also heat the barrel more quickly. You'll also need a basket to hold the growing medium. Stack a few bricks in the bottom of the barrel, building enough of a platform to keep your basket suspended about 8 inches above the bottom. Add about 6 inches of water, and begin heating. That little puddle of water will heat quickly and also evaporate, so every thirty minutes or so, dip a wooden dowel down to the bottom and take depth reading. As needed, add enough water to keep the depth at 6 inches, and replace the lid to build steam and pressure. Monitor the temperature as well, aiming for a range of 180 to 190°F (82–88°C). Once you see live steam escaping from beneath the lid consistently, the barrel is ready for your growing medium.

Lower the basket full of wet, unpasteurized growing medium into the steaming barrel, replace the lid, and start timing the cook, which will last for twelve hours. Reduce the heat; you now only need a small flame to maintain the steaming process. Check water levels with a stick to make sure there is at least a few inches of water actively producing steam, adding small amounts of water if necessary.

After twelve hours, remove the basket and continue the process as you would in a hot water immersion, transferring the hot medium to a clean, sanitized tarp or table to allow it to cool down. Since this medium was steamed and not submerged in hot water, it will not need to be drained and will cool quickly, typically within an hour. Once the medium has dropped to below 100°F (37.8°C) throughout, it is ready for spawning.

SOLAR PASTEURIZATION

Especially in remote areas of the world where fuel is a limited resource, a simple and inexpensive solar hot water system can be used to power a hot water immersion pasteurization setup. Solar heating is quite adequate for bringing water up to the pasteurization temperature of at least 165°F (74°C) and keeping it there for one to two hours. Since the water is not going to be used for drinking, it's not critical that the system kill all of the waterborne pathogens. In fact, once the medium is inoculated with mushroom spawn and the mycelium begins to colonize, the mycelium will filter any water remaining in the medium, transferring clean water into the fruiting body, providing a safe edible food that is 90 to 95 percent filtered water. (But note: As a filter, the mycelium will efficiently break down most chemical pollutants, such as pesticides, and destroy biological pathogens, but it hyperaccumulates heavy metals such as lead and mercury.)

You can also use solar power to pasteurize growing media by placing moistened substrate in shallow bins or trays, covering them completely with plastic, and letting them sit in a sunny spot on a hot day for six to eight hours. Reflective fabric or Mylar panels can hasten and improve the process if you place them strategically around the setup as you would with a parabolic solar cooker. (Old satellite dishes, the extremely large ones, can still be found and converted to parabolic lenses to help focus the light energy more efficiently to one particular spot.) You can further enhance the process by attaching your bins or trays to an automatic solar tracker, a unit that constantly follows the sun, keeping the most direct concentration of rays focused on the media.

CHEMICAL TREATMENTS

If you have limited options for heat-treating your growing media, you might consider chemical treatments. These heatless methods can be helpful in remote or developing areas of the world where fuel resources such as firewood are limited.

You can use a peroxide soak to treat growing media by making a 10:1 dilution of over-the-counter hydrogen peroxide, which is typically 3 percent and will make a 10-part water to peroxide solution ending in .3 percent. Commercial cleaning companies sell more concentrated solutions that can be diluted accordingly to achieve the .3 percent end solution needed to treat the media effectively. Submerge the growing medium in the solution and let it sit for twelve hours. Remove the medium and allow it to drain. Once the medium has drained thoroughly, you can proceed to spawning.

Hydrated lime treatments have also shown some success. Mix 4 cups of hydrated agricultural lime (calcium hydroxide) into a 55-gallon drum filled with clean water. (Burnt wood ash from fires and stoves, which is high in calcium carbonate, also can be used to treat the water if no hydrated lime is available.) Submerge the growing medium in the solution for six hours. Once you remove and drain the media, you're ready to spawn. One drawback of this method is the lime solution waste. I encourage you to neutralize or dilute the liquid before disposing of it since the pH is around 9.0, which can be hazardous to the environment. You can use this diluted liquid to balance the pH of your garden, as well as for watering open-topped surfaces of mushroom trays and containers; acidifying mycelial enzymes will naturally lower the pH, while a higher pH from the lime solution will limit mold proliferation.

STEAM STERILIZATION

Sterilization requires heating to a temperature of 250°F (121°C) for one to two hours, which is sufficient to kill all microbes in a growing medium. Most large commercial mushroom farms sterilize their growing media (usually highly supplemented sawdust) under high pressure in large industrial autoclaves, also known as retorts, or steam sterilizers because it's faster and more efficient than prolonged pasteurization. If you are thinking about getting into commercial production of species that grow on sawdust substrates (shiitake, maitake, lion's mane, et cetera), a commercial autoclave is currently your best option. Our lab uses only well-reinforced commercial autoclaves that are tested and certified for safety, since we are operating at extremely high temperatures and pressures.

If you are using an autoclave and you have a high-nutrient growing medium, you will want great air filtration controls throughout the process to avoid contamination of the sterilized medium. At the least you'll need a clean room with HEPA filtration.

The limiting factor is cost—the average autoclave can run in the range of $10,000 for a used machine

CULTIVATING MUSHROOMS
ON INVASIVE WEEDS AND TREES

Invasive plants can be found all over the planet, primarily as the result of ecosystem disruption from human activity. In recent years, land managers and conservationists have focused on aggressive eradication efforts using, in many cases, hugely toxic herbicides. In fact, according to a 2011 special investigation by Truthout, "some biologists and conservationists are now questioning the idea that every species deemed 'invasive' is actually bad for ecosystems. Others flat out say that the invasion biology 'greenwashing' myth was created by the land management industry to sell pesticides and generate an endless stream of government contracts" (Ludwig, 2011). Still, it's hard to deny that these aggressive species take a toll on ecosystems, suppressing or preventing the growth of native plants—even by changing the composition of the soil so that it becomes increasingly unsuitable for native species and conducive for the invasives, and inhibiting the mycorrhizal relationship between native plants and their symbiotic partners. Although views differ on the best way to manage invasives, there's one thing we know: They're here (meaning everywhere), so we might as well use these plants as a growing substrate for mushroom cultivation.

Needless to say, if you're using invasives as a mushroom growing substrate, you need to be extremely careful not to further facilitate their spread in the process. You'll need to educate yourself about the invasive you plan to use (understanding its ecology and life cycle just as you do for your mushrooms) and plan to harvest it in a way, and at a time in its life cycle, that reduces to possibility of propagation to the greatest degree possible. You will absolutely need to heat-treat it, by either pasteurization or sterilization, to destroy any reproduction potential within the growing medium.

If you don't know what plants are invasive in your area, lists are available from most state departments of natural resources and USDA offices, agricultural colleges, and native plant societies. Since we all have our unique problem plants depending on locale, I'll touch on just a couple of notorious examples from the southeastern United States. You can take these ideas and apply them to your own problematic plants. Bear in mind that research and trials with this kind of mushroom cultivation may be exceptional candidates for grant funding.

Kudzu

Kudzu (*Pueraria lobata*), also known as Japanese arrowroot, is a large-leaved, aggressive crawling and climbing vine in the pea family that is capable of completely strangling and shading out very large areas, extending up into the canopies of nearby trees. Originally introduced in the American South from Asia as a form of erosion control, it has now become the focus of its own control, since it is much more aggressive in its new home than in its native ecosystem, where it faces natural balancing pressures.

up to $100,000 for a high-end model capable of producing enough growing medium for 1,000-plus pounds per week of shiitakes. Although the yields can be extremely high with an autoclave, so is the energy input (in addition to the cost of the machinery), so I would evaluate whether this kind of investment is necessary for your goals. It is better to start with the no-tech/low-tech methods of cultivation before thinking about upgrading to a high-energy system with a big environmental impact that goes with it.

It's possible to sterilize media using small professional sterilizers, large pressure cookers, or homemade steaming units, but be aware that pressurizing any container not designed specifically for this purpose can be dangerous. If you are able to construct a vessel

Farmers have been quick to discover that kudzu is great livestock forage, however. It is more than 16 percent protein, and up to 60 percent of the plant is digestible nutrient content. Given its nutrient profile, kudzu is a perfect candidate for a substrate for the cultivation of many edible mushrooms such as oyster (*Pleurotus* spp.), paddy straw (*Volvariella volvacea*), and even the tropical giant milky (*Calocybe indica*). Simply harvest, dry in the sun, shred, and pasteurize or sterilize. Composting these cut vines can also produce a nutrient-rich compost capable of supporting secondary decomposers such as almond portabellas (*Agaricus subrufescens* and related species).

Water Hyacinth

When water hyacinth (*Eichhornia crassipes*) was introduced to Florida in 1884, it quickly choked up waterways and is now considered one of the fastest-spreading and fastest-reproducing plants on the planet. This flowering aquatic plant with global distribution in the tropics is easy to identify by its large, inflated, bulbous stalks. It is extremely buoyant and floats around in large, connected mats that can be dragged out of rivers and streams. (Remaining smaller pieces will continue to colonize.) Land managers sometimes use floating harvesters to clear waterways, bundling up the plant material and dragging it ashore, where it is typically composted or otherwise disposed of. More often managers use aquatic herbicides,

Water hyacinth can be sun-dried and used to cultivate many mushrooms in warm climates around the world.

which are less expensive and easy to apply compared to the physical labor and expense of harvesting—but also pollute the water with chemicals that damage the ecosystem more than the water hyacinth itself.

Water hyacinth has high levels of cellulose and nitrogen, which is desirable for mushroom growers. On my farm, we have sun-dried the plants first on land and then used them as a substrate for cultivating heat-tolerant species such as paddy straw (*Volvariella volvacea*), which is primarily a cellulose degrader and not fond of lignin-rich substrates.

that builds steam pressure (whether by injecting live steam or heating a bit of water in its base, much like in steam pasteurization) and is fitted with a pressure gauge and safety release valve (available from boiler companies and easy to install with rubber gaskets), you can create a super-pasteurization unit, a higher level of pasteurization that rivals sterilization without pressure but takes longer exposures, that can reach 212°F (100°C) for at least six hours. Maintaining the pressure at somewhere between 2 and 3 pounds per square inch is enough to drive the temperature up even higher while maintaining a level of safety. Vessels or containers can be fitted with low-pressure release and safety valves to drive temperatures higher without risk or danger, while shortening the heat treatment time.

Shiitake mushrooms (*Lentinula edodes*) fruiting from supplemented sawdust sterilized in autoclavable bags using a standard pressure cooker.

Loading a pressure cooker. Be sure to leave air space above and around each layer to allow for steam penetration.

Small, professional sterilizers are similar to home pressure cookers for canning except they have metal-to-metal seals and many safety features to allow for pressure release.

A 55-gallon drum can be modified to hold a few pounds of pressure using a release valve and pressure gauge. It can be used to super-pasteurize media for six hours.

CHAPTER 8

Cropping Containers

You can grow mushrooms in just about anything, so long as you understand a few basic concepts, including the "mushroom math" of substrate volume versus the exposed surface area of your cropping containers. In this chapter, we'll look at which containers are best suited for particular growing media, including different options for colored bags, plastic columns, reusable buckets, nursery pots, trays, and untreated wood for making outdoor raised beds—and how your choice of containers can minimize environmental impact, produce the greatest yield, and make the best use of your fruiting space. Everyone's growing situation is unique. Armed with information and understanding of some key concepts, you can make the best of yours.

There are two separate issues around the concept of exposed surface area that come into play when you are choosing containers: gas exchange and photosensitivity. You'll want to consider them both when you're choosing containers.

A WORD ABOUT ENVIRONMENTAL STEWARDSHIP

Disposable plastic bags and columns are the norm in the mushroom production industry, but that doesn't mean that's what you have to use. There's nothing wrong with salvaging or recycling containers; in fact, the opposite. You should get in the habit of using recycled materials and containers (while being mindful of the time and energy required to clean a given container). Although biodegradable containers might seem like a good idea, the mycelium will use them as food, turning your container into a fruiting substrate and dramatically decreasing yields. That said, as a last resort, you can use burlap bags or cardboard boxes.

EXPOSED SURFACE AREA AND GAS EXCHANGE

The total volume of the substrate in the container versus the surface area that is exposed to the air is a critical factor for mushroom formation and yields. If there's not enough substrate volume relative to the exposed surface area, the mushroom beds will "abort," much like a fruit drop. Mushrooms may initiate or form, but they'll quickly dry up and never mature. Although this is very disappointing and a waste of time and resources, it's easily preventable if you give consideration to the dimensions of your cropping containers and limit the exposed surface. For example, you could use a container that completely encloses the substrate, exposing surface area simply by punching a few holes or slits in the container. Or you could

73

grow mushrooms in open trays that are relatively deep proportional to the exposed area. But there is a limit to this sort of setup, because over a certain depth, you will be wasting media. In other words, as the cultivator you must find the balance: building enough substrate volume to support mushroom production, but not so much that you're wasting substrate.

A similar balance applies to the concept of exposed surface area. In an enclosed plastic container with holes punched in it, for example, the holes must be kept to a size and a number that allow for maximum mushroom production. Not enough holes and the medium cannot breathe and will become anaerobic, creating large, dead marbled areas inside the biomass that will compromise yields. But too many holes overstimulates the colonized substrate, signaling to the mycelium to produce more mushrooms than it can support, causing a partial or complete crop failure.

In general, there is a set weight of mushrooms capable of fruiting off a given volume of colonized substrate. Without knowing how much exposed surface area to give a mushroom, you may need to perform some trials and see how they fruit. Let's consider a hypothetical series of plastic bags, each packed with an identical volume of substrate, inoculated with oyster mushroom spawn, kept under optimal conditions, and capable of producing 10 pounds of mushrooms—but with a different number of holes punched through the plastic.

Scenario 1: One Hole on One End

Depending on where you put this hole, chances are there will be a single mushroom or one massive cluster. The mushrooms in total will weigh considerably less than 10 pounds; their weight will instead be proportional to the amount of the growing medium that the mycelium was able to colonize. In this arrangement, the carbon dioxide the mycelium produces as it colonizes the substrate has to travel an ever greater distance to escape the plastic bag, while oxygen

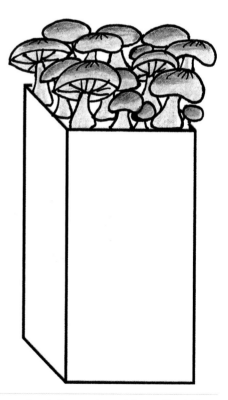

Depth versus exposed surface area on two trays of equal volume. Limiting the exposed surface area, such as in the cropping container on the right, can increase yields by reducing exposure of the mycelium to environmental stimuli such as oxygen and light.

backstreams inward; like a two-lane highway at rush hour, the gas exchange pathway becomes congested and cannot supply the mycelium with enough oxygen to allow it to stretch into the farthest reaches of the substrate. The mycelium responds by cutting itself short, focusing only on the small volume of medium capable of supporting adequate gas exchange. If we estimate the uncolonized growing substrate as 40 percent of the total, and a yield of 60 percent of 10 pounds, we achieve a maximum of 6 pounds.

Scenario 2: One Hole in the Middle

Since the distance the gas has to travel through the substrate is not as great here as in scenario 1, the mycelium may be be able to colonize almost the entire substrate, leaving only the extreme ends of the plastic bag uncolonized. This hole can most likely produce a cluster that weighs 8 to 10 pounds, but I wouldn't recommend this setup, since that cluster is likely to comprise just a few or even one mushroom—and not many people want a 10-pound mushroom (unless you are going for a world record).

Scenario 3: Two Holes, One at Each End

Assuming complete colonization, having a hole placed at each end of this plastic bag will theoretically produce two clusters that weigh 5 pounds apiece, for a total yield of 10 pounds. Again, the size of these clusters may not be what you are hoping for, but now you can begin to see how to orchestrate the size you need to suit your production and, if you're growing commercially, marketing strategies.

Scenario 4: Four Holes, Evenly Spaced

Once again assuming complete colonization, four holes should produce four clusters with an average weight of 2.5 pounds each for a total yield of 10 pounds. From here, you want to fine-tune for the mushroom size you want while maintaining your 10-pound yield, bearing in mind that adding more holes can decrease

A beautiful flush of pink oyster mushrooms (*Pleurotus djamor*) with just the right number of holes poked in the container for the volume of substrate. This ensures that all the mushrooms have enough energy and water to mature properly.

the size of the mushrooms up to the point that you have too many mushrooms that will never mature.

Scenario 5: Eight Holes, Evenly Spaced

Eight holes should produce clusters averaging about 1.25 pounds apiece for an overall yield of 10 pounds. This is my recommended target size if you are selling your oyster mushroom clusters, since 1 pound of mushrooms is easy to pack and affordable for customers to pick up and take home.

Scenario 6: Fifty Holes, Evenly Spaced

Now we have a problem. I jumped to fifty holes to get straight to the point. Doing the math of 10 pounds divided by fifty holes should give us an average cluster weight of 0.2 pound each, right? Not quite. An overabundant fruiting of small mushrooms takes so much energy for the mycelium to produce that a lot of the small clusters abort, either rotting or shrinking during the pinning stage, leaving the mycelium able to redirect its energy to the remaining clusters. This loss of primordia equates to a loss in yield that you will soon learn to calculate based on the immaculate production notes that you should be keeping. I would expect that you could lose approximately 10 to 30 percent of your yield this way.

Having too many holes and, as a result, clusters can also create competition among the clusters themselves. Mushroom clusters that initiate earlier than the rest of the pinset can hog the nutrients and water. Like piglets fighting over the mother pig's milk, there will be fat babies and skinny runts.

So try to minimize the runt clusters by keeping the hole counts or surface area to a minimum, so all the developing mushrooms have enough energy and water not only to form simultaneously, but to form into uniform clusters of equal size and weight. If you do happen to poke too many holes and your first flush is ruined, go back and tape most of the holes shut; the next flush should come out okay. Similarly, if you're cultivating in an open bed or tray and you notice too many mushrooms initiate, you can reduce the pressure load by cutting or scratching off a few primordia to reduce the numbers.

EXPOSED SURFACE AREA AND PHOTOSENSITIVITY

If you recall the importance of light to mushroom formation from chapter 1, you'll realize that the opacity of the container you choose can affect the timing and location of mushroom formation, or initiation, based on photosensitivity. Some strains are more triggered by light than by gas exchange. If you're using clear plastic bags, as an example, this could be a disaster, since mushrooms will form wherever light conditions are right (everywhere), instead of the spots where gas conditions are right (the ventilation holes you've punched in the bag). In this case, small mushrooms will form all over the biomass under the plastic, the mycelium will direct energy and resources toward each trapped mushroom or cluster, and they will all soon become stifled. (If this happens, resist the urge to poke more holes into your bags. Instead, slip a dark bag tightly over the clear one, and poke holes through both bags to salvage the remaining flushes. And then switch to opaque bags for future batches.)

I prefer the use of opaque containers for production, with a few smaller transparent "spy bags" for

Photosensitivity can cause mushrooms to become trapped under the plastic of a clear bag. If this happens, resist the urge to poke new holes and instead slip a dark bag tightly over the clear bag, poke holes through both bags to salvage what production you can, and switch your future production to darker containers.

tracking the contamination and spawn run of larger batches. The mycelium generates very high levels of carbon dioxide in a container, so in an opaque container mushrooms naturally form directly near and around the ventilation holes, forming where oxygen is available. This places them in the perfect position to escape and mature into healthy clusters, without the mycelium wasting resources on trapped pins.

PLASTIC BAGS

Widely available and easy to use, plastic bags come in many shapes and sizes, and once filled with growing medium they can be stored neatly on racks or shelving. They are often used to cultivate oyster mushrooms indoors. The thickness of the bags is critical; thin-mil plastic stretches easily and allows you to pack the spawned growing medium tightly. If you tie off the bags while they are compressed, they'll become springy with pressure—which is your goal. As the mushrooms form and the biomass shrinks, the plastic relaxes and maintains a tight fit, which will limit air pockets and improve yields. Although most bags are not sterile, if you keep them packaged and clean, they can be used directly from the manufacturer with pasteurized media. After you fill and tie off the bags, you need to immediately poke holes to allow gas exchange. These holes, however, are one reason why plastic bags, although cheap and useful, are hard to reuse, since they are likely to tear from the holes and the stretching, or else they become contaminated by the substrate.

If you are planning on cultivating mushrooms on sterilized substrates (such as supplemented sawdust formulas for the indoor cultivation of shiitake, lion's mane, and maitake), you should consider using autoclavable (heat-tolerant) bags, which can withstand sterilizing heat treatment in a pressure cooker or autoclave for several hours. A micron filter "breathing" patch on the bags allow for contaminant-free gas exchange during colonization. These bags are also used in laboratory settings for making larger quantities of spawn. Once they're inoculated with spawn, you can pack sterilized sawdusts and other growing media

inside these bag for colonization. When they are fully colonized, you can either poke exit holes for a few mushrooms to emerge (as is done for lion's mane) or cut the very top of the bag open and roll down the sides so the bag is not destroyed. The autoclavable bags are reusable for several sterilization runs; after you have used their contents, rinse and dry the bags, then refill with growing medium and sterilize—but you should never reuse them if they have ever contained contaminated media or if you have poked holes in the bags.

PLASTIC COLUMNS

Many mushrooms can be grown in cut-to-length plastic (polyethylene) columns, also known as poly tubing. Species that can be grown on vertical surfaces (primarily all of the oyster mushrooms) benefit from this naturalistic, standing-dead-tree configuration, and the majority of them prefer pasteurized straw or other dried vegetable waste as a growing medium, which is easy to pack into the columns.

One major advantage of using columns is it makes excellent use of vertical space and eliminates the need for shelving and racks—which can save you a lot of money. On the other hand, poly tubing is intended for onetime use only. As with plastic bags, you could conceivably rinse and invert them between uses, but it's difficult since they're likely to tear or become contaminated from the previous substrate. If you plan to try to reuse them, once the columns stop fruiting, empty them by opening one end and pushing the spent media out, rinse the plastic in a weak bleach solution, and then dry them before restuffing with freshly spawned media.

The tubing is sold in long, continuous rolls that can be cut to length. The diameter should be at least 8 inches, and can be up to 16 inches. Anything larger creates an anaerobic core that doesn't colonize and overheats, especially in warmer climates. We switch from larger-diameter columns in the winter to smaller ones in the summer to compensate for the thermogenesis that accompanies the colonization process. The length at which you cut the plastic tubing depends

Cultivating in columns can make great use of vertical space in a fruiting room.

HARD COLUMNS

Hard columns, such as metal or PVC culverts available from commercial landscaping and building companies, can be a great alternative to the plastic columns, without the waste of onetime-use polyethylene. The only mushrooms I would recommend for them are oyster mushrooms; since they are going to be filled presumably with spawned, pasteurized substrate, this limits the types of mushrooms you can grow. The biggest obstacle for cultivating in this type of container is the variable cost. Recycled plastic culvert pipes or other tubular nonbiodegradable materials will also work if you can find them. Whatever you use, you'll need caps that fit snugly over the ends of the tubes.

To begin, cut the columns to length to fit your growing space (6 feet is common), allowing for a cap on each end. Predrill the columns with ½-inch holes staggered about 6 to 8 inches all around the column surface to allow for gas exchange and mushroom formation. To make your own reusable columns, install removable caps on each end, fill with spawned fruiting medium, and pack tightly to remove air pockets. The columns can sit upright on the ground, held straight by an overhanging tether or hook, or arranged in frames that align the columns sitting on top of each other like creating a wall. For this configuration I recommend only drilling fruiting holes on the sides of the column facing your aisles so mushrooms are only fruiting where they will be easily picked. Spent columns can be emptied by using the same type of packing device that fits the diameter of the column.

BLACK NURSERY POTS

I started using nursery pots for mushroom production out of curiosity in 2006, testing different containers that I had available. I saw it as an opportunity to use an existing container that was already designed to be filled with mixed media, conveyored around plant nurseries, and watered—all with automated machinery. The fact that using these pots wouldn't require any custom engineering to automate an operation, and that I could acquire the pots used from nurseries

on how much weight you are capable of lifting comfortably (and repeatedly). I can carry a filled 6-foot column—which can weigh over 100 pounds—but I wouldn't recommend that on a daily basis, both because you risk injuring yourself and because you risk breaking the column. Cut the plastic 2 feet longer than the length you want to give yourself enough room to tie off both ends. Tie one end, pack it tight with substrate, and then tie the other end to make a giant "sausage." You can also make smaller "sausages" and link them together with S hooks. Move the columns into your fruiting room. The best way to carry a long column is with two people, one on each end. Hang them in place, and then poke your holes. (If you poke holes prior to moving the columns, you risk tearing and breaking them in half.)

liquidating their inventory, appealed to me. When they're stacked, it's easy to water the pots between flushes, unlike enclosed containers such as columns and sealed bags. Even stuffing them by hand is much easier than stuffing bags. Before filling the pots, be sure to sanitize the pots with diluted bleach, a 10:1 water to bleach solution, and rinse with clean water. Having two barrels in sequence with these solutions prepared makes it fast and easy.

Nursery pots in varying sizes are easy to find. You can usually find used pots available for free, and as a last resort you can buy them new, but trust me: They are everywhere if you know where to find them. Landscaping crews seldom recycle their pots, opting

to dump them or sell them back to the nursery where they bought them for a fraction of the price. (Returning the pots is a good deed on the part of the landscapers, since they are returning them for a loss after the labor is factored in for delivery.) So contacting local landscapers is a great way to secure a supply of these pots; offer to pick them up or pay the landscaper a small fee to cover the redemption cost. (Before offering payment, check with a local nursery to see what they are worth.)

Vertical Growers in Pots

For vertical-growing mushrooms, such as oysters, you can stack the pots in columns. Make holes or slits in

The plastic nursery pots commonly used for landscaping are great for cultivating many species—and they can be washed and reused. Shown here are golden oyster mushrooms (*Pleurotus citrinopileatus*) fruiting from stacked pots.

the sides of the pots at offset intervals, staggering them to create more exit points for the fruiting mushrooms. Stacking the pots three or four high may be the limit for a freestanding column, since they are prone to leaning, but you can set up a wooden or metal brace to increase the number of pots you can stack. Another option, since the pots have holes in their base, is to pound rebar stakes into the ground, fill the pots with freshly spawned media, and slide them down over the rebar, stacking them to a stable height.

To minimize the exposed surface area of the uppermost pot, simply top it with a plastic bucket lid or similar circular, nonbiodegradable cap. Once colonization begins, water sparingly, just to keep the growing medium moist. You can check the progress periodically by lifting a pot or two here and there to see the spreading mycelium and to monitor for contamination. One of the best revelations I ever saw at my operation was that when we stacked pots of growing medium all inoculated with the same strain of spawn, they fused. The entire stack of pots become one organism. Since such an organism is sharing water and fruiting information throughout the pots, once it is close to becoming fully colonized, do not disturb or separate the pots or you could delay the fruiting cycle. If you need to move the pots, separate them only immediately after a harvest; if you time it correctly, the pots will quickly fuse back together and not skip a beat.

Horizontal Growers in Pots

If you're using the pots for horizontal-growing mushrooms that fruit on the surface only, such as *Stropharia* or *Agaricus* species, you need to pack the growing medium into the pots tightly, but leave about 2 inches of space at the top of each pot for casing soil. (See chapter 20 for more information on casing.) You can stack the pots after spawning to save space and speed colonization, but as soon as colonization is complete, separate the pots, case each of them with a pasteurized, nonnutritive soil, and place them individually on the ground or on shelves. The mycelium will drive upward into the soil, creating a dense mat just beneath the soil surface. Mist the pots to keep the casing soil moist and ideal for primordial formation.

REUSABLE BINS AND BUCKETS

You can use buckets or storage containers much like nursery pots, drilling holes along their sides and stacking them to a safe height. This method is suitable for vertical-growing species that do not require a casing soil, such as oyster mushrooms; the primordia will escape through the small holes in the sides. For horizontal-growing mushrooms that fruit from the upper surface, such as *Stropharia* and *Agaricus* species, do not drill or stack the bins or buckets; simply crack the lid slightly for gas exchange and periodic misting.

As with any cultivation decision you'll need to weigh the pros and cons of using bins and buckets. In this case, you'll need to wash and rinse them with diluted bleach each time you use them. But buckets and bins are great for urban mushroom cultivation because they take little space, do not have to be supported like a plastic column, and can produce oyster mushrooms year-round. Anyone with a stove, a large pot, and a few buckets can set up a highly productive cultivation system using these containers in an extra bathroom, closet, or basement, scaling the number of buckets to their individual harvest needs.

WOODEN TRAYS

Wooden trays are similar to enclosed bins, but they're more suitable for horizontal fruiters that require an open top and no side holes, like *Agaricus* (almond portabella, white button) and *Stropharia* (king stropharia) species. The exposed surface area is ideal for casing with a nonnutritive potting soil mixture to promote mushroom formation and to introduce bacteria that stimulate primordia (see chapter 20 for more discussion of casing soils). Of course, because the top surface area is exposed, you'll need to calculate a sufficient depth and volume of substrate to maximize your yield. In general, with this setup, for every cubic inch of growing medium, allow ½ square inch of exposed surface area.

Use untreated wood, preferably a conifer to delay the rapid degradation of wood by most fungi. We have used untreated, rough-cut pine planks from our sawdust supplier to build outdoor beds, and these would be perfect for indoor trays as well. The beds have lasted for three years and show no signs of deterioration. And buying direct from the mill ensures that we are getting a pure product at a reasonable price.

EVERYDAY ITEMS THAT YOU CAN "FILL AND FRUIT"

As I stated at the beginning of the chapter, you can grow mushrooms in just about anything. Don't limit yourself to the cropping containers discussed above; use what is available. Scavenged items can easily take the place of store-bought containers or bags as long as they respect the "mushroom math" rules of surface area versus volume, discussed above. If the exposed surface area of your scavenged item will be too large, simply cover it or stuff the openings with a plastic bag to make it work.

Ideally the containers should not be clear (to limit light) or biodegradable (or else the fungi will "eat" it), but whatever you can find will work, including large plastic soda bottles, terra-cotta pots, garbage bags, large-diameter bamboo, dried gourds, woven baskets, livestock feed bags, and more. Dumpster divers take

Containers are everywhere! Here we have salvaged items such as lidless coolers, gourds, old pots, and even large PVC elbows. Make use of whatever is available if the need for production exceeds aesthetics.

heart: Just about anything you can call a container will work, and in a time of emergency—after a disaster, for example, whether natural or man-made—a little creativity can lead to great results. Note that if you are scavenging what appear to be cast-offs from local businesses, you should ask the store staff for permission first; many stores recycle their waste products for credit, and they may take exception to you helping yourself to their recyclables.

CHAPTER 9

Natural Pest Control and Disease Management

Whether you are a small or large operation or growing indoors or out, there will always be pest pressure. In a commercial operation, where the stakes are higher, your management of these issues can make the difference between staying in or going out of business. Your main goal should be to prevent outbreaks and contamination before they happen by following protocols for proper media treatments, cleanliness, and tool sanitation. The cultivator is the most frequent vector of contamination during media preparation and spawning, so remember that the effort you put into making intelligent decisions about your daily schedule, hygiene, and sanitation can be a big factor in your success and failures.

Pests can be defined as any living organisms that will compete with and cause harm to your mushroom growing operation. They can be other fungi, such as molds, or insects that feed on or lay their eggs in the developing mushrooms. Insects themselves can be a vector, spreading contaminating spores from older columns or cultures to new ones—and rapidly throughout a fruiting room. If you can control flies and other insects, you will have a much lower contamination risk. If you can eliminate molds at the same time, you will drastically reduce your contamination risk, since mold-contaminated cultures draw flies, and the flies

spread the fungus. This cycle can magnify throughout your fruiting room and quickly become a nightmare. Outdoor cultivation makes it even harder to control infestations, so consider covering your growing medium with insect screening or fine netting, such as the row cover cloth that vegetable farmers commonly use to protect their crops from frost damage. These fabrics "breathe" (allow gas exchange) and may serve double duty by providing shading or light reflection.

Never spray pesticides or synthetic fertilizers on your fruiting mushrooms. Mushrooms absorb gases, liquids, and sometimes heavy metals from their growing environment, which is why it is so important to avoid the use of pesticides, herbicides, and any other toxic chemicals that could end up in the fruiting body. Although your growing medium may contain trace amounts of herbicides and insecticides, many mushrooms are capable of breaking those down using their extracellular enzymes, so that the herbicides and pesticides in the substrate do not translocate to the fruitbodies. That doesn't mean that it's a good idea to use chemicals on your substrate as long as you don't spray the fruitbodies. Some large commercial farms spray chemicals directly on their beds as the mushrooms emerge as a way to protect their crops; as a result, pests develop resistance, forcing cultivators to increase application rates, or to mix even more toxic chemical cocktails, which they drench onto and into the growing medium. (Larger commercial farms actually applied for and received a "special use" permit from the EPA for a banned fungicide, having claimed that they had to use it or their industry would fail.) Mushroom fruitbodies that are directly sprayed with pesticides are not capable of breaking down those chemicals, so be aware of this when you buy them. The good news is that there are alternatives that work as well as or better than pesticide sprays.

CONTAINER CONSIDERATIONS

Insect and mold infestations can be better or worse depending on their movement and contact with growing media. Containers, bags, and trays that are relatively open allow greater opportunities for pests and mold. To minimize the risk of flying insects, you can cover your mushroom cultivation trays or outdoor beds with screening. If you're working with enclosed containers, such as plastic bags, in which you must make ventilation holes, after inoculation poke just minute holes—small enough that flies cannot enter, deposit their eggs, and track any mold spores from module to module. After full colonization, you can enlarge the holes to encourage mushroom formation. At our farm, we created a custom paddle with over a hundred pin needles; we use it to slap plastic bags and columns on the day we fill them with inoculated substrate, making ventilation holes. We then go back through approximately a week after full colonization and poke larger holes for the mushrooms to emerge through.

Whatever kind of container you are growing in, be sure to inspect your spawned growing medium daily, monitoring the progress of the mushroom mycelium. Once the bags are fully colonized there is less risk of mold contaminants being spread by flying insects.

GROWING SPACE DESIGN

If you are cultivating indoors, the design of your growing space can be a big factor in the challenges you'll face with pest and disease control. To begin, it's important to partition your space. Large, open spaces where workers, air, insects, and spores can move freely among different stages of the mushroom cultivation operation facilitate the spread of pests and diseases. And once an outbreak occurs (and it will, even with the best sanitation measures, from time to time), it will be much harder to control in a large, open space than if you partition your space into smaller rooms or sections.

Examine your growing space with an eye toward its total holding capacity, the different stages of mushroom cultivation (see chapter 2), and the particular requirements of the species you will be growing. Then you can figure out how to divide up your space. For example, in my operation, my shiitake strain colonizes for two weeks, goes through a browning or curing phase for four weeks, then is transferred to a fruiting room for the remaining twenty-one weeks (five flushes twenty-one days apart). Let's say that I was considering installing this shiitake operation in a space that was 100 square feet (to make the calculations easy). Knowing that I'm going to keep mushroom production going pretty much year-round, with containers of shiitakes accumulating for every stage of production, I have to decide how much space to allocate to each stage. The math computes something like this:

Total number of weeks needed: 27

Colonization room:
2 weeks divided by 27 = 0.074
= 7.4% of the space, or 7.4 square feet
Curing and browning room:
4 weeks divided by 27 = 0.148
= 14.8% of the space, or 14.8 square feet
Fruiting room:
21 weeks divided by 27 = 0.778
= 77.8% of the space, or 77.8 square feet

You can partition a space using curtains, movable partitions, or permanent walls. Many larger commercial operations are now using long "tunnels," similar to shipping containers linked end to end. They fill the tunnels with a single batch of mushrooms and keep them in there for the duration of the spawn run and fruiting cycles. After a few flushes they empty and sanitize the entire tunnel to keep pests from accumulating and creating contamination problems. This practice can reduce or eliminate the "need" for pesticides. Since the individual batches are isolated from each other, there is little to no risk of them cross-contaminating unless workers mistakenly drift from one to the other. In the event of an outbreak, a tunnel can be shut down and dealt with independently of the other tunnels.

Cleanliness is also important. Keep the floor well swept of organic debris such as bits of mushrooms and cut stems; insects will treat these as breeding grounds, and they can also rot and attract bacteria and molds. Similarly, never leave mushroom remnants (cut stems or torn caps) on your growing medium. This debris quickly softens and rots, drawing in flies and bacteria that can ruin future flushes, as well as contaminating neighboring mushroom batches. Organic debris can also concentrate in the plumbing, spoiling and offering an ideal place for breeding flies and gnats. Once a week or so you should clean out any drains and plumbing traps and sanitize them with a small amount of bleach. And under no circumstances should you allow standing areas of water in your growing space; they'll attract and breed pests, bacteria, and mold.

Approach pest and disease control proactively to improve your overall risk reduction. By focusing on cleanliness and isolation of different batches and stages of mushrooms, you can eliminate the majority of outbreaks that lead to total crop loss. Remember that every insect can multiply by the hundreds, even thousands, so capturing or eliminating them even in small numbers is an achievement. Treat a contaminated module or area like fire: Lock it down and deal with immediately, as soon as you see signs of contamination or insects. Figure out what works best for you and keep up the protection; otherwise another bloom of insect, bacteria, or mold problems will arise quickly.

INSECT CONTROL

Common problem insects vary depending on where you live, but they all share a few things in common with regard to a mushroom growing operation. They can inflict a great deal of damage, chewing through flesh and laying eggs inside and in between the gills, and their larvae quickly emerge to devour mushrooms from the inside out. The adults, usually winged, move from mushroom to mushroom and spread contaminants. In the same way that you should understand the ecology and life cycle of cultivated mushrooms, the more you can understand about the ecology and life cycle of common insects, the better positioned you'll be to recognize solutions to their infestations, many of which can capitalize on vulnerabilities in their life cycle.

Fungus Gnats and Fruit Flies

The most common insect pests that indoor mushroom cultivators encounter are fungus gnats, which include members of the Sciaridae family and can include fruit flies in the genus *Drosophila* spp. Fungus gnats and fruit flies fit perfectly into the gills of many mushrooms, especially oysters. As mushrooms mature and age, the gills part, making room for intruders and egg layers. Pick your gilled mushrooms before their cap edge turns upward or ruffles to help reduce the risk

Adult fruit flies fit perfectly into the gills of many mushrooms, especially oysters. As mushrooms mature and age, the gills spread apart in preparation for sporulation, also making room for intruders and egg layers. Pick your gilled mushrooms before the mushroom's cap edge turns upward or ruffles to reduce the risk of these pests taking over.

of these pests. And if you have fungus gnats or fruit flies, chances are you may also have worse problems, such as *Trichoderma* molds, so inspect your substrates to see if there are any visible traces of molds that need to be removed.

One solution to a fungus gnat or fruit fly infestation is a diluted concentration of *Bacillus thuringiensis* var. *israelensis* (Bti), a bacterium that kills the larvae before they become reproductive adults. Larger growers introduce this beneficial bacteria via overhead sprayers in the substrate preparation and inoculation room. Home or small-scale growers can use a small hand sprayer to accomplish the same task; once your growing medium is cool, mix this solution into the substrate during inoculation to reduce future infestations. The bacteria have no effect on the adults, though, so it will take a few days for the population to drop. I have also mixed Bti into a small pump sprayer and injected a few heavy streams into the holes of oyster mushroom columns during severe fungus gnat outbreaks, which helped greatly, since these holes were the insects' way in and out,

meaning the larvae were not far. Do not spray Bti on mature mushrooms; it's wasteful, since the larvae are typically inside the growing medium or mushrooms, and also leaves Bti residue on your mushrooms.

Small, shallow pans of apple cider vinegar or milk mixed with a little dish soap attract and drown many adult flying insects (but not pleasing fungus beetle or Sciarids). This method works well for fruit flies and is highly effective when used in conjunction with Bti. To make the fly attractant, pour ½ cup of apple cider vinegar or milk and a few drops of liquid dish soap into a small bowl. Mix by swirling gently with a fork, so no bubbles form; don't froth the soap. The soap creates a surface tension that pulls the flies in when they come to drink the solution and drowns them quickly. Each small bowl can capture and kill hundreds of flies in just under an hour, so place these containers near the developing mushrooms wherever you see populations of gnats. If you want to hang these traps, cut a few small windows in some plastic beverage bottles and pour some of the solution into each bottle. Tie a string

A mixture of soap and vinegar will attract and trap airborne insects (although not beetles or Sciarids) feeding on mushrooms.

to the top of each bottle and hang them in strategic places where the flies are gathering.

Yellow adhesive ribbons or dangle tags, sometimes sold under the name Tanglefoot, are nothing more than glue or sticky traps that you can make yourself if you do not want to buy them. They are bright yellow, a color that attracts the flying insects, which land and then never leave. These sticky mats in a fruiting room can help you monitor fungus gnat and fruit fly populations. Knowing whether they are on the rise or decreasing can help you stay alert to what strategies are working.

"Pleasing" Fungus Beetles

Anything but "pleasing," these rapidly reproducing beetles will wreak havoc if they enter your mushroom fruiting room. They dig deep into mushroom gills and bore tunnels into the cap and stem from underneath, eating their way through the mushroom's flesh, while breeding and raising larvae. As with fungus gnats and fruit flies, harvesting mushrooms before the gills have fully spread and extended will help reduce the beetles' ability to damage your crop. If you see adults flying near or in the gills of your mushrooms, crack open and inspect a few mushrooms at different maturity levels to see how bad the infestation is.

When you're picking mushrooms that are infested with beetles, pour 1 to 2 inches of a strong bleach solution into a bucket with a snap-on lid, and gently swirl in a tablespoon of liquid dish soap, taking care not to make many bubbles. Place the bucket near the fruiting spot and open the lid. As you pick the mushrooms, hold them briefly over the bucket and tap the caps; the beetles will fall down into the bleach mixture. Replace the lid immediately in case the beetles do not die right away and to preserve the chorine content of the solution. The bucket can be used repeatedly for up to a month or so if the solution remains strong. To discard it, strain out the dead insects and dilute the bleach for use in sanitizing the floor, inoculation tools, or other surfaces in the fruiting room.

Mites

If the mycelium during the spawn run seems blotchy and does not fully colonize the substrate, you may

"Pleasing" fungus beetles can be extremely damaging to mushroom production. Treat them like the plague, reducing their numbers using different strategies until they are gone.

have mites. Inspect any fruitbodies with a hand lens and you may discover little bodies that resemble crab-like spiders. If you cannot see any mites but suspect an infestation, try using a filtration method. Wash a liter of fluid through the growing medium or casing soil and capture the leachate in a container. Strain the liquid through a coffee filter and look for movement in the debris left behind, using a hand lens or dissecting microscope.

VACUUM PEST CONTROL

My last resort for flying insects (including pleasing beetles and fungus gnats) that are embedded in the gills is an inexpensive wet/ dry vacuum that I leave in the corner of the fruiting room for emergencies. Fill the vacuum canister about one-third full with soapy water. Holding the nozzle with one hand, and pointing it upward, gently rake the bottom of the gills with the nozzle, while lightly tapping the top of the cap with your other hand. Repeat for all the mushrooms in the room. This technique quickly reduces adult populations that are in their egg-laying state before they create a new generation. It has saved many fruiting rooms and can turn around quality control if done repeatedly every day for one week in conjunction with Bti and the sticky mats.

Leopard tree frogs patrol our greenhouses. I collect their tadpoles from around our farm and raise them in aquariums inside the fruiting rooms. They eat insects all night long, reducing contamination vectors. They also fill the night air with an earthy, grainy song that I love to listen to from my back porch.

Mites can wreak havoc in a laboratory, especially if they invade the process at an early stage such as when you are using agar plates, so beware. Since the mites are so small that they're difficult to see, it is best to visually inspect the plates prior to grain spawn expansion, especially if you notice any irregularities such as minute pits or zones of mycelium that look abnormally eroded for no reason. These could be signs of mites feeding on your fungi. Mites are allergic to cinnamon, and there are a few manufacturers that use a cinnamon extract as the active ingredient in their mite treatments to suppress them. I recommend making a solution of cinnamon and incorporating it into your daily rubdown and the disinfection solutions you use on laminar surfaces; the cinnamon leaves a trace residue that repels mites in your working area.

Slugs and Snails

As outdoor cultivators or anyone growing in primitive humidity tents and greenhouses can tell you, slugs and snails always seem to slime their way up and onto fruiting mushrooms. These critters love alcohol and are attracted to the gases produced by both yeasts and mushrooms as they mature—a proclivity you can use to your advantage. The most common remedy for slugs is to place shallow pans of inexpensive beer around the bases of logs, racks, or trays or on the surface of beds. The slugs and snails can locate these dishes from many feet away. They will stream directly toward them, wade in, and drown in the malty brew.

MOLD CONTROL

Molds can be a huge problem when you're growing mushrooms on pasteurized or sterilized substrate. Unlike most cultivated mushrooms, these fungi produce spores prolifically on their mycelium as it grows and encounters oxygen. And some molds can produce spores even under very high carbon dioxide concentrations, such as are found in the growing medium of cultivated mushrooms (manure-based substrates and underpasterurized bulk substrates), making them difficult to contain as they thread away through the substrate.

Many different kinds of mold can compete with your mushroom cultivation efforts; the most common in growing media include *Aspergillus*, *Rhizopus*, and *Trichoderma* species. (*Penicillium* and *Aspergillus* species are very common in agar culture.) Some *Trichoderma* species are mycoparasites, meaning they parasitize mycelium, which actually makes them easier to identify: You will see a dark green mold parasitizing your mycelium.

Monitoring for and removal of mold should be regarded as a high priority both in the fruiting room and (if you have one) in the lab. If you discover mold in your fruiting room, apply a mist of full-strength bleach (typically 5.25 percent sodium hypochlorite) onto the mold surface (obviously never on the fruitbodies), and then remove any contaminated materials. Commercial button and portabella mushroom farmers commonly use spot treatments with powdered lime and bleach on casing soils, and this treatment is also fine for any growing medium. To spot-treat a mold-contaminated area, first sprinkle it with agricultural lime. Fill a small bottle sprayer with a 10:1 dilution of bleach, put the nozzle directly over the spot or into the hole (in a column or bag) where the infection is occurring, and spray one or two hard blasts into the moldy area. Use just enough to coat the contamination. If the contamination site is a hole in a column or bag, also wipe the surface of it with a cloth or paper towel saturated with 70 percent isopropyl alcohol. Allow it to dry for a few seconds and then tape up the hole with clear packing tape so you can see if the mold recovers or is contained. Even just wiping and taping the hole is a good way to "shut the door" on the contamination spot.

PREDATORS IN THE FRUITING ROOM

Predators are everywhere in nature and help keep populations in balance, so why not use them in your fruiting room? Compared to monoculture, biological diversity can reduce problems in a sterile growing room, filling the ecological niches that naturally slow the growth of highly reproductive or invasive species. That doesn't mean you need to add a bat house in your fruiting room (which would be very cool, but not good on account of the guano), but there are some other organisms that I encourage to set up homes in my fruiting room. My greenhouse isn't exactly airtight, so they can come and go as they please, but they usually stay due to the nice weather and free fly cocktails.

Spiders dominate the upward region of the rafters, and at night they descend and weave their nets in between the oyster mushroom columns, which are only about 1 foot apart and densely arranged. These spiders are great allies, fishing for fungus gnats every day and reducing numbers, as made evident by the plethora of tiny corpses in their webs. Anole lizards too, here in South Carolina, patrol the racks and tables like miniature bulldogs, jumping at beetles and snapping them up when they see them. I have a few leafy potted plants scattered about to help them blend in.

The third predator I bring into the room is our native leopard tree frog. I collect the tadpoles—from, among other places, an abandoned bathtub out in the nearby woods where the adult frogs like to lay eggs—and transfer them into a few aquariums in my fruiting rooms. I feed the tadpoles with pellets of tilapia food and the occasional crisp lettuce leaf. Floating a few small aquatic plants or even a wooden raft gives them something to crawl up onto once they develop their legs. Once fully matured into frogs, they roam freely about the fruiting room, positioning themselves perfectly near the exit holes of my oyster mushroom columns, waiting for the next victims, which are usually fungus gnats.

A cloned specimen is not always sterile. *Aspergillus* and other molds can quickly outcompete mycelium in sterile conditions, and any mold-contaminated plate or container should never be opened or exposed in the lab. Isolate mycelium away from contaminants and be prepared to get rid of any contaminated media.

The "greasy spots" in incubating grain spawn are indicators that the grain is covered with bacteria or yeasts. Never expand these. Start over with cleaner cultures and longer sterilization periods.

In the lab, remove all agar plates, bags, or other containers that show signs of mold contamination. To rescue fungal cultures that show signs of mold contamination, you can use any of the techniques described in chapter 18.

Bacteria Control

Bacteria are most commonly encountered in the laboratory on agar media and when making grain spawn, but they can sometimes become a problem in the cultivation process as well. Thankfully, you can control most of the cultivation-related bacterial problems fairly easily.

First, the growing medium may be the cause. If you are using a super-supplemented formula containing high nitrogen levels as your growing medium, you are at risk of promoting bacterial growth and not mycelial growth, which creates excessive heat above 100°F (38°C), which cooks and damages the mycelium, turning the growing medium into a bacterial playground.

Once the bacteria have colonized an area, they exude metabolites that prevent fungal colonization (unless the mushroom is one that attacks and kills specific bacteria). With their rapid life cycle, thermophilic bacteria outcompete the mycelium and exploit the substrate, leaving a marbled and uneven colonization, which you'll discover if you cut open and inspect your containers. The more nitrogen you use, the more you risk losing entire batches to bacterial contamination and having insect vectors spread the disease. If you are having problems, lower the temperature to see if this makes a difference.

Second, if you haven't provided adequate gas exchange or drainage in the growing medium, you'll create a dead core or anaerobic zone, populated by anaerobic bacteria, which will result in a similarly marbled and uneven colonization by the mycelium. So be sure that you poke holes early enough, provide adequate drainage, and keep the temperature of the spawn run below 85°F (29°C).

PART II

Mushrooms for Life:

Innovative Applications and Projects Using Fungi

CHAPTER 10

Recycling, Composting, and Vermicomposting with Mushrooms

In the United States, an average of 35 percent of home waste and 60 percent of business waste is suitable for use as a mushroom growing substrate. Mushrooms can be grown on toilet and paper towel rolls, egg cartons, newspapers, magazines, coffee grounds, tea bags, old cotton clothing, tissue boxes, shredded paper, cardboard boxes, and many other common materials. In addition to yielding a bountiful mushroom harvest, these products can also be used to expand mycelium into a biomass that could conceivably be used to inoculate larger waste streams or substrates for a wide spectrum of applications, including composting, mycoremediation projects, and creating value-added consumer goods such as insulation or living paper products, which are made of recycled mushroom growing media, such as spent oyster mushroom substrate, that are pressed into forms, and only need water to begin the composting process.

To recycle and compost with mushrooms, start by simply identifying your biodegradable waste. Separate your weekly garbage for a few weeks to determine exactly how much waste of each type—paper, cardboard, glass, plastics, food—you are generating. (This will also help you determine where you can improve consumer packaging decisions, reducing your plastic and Styrofoam purchases as you shift to packaging that can be put to better use with mushrooms.)

Pizza boxes aren't usually recyclable due to the oils and food remnants they contain. Oyster mushrooms, on the other hand, can eat them up and then fruit, plus produce a beautiful worm composting feedstock after two flushes.

Open your cupboards, look in your refrigerator, and peek in your cellar for anything you are consistently producing as waste. Check with local businesses about the waste they pay to get rid of, and you may be surprised to find them willing to let you cart off some of their trash. Dumpsters and other sites where debris is often piled up on street corners or behind restaurants and businesses are also great areas for collecting recyclable debris. Smaller companies that cannot afford or don't have the space for a recycling Dumpster often just flatten boxes and stack them up for trash removal; those boxes can be gold for a mushroom cultivation operation.

I try to think of treating my home and life like a "space bubble," attempting to minimize the nonrecyclable goods I bring in, pretending that landfills do not exist. Thinking this way involves a shift in consciousness where you start to look at everything available in terms of its potential to be recycled and its potential as a cultivation resource. Here are a few of my favorite mushroom composting and recycling projects.

CULTIVATING OYSTER MUSHROOMS ON SPENT COFFEE GROUNDS

Cultivating oyster mushrooms on spent coffee grounds is a simple and enjoyable home activity for all ages, resulting in some good edible mushrooms to boot. If your home brewing doesn't provide enough grounds, try asking your local coffee shop or roaster if you can leave a bucket for them to toss their grounds into, especially if they would otherwise go into the trash. If you're not able to inoculate your grounds with spawn right away, freeze them until you're ready to do so; otherwise molds will form within days.

Although the yields you'll get from this method are not as high as when you use commercial oyster mushroom formulas, such as pasteurized wheat straw or cotton waste, if you factor in the production costs, the lower-yield coffee grounds method becomes as economically viable as the more sophisticated cultivation. If you simply recycle your own

grounds you can expect to produce a few pounds of beautiful oyster mushrooms a week—at which point you'll need to create an oyster mushroom dressing, sautéing your harvest in a balsamic vinaigrette and tossing it over fresh greens crumbled with feta cheese. (Please note: although I have primarily used this process for cultivating oyster mushrooms, some European growers have successfully fruited parasols from coffee grounds.)

Step-by-Step Cultivation on Coffee Grounds

To begin, you'll need a container with a lid, a steady supply of coffee grounds (with or without paper filters), and grain- or sawdust-based oyster mushroom spawn.

Step 1. Carefully collect the cooled and spent coffee filter, grounds and all, and place it into the container faceup. If using a press or strainer just add the grounds to your container once they are drained well.

Step 2. Massage your mushroom spawn bag to separate the grain or sawdust into individual bits to maximize the spreading capability.

Step 3. Sprinkle the mushroom spawn sparingly over the surface of the coffee grounds. You only need a small amount. Crack the container lid so it can breathe. The container can be located anywhere, such as a kitchen counter, garage, or any other space where there is indirect light, never direct sun.

Step 4. Add coffee grounds and filters daily, sprinkling spawn sparingly over each layer as you add more. After just a few days, mycelium will start to be visible as white threads growing together.

Step 5. Fill the container almost to the top, leaving just a few inches of space to make room for developing mushrooms. When you stop adding filters and coffee, the mycelium will finish colonizing.

Step 6. Once the container is completely colonized, expose it to diffuse natural or fluorescent light at room temperature. (If it gets direct sunlight the mycelium and mushrooms will dry up and you won't get a harvest!) Keep the surface misted lightly and the lid just cracked, to preserve moisture. If you have filled a 5-gallon bucket or similar large container, you can drill ½-inch holes around the sides, every 10 inches or so, where the mushrooms can also emerge, but you will need to either mist the holes several times a day indoors or cover the container with a large, clear bag to make a humidity tent until the primordia have safely emerged and are no longer at risk of drying out.

Step 7. Two to three weeks after the colonization is complete, mushrooms should begin to form. Remember that mushrooms only form when they run out of food or space, at which point they recharge their battery and fruit. Baby mushrooms will appear overnight, so check your buckets at least once a day and keep the surface misted, though not underwater. The mushrooms should double in size every day. Harvest them when the fruitbodies' growth slows. You may notice a powdery spore deposit forming underneath the caps when they are ready to harvest.

Step 8. After you've harvested the mushrooms, allow the mycelium to rest by not watering or adding any additional growing media, and it may fruit again in a few weeks. During the rest period no light is needed if you need to move the container. Soaking the coffee grounds with a generous amount of water after a few weeks of resting can help shock the mycelium into fruiting more prolifically. Once rehydrated, the biomass will respond with additional fruitings.

Step 9. After the second flush, your coffee grounds substrate will be pretty much spent as a mushroom growing medium. However, being full of fungal life, it has now become a living compost starter and can be mixed into your outdoor compost pile to help with the decomposition, or you can use it to inoculate cardboard cultures (see chapter 12). Worms also love this spent media, so adding the grounds to your vermicomposting bin could possibly start a worm revolution.

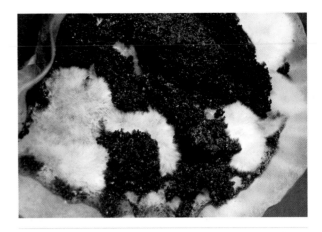

Oyster mushroom mycelium colonizing spent grounds and filters from a local coffee shop.

Oyster mushrooms fruiting on spent coffee grounds and filters in a 1-gallon container. This cluster stripped away the threading and pushed off the lid to escape—reminding me who is in charge here.

Cultivating Mushrooms on Cardboard

There is cardboard all around us. The only thing missing is a little water and a mushroom starter culture, which you can either purchase or make yourself (from the coffee cultivation system above, if you like). This method is best for fruiting oyster mushrooms, but it works well as an expansion method for generating pounds of mycelium from other wood-loving saprophytes. (See chapter 12 for information on isolating and expanding spawn on cardboard.)

Step-by-Step Cultivation on Cardboard

Step 1. Locate a large box or bin in which you can stack sheets of cardboard. A plastic bin will work best, helping to maintain humidity and promoting mushroom formation only on the inner top layer of cardboard. As a last resort you can use a cardboard box as a container; you will need to water it more often, since cardboard is prone to drying out, and mushrooms may form all over the outside, which can reduce your overall yields.

Step 2. Stack all of your cardboard in the bin, and add enough water to cover it. Let the cardboard soak until completely saturated. This may take an hour or so. After soaking, drain the excess water (into your garden) and remove the soaked cardboard from the bin.

Step 3. Layer the bottom of the bin with a few sheets of cardboard, then sprinkle a small amount of spawn across the surface. Repeat, layering cardboard and spawn, until the container is full.

Step 4. Set a lid on top of the bin, leaving it just cracked open, or lay a plastic bag over it. You want the cardboard inside to remain humid but also to allow it to breathe.

Step 5. Monitor the moisture level inside the bin. You want to keep the cardboard moist but not too wet. If it happens to dry out, you can soak the mass overnight and pour out the excess water the next day. Most

Sprinkle spawn onto soaked cardboard to create a mother culture that can be expanded almost indefinitely into additional cardboard and coffee grounds.

A few weeks after spawning, inspect the cardboard to see if it is ready to expand further.

Spawning a pair of jeans. After inoculating the jeans with spawn, roll up each pant leg and let the pants sit for several weeks to allow the mycelium time to colonize. At that point, you can either leave it to fruit or unroll the pant legs and layer them with additional clothing to further expand the mycelium.

mycelium colonize best at a warmer temperature, between 65 and 85°F (18–30°C). Once the bin and cardboard is fully colonized and completely white, you have created a mother culture of cardboard spawn capable of making many more.

Step 6. At this stage, you can expand the mycelium in new bins; just separate the layers of spawned cardboard and shuffle them into additional layers of fresh cardboard in new bins. Or you can leave it alone to fruit. Fruiting will usually occur around the perimeter of the bin. You may also consider drilling ½-inch holes every 8 to 10 inches around the bin. These holes can provide ventilation and a site where mushrooms can emerge, but you'll have to mist them several times a day to keep the primordia from drying out.

Step 7. Once the growing medium in the bin is completely colonized, expose the bin to diffuse natural or fluorescent light at room temperature. Keep the lid just cracked, and mist the surface and any holes regularly. Mushrooms will fruit, most of the time at least twice, and you can then use the spent waste as spawn or you can compost it for an excellent soil amendment. Following complete colonization, it can take several weeks to fruit mushrooms—typically slower than commercial fruiting media, such as pasturized agricultural waste. Still, the easy low-tech approach is a winner when no other options are present.

CULTIVATING MUSHROOMS ON CLOTHING

I started growing mushrooms on clothes when I first became interested in mycoremediation of waste dyes and pigments. There was a textile mill near our farm that manufactured denim for the production

White oyster mushrooms (*Pleurotus ostreatus*) fruiting on old jeans. Scraps of unused or unwearable cotton clothing can be collected and used to produce edible protein in just a few weeks.

of jeans and other clothing. My wife Olga and I went to the mill one day and were greeted by a few friendly folks. I told them I was interested in remediating the indigo carmine they were allowed to release into the waterway based on EPA daily allowable standards. They looked at me a bit nervously, as if I were a whistle-blowing undercover environmentalist; picking up on that, I quickly told them about my mycoremediation research and passions. The man I was speaking to happened to be the owner, and he was excited to hear about the prospect of lessening the mill's environmental impact. The following week I decided to grow mushrooms on old jeans to see if they could decolorize the indigo carmine that makes them blue.

My first experiment was a success, with oyster mushrooms colonizing and fruiting very well on old cotton jeans, but the decolorization of the indigo carmine that I expected was not evident. Turkey tail mushrooms and a few other species are more efficient

at the decolorization process, but what I learned is that old cotton clothing can support fruiting oyster mushrooms. (This could be potentially valuable survival information for anyone directly impacted by a natural disaster, where there is a huge amount of debris, but food is scarce.) Old cotton shirts, bits of rugs, hemp and sisal rope—any material composed of natural plant fibers, including cotton, hemp, and bamboo, can be used to cultivate mushrooms. It only needs water and a bit of oyster mushroom mycelium to get started.

Step-by-Step Cultivation on Clothing

Step 1. Soak the clothing in fresh water. The water does not have to be sterile or clean, only free of heavy metals.

Step 2. Flatten the clothing on a surface. Sprinkle the mushroom starter culture over the surface sparingly. Remember, more spawn will speed the process, not necessarily produce more mushrooms.

Step 3. Roll the clothing tightly, or if you have more than one article of clothing, stack it in spawned layers. Place the clothing in a plastic bag or an enclosed container with a few holes.

Step 4. Check the moisture content of the clothing every few days during colonization to make sure the fabric does not dry out; mist or water it as needed. Room temperature or cooler is perfectly fine for colonizing clothing scraps.

Step 5. When the entire mass of clothing seems to have been completely colonized by the mycelium, increase ventilation by adding more holes or cracking the lid of the container, but not enough that the clothing will quickly dry out. Keep the surfaces misted slightly to induce mushroom formation. The colonization process can vary from one to two weeks depending on how much spawn you use. At this point the mushrooms are not interested in fruiting so no light is needed to promote primordia formation.

Step 6. Once mushrooms begin to appear, which can occur a few days to weeks after colonization depending on temperatures and spawn amount used, they will double in size every day. Mist as frequently as needed to keep the mushrooms from drying out at a young state. When the mushrooms stop growing, they are ready to harvest.

MYCOVERMICOMPOSTING

Composting with worms, known as vermicomposting, is a lucrative business, providing worms for bait shops, feed for chickens and other poultry, and some of the best compost additives for creating organic soil amendments that recharge and revive soils. Worm castings are, ounce for ounce, one of the best and most effective natural fertilizers you can use, and they can easily be generated on home waste, including spent mushroom growing substrate. Since the end result of mushroom cultivation is the creation of soil, and since red composting worms (*Eisenia fetida*) are extremely fond of mushroom mycelia, mycovermicomposting is

a great way to manage the end product of your cultivation efforts. In this way, your mushrooms can reach their full yield potential while also producing a rich and valuable compost.

Spent mushroom growing medium is essentially fully colonized with mycelium, a worm's favorite natural food source. The sweet-smelling metabolites in the spent medium lure worms from afar to make a home and breed in the nutrient-rich substrate. Red composting worms, also known as red wigglers, are vertical migrators, which are the best option for mycovermicomposting because they are able to penetrate every cubic inch of the growing medium. Earthworms are less effective because they tend to compost horizontally, at the soil level where the fresh medium meets the old, operating in a very thin layer. You can find commercially raised composting worms online, or you may be able to find them from local worm farmers. If you have trouble finding the worms, it's possible to collect and raise worms that are native to your area. Although they may not be as efficient as red wigglers, it's better than not having any worms at all.

Although spent mushroom substrate or mycelium-colonized cardboard is a superfood for worms, there is a missing component that is critical for worm health and for your composting success. Worms need fine sand or grit to process the food they eat, grinding it

This sign says it all. For a worm, spent mushroom substrate is where the party is at. "Have Wine Caps, Will Travel!"

The "last gasp," or effort to fruit after multiple flushes, by these little white oyster mushrooms (*Pleurotus ostreatus*) just before we add them, substrate and all, to our worm composting bins.

The worms quickly invade the entire pile, feeding on the sweet mushroom mycelium and growing medium.

This is the same pile eight weeks later. If you don't turn the compost, your substrate will be fully composted into worm castings in about three to four months.

up in their gizzard much as birds do. A small amount of your native soil mixed in with the substrate or cardboard will improve the health of the worms and thereby the speed the rate at which they produce castings. Using your native soil can also supply a dose of beneficial microbes that the worms' gut bacteria need, giving your worm castings rich, beneficial properties. Worm castings are loaded with beneficial microbes that perform essential functions for plants. A seedling selects the beneficial bacteria it needs, and those bacteria form a symbiotic biofilm on the plant's developing root system. The plant responds by cultivating the bacteria, allowing them to multiply and spread with the growing root system for the entire life of the plant. Using native soil for your worm composting additive ensures that the many symbiotic nuances are preserved and perpetuated throughout your operation.

Mycovermicomposting with Spent Columns

If you're cultivating mushrooms in colums, when your substrate is spent and your mushrooms are no longer fruiting, there's no reason to slice open the columns and discard the spent growing substrate. Introduce some red wigglers, and let the worms migrate from

REVERSING MICROBIAL DESERTS

Synthetic fertilizers are killing our soil microbes. When presented with the choice between synthetic and organic compounds, plants will opt for the synthetic, since they do not have to reciprocate any energy loss as a trade. With the synthetic fertilizer as its "fast food" source, the plants roots are no longer nterested in forming a lifelong relationship with soil microbes, which include bacteria and mycorrhizal fungi. The root systems then become microbial deserts, and once the temporary fix of synthetic fertilizer is consumed and gone, the sterile soil has lost connection with the plant host and can offer no support. Then the cultivator must reapply fertilizer, making matters worse. On the other hand, when you remove synthetic fertilizers and provide your garden a more complex source of nutrition such as worm castings, it helps to restore the relationship among plants, bacteria, and fungi, allowing the soil to return its natural state of balance.

Once oyster columns are finished fruiting, you can lay them down and introduce red wiggler worms to the holes. In six to eight weeks, you can add rooted plant starters, such as these sweet potato vines, which will fill the column with potatoes, making the harvest extremely easy!

hole to hole. If you stack the columns, eventually all of the columns will be threaded through with a worm population that is composting the spent media in situ. Mycovermicomposting combines two stages of handling—removing the spent substrate and composting it—into one, which reduces your work and simplifies your operation.

Then the spent column can even double (actually, triple!) as a planter. Six to eight weeks after introducing the composting worms, plant the columns with vegetables and herbs, inserting rooted plugs or starters, for a bonus crop of edibles. You can then recycle the dried vegetative waste back into your mushroom production system.

You can even incorporate the columns into a hydroponic system. Add drip irrigation emitters along the plant openings, directing water from the biofilter into the columns. Arrange the columns at a downward angle to drain. Recapture the excess irrigation water at the lower end, and either route the effluent back through the filtration system or use it to drip-irrigate yet another level of crops.

Mycovermicomposting in Pots or Bins

Worm composting is also compatible with a mushroom cultivation operation using nursery pots—with no modification needed. In fact, commercial mycovermiculture systems are essentially themselves a series of nursery pots. Nursery pots have holes in the bottom, which allow worms to migrate, so you can simply introduce composting worms to pots of spent growing medium and then stack them, placing a tray at the bottom to keep the worms from escaping. (You could do the same with any stackable containers, so long as they have holes in their bottom.) Make sure there is adequate bright light and even some dappled sunlight to keep the worms from wanting to crawl out and away. Add newly spent pots every few weeks to the top and remove the bottom ones, which will be filled with worm compost and castings. Once the worms have exhausted the resources of one pot, they'll migrate to the new food source, so you'll want to time the addition of new pots with the worms' progress through existing media. When the substrate in the pots with worms looks like finely ground wet soil, add another pot on top. This encourages the worms to migrate, which takes just a few days. As a bonus, this means that when you remove the bottom pots and harvest the worm castings, you don't even have to sift out the worms, making this system a potentially flawlessly flowing addition to your cultivation efforts.

CHAPTER 11

Urban Mushroom Cultivation

No forest? No spare garage? With a little creativity, mushrooms are easier to grow in tight places than you might think. You can grow a substantial amount of mushrooms by incorporating them into community or rooftop gardens, or even by growing them indoors in closets and spare bathrooms. Of course, the amount and type of space you might have can vary considerably. Some people have horizontal space; some have vertical space; some may have both. The key is to evaluate your situation—with a site analysis or, if you are indoors, a walk-through—and choose the methods that will help you maximize yields for your given situation.

INDOOR SMALL-SPACE CULTIVATION

The most common and efficient mushrooms for fruiting indoors in small spaces are oyster mushrooms (*Pleurotus* spp.). You'll be surprised by how little space they take up. You can pasteurize small batches of growing medium on your stovetop. You do not need a humidified room; an extra bathroom or closet works fine as long as you provide a simple humidity tent over the fruiting cultures, so that the primordia don't dry out and abort.

You can house as many as thirty 5-pound bags of inoculated fruiting substrate—enough to produce 8 to 10 pounds of oyster mushrooms a week—on a five-tiered rack placed near a window, which typically takes up about 6 square feet of floor space and rises to a height of about 6 feet (which coincidentally is very close to the size of a small closet, if you have an extra one you would like to devote to fruiting). Or you can cultivate them in buckets on spent coffee grounds, as described in chapter 10. Wherever and however you grow in your small space, if you're indoors be sure to provide ventilation to allow for gas exchange, add a fluorescent light if your setup isn't near a window, and cover the rack or containers with a humidity tent. (I would line the floor and walls with plastic if you are experiencing excessive moisture buildup. The object is to provide extra humidity to the mushrooms, but you also need to protect your structure from excess water to avoid rot.)

The 4×4 Indoor System

This 4×4 system will require about two hours of work per week to maintain, but the returns are worth the effort. You can locate your containers in a closet, spare bedroom, or bathroom, or even outside during warmer months. The system is scalable depending on the container size. I call for 5-gallon buckets here, but you could also use pails from a local restaurant or stacking plastic bins in the 5- to 13-gallon range. Just pick a size that is appropriate to the space you have available and that you can easily fill with the amount of growing medium you can prepare on your stovetop. Oyster mushrooms fruit about three weeks after spawning and can flush at least three times over the course of thirteen weeks, so to keep your operation going consistently, you'll use as many as fifteen containers. Drill ½-inch holes evenly spaced around the sides, about every 6 to 8 inches. Make sure all the containers have secure lids and can be stacked several units high without any danger of tipping.

To begin, procure your growing medium. This can be spent coffee grounds or any kind of pasteurized agricultural by-products, such as grasses and cereal straws. To pasteurize a substrate, heat a large pot of water, three-quarters full, to a near boil, then add dried plant debris such as shredded garden plants, chopped cereal straw, plant-based kitty litter, or any other organic material that can be used to grow oyster mushrooms. Push the floating medium down to submerge it, heating for one to two hours on low heat with the pot covered. Remove from the heat, drain all the water, and allow the medium to cool completely. You may have to pasteurize two separate batches to have enough substrate to fill the container completely; if you want to make it a onetime cooking event and you don't mind having a smaller harvest, just use smaller containers, such as 2-gallon buckets. If you are having a hard time finding growing media for oyster mushrooms, try pet or livestock feed stores to see if they have any bulk shredded straw pellets (such as the Streufex type); or you can buy bags of wood pellets (for use in pellet stoves). Both shredded straw and fuel pellets are good for oyster mushroom substrate when you mix them with a little shredded alfalfa (from a pet supply store) as a nitrogen supplement.

Mix the growing medium with your spawn. You can do this right in your growing container, but to make things easier I generally mix the substrate and spawn in a larger tub and then transfer the mixture to the growing container. Fill the container, and then label it with the date, the substrate, and the type of spawn you used. (Keep a log and record what you are doing if you wish to improve your yields.) Snap the lid onto the spawned container, and move it to your growing space. Keep all your newly spawned containers under a loose layer of plastic; this forms both the humidity tent and the fruiting chamber.

Once you have a series of container under way, you can organize them by the order in which you expect them to fruit. Oyster mushrooms will generally pin (begin forming mushrooms) in three weeks. Keep the pinning buckets in the front and any resting or colonizing buckets in the back. Once the mushrooms flush and you harvest them, you can rearrange the containers to position the buckets that will fruit next in the front. If you want to stagger the harvest, once the mushrooms are producing and harvested, let some of the buckets rest and dry out a little, which means pulling them out of the humidity tent and reducing misting for at least two weeks, then return them to the humidity tent and resume misting and watering.

The second flush will typically produce half the weight of the first, and the third will produce half of the second (5 pounds, 2.5 pounds, and 1.25 pounds, for example), so if you are making a container a week, all of the containers will have overlapping flushes producing different amounts. Weigh and add up the yields of each flush every week to see if you are producing too much or too little for your goals.

For calculating yields, after the first thirteen-week cycle, when your system is up and running and you have mushrooms in all stages of cultivation, I would use a starting estimate of 1.75 pounds of oyster mushrooms for every gallon of substrate you prepare every week. So my 5-gallon-bucket system should an average of 8.75 pounds a week.

The up-front costs for this system would probably run about $170: $65 for buckets, $10 for a shallow tub in which to mix the growing medium with the spawn, $20 for the growing medium, and $75 for three bags of spawn. Of course, it will be cheaper if you use secondhand or salvaged items. During your first thirteen-week cycle you will basically be paying off the cost of any purchased materials, but the return on your investment will only get better after that. Thereafter, your costs for every thirteen-week cycle will be for the spawn and the growing medium (approximately $95, or $7.31 per week).

Given potential yields of 8.75 pounds per week and a value of $10 per pound, this production system can be a good investment. Aside from preparing the growing medium and filling the weekly container, the only maintenance it needs will be to rotate the containers once a week, to mist frequently, and to harvest the wonderful, protein-rich mushrooms—enough for a family of four to enjoy year-round. What is amazing is that this entire system takes up only about a 4-foot by 4-foot space, or 16 square feet, but it can be scaled to produce as many mushrooms as you, your family, and your neighbors can use. And after the buckets are finished fruiting, you can add composting red wiggler worms to produce soil that can be used to grow greens and other vegetables on sunny balconies and rooftops!

ROOFTOP AND BALCONY GARDENS

There are a lot of advantages to cultivating mushrooms outdoors on rooftops and balconies rather than indoors. The fresh air and higher humidity outdoors—rather than the air-conditioning and dry heat inside—will benefit your mushrooms, plus these areas account for a large amount of vacant square footage not being used in cities. And when a rooftop space is being converted to a garden, you can create a hybrid system combining vegetables and mushrooms in the same growing environment to establish a circular flow of nutrients, such as using dried plant matter for the growing substrate and then cycling the mushroom compost back into the vegetable beds.

If you're cultivating on a balcony or rooftop garden, you'll want to be sure there is adequate shade available so that the sun doesn't dry and damage your developing mushrooms. Rooftop gardens especially, with their vast open spaces, can be too sunny for mushrooms to develop. I recommend using reflective shade cloth in the range of 80 to 90 percent shade during the summer months or year-round in tropical climates to reflect thermal heat transfer to lower the temperatures of the fruiting area. Your shade cloth, if tightly woven, can double as insect netting, which you'll also need since your mushrooms will probably attract fungus gnats.

Whether you are on a balcony or on a rooftop, you can design a simple horizontal pole structure (or something with a few posts and crossbeams) with tightly woven shade cloth draped over the top and down the sides to the ground. Some shade cloth comes with sewn-in grommets that will allow you to tie it in place or reposition it as needed. Overlapping flaps can serve as an entry point, unless you want to add a screen door.

If you are growing mushrooms in rooftop beds you can simply cover them with row cover fabric, providing both shade and insect protection. If no commercial shade cloth is available, you can use other materials such as bamboo screens, palm thatch roofing, a living shade cloth made of vines, or even a tarp or sheet. In this case, be sure to add insect screening or row cover cloth to the structure to prevent insects from entering the structure as much as possible.

Remember that these structures cannot be airtight, unless you are cultivating in an enclosed system such as a rooftop greenhouse, in which case you'll need to make accommodations for managing temperature, humidity, and gas exchange. (These accommodations are in many ways similar to those you'd make if you were cultivating mushrooms in an indoor environment; see the discussion of infrastructure in chapter 3.) Otherwise, if you are in the open air, your shade cloth should allow for the passage of fresh air and, ideally, for rainfall to permeate the growing space so you don't have to hand-water as much. It can also double as a raindrop dispersant, especially during downpours; large raindrops and volumes of water that strike the shade cloth are filtered into smaller water droplets, sometimes like a descending mist that lands gently on your mushroom growing beds or containers.

Whenever you're cultivating outdoors, extreme heat or cold is a concern. Although it's an option, moving containers back indoors during either short- or long-term temperature extremes may not appeal to you or be very realistic. Again, consider starting with oyster mushrooms, which can be grown outdoors almost year-round in many temperate and tropical climates. Look for strains whose fruiting temperature window matches your climate. Phoenix, warm blue, golden, and pink oysters are often good choices in warmer months, and elm, brown, cold blue, and white are often good choices in the cooler months. Or it may be that growing seasonally is right for you; you can begin your outdoor cultivation when the weather warms and continue production until it begins to get too cold for outdoor fruiting, which is typically around the freezing point for cold-weather strains of oyster mushrooms. At that point, if you have the space, you might want to shift cultivation indoors, or cease production and start up again the following year. Since mushrooms can be dried and store very well, I would recommend overproducing for your needs, so that in the off-season you will have a stockpile that you can rehydrate and consume.

COMMUNITY GARDENS AND PUBLIC SPACES

In urban environments, open spaces with healthy soils are real community assets, and mushroom cultivation can help optimize the use of that valuable space. In community gardens, for example, where vegetable cultivation tends to be the primary focus, mushrooms can integrate nicely with and complement cropping cycles. To begin, many vegetable plants are tall and sun-loving, and many species of mushrooms will do well in the shade they create. And mushroom mycelia unlock nutrients in the soil, enhancing its overall fertility and encouraging those all-important mycorrhizal relationships.

The ideal mushrooms for community gardens are those that prefer to grow in the substrates used there, which are commonly wood chips and composts. King stropharia, for example, will do well in the hardwood chips that are often used as mulch and pathway materials, and blewits and almond portabellas thrive in manure-based composts. Log cultivation is one of the easiest and most attractive methods for incorporating mushrooms into a community garden, since it eliminates the need for building raised beds out of blocks or lumber and the edging will fruit for several years.

You can also build raised beds exclusively for mushrooms using masonry blocks, stones, or recycled concrete. Raise them to a height of 1 to 2 feet and fill them with manure-rich compost. This method works well for cultivating *Agaricus* species such as almond portabella in warmer months. The following season, you can rotate the beds and plant vegetables directly into the composted mushroom medium. This cycle—creating heaps of organic material in prepared beds to fruit mushrooms first, and then rotating in vegetable crops the next season—is an excellent strategy for building healthy soil. And at the end of the season, spent vegetation—squash leaves, cornstalks, and tomato vines—can be used as mushroom growing substrate. Sun-dry it for several days, and then shred it into smaller bits. You can now pasteurize it and use it as a growing medium for oyster mushrooms, or you can add it to your compost and use it for secondary decomposers such as blewits and almond portabellas.

Whether you want to add mushroom cultivation to an existing community garden or are proposing a new site, I recommend meeting with the garden manager or the city officials who oversee the site to discuss the benefits of mushrooms—soil creation, mycorrhizal relationships with plants, a high-protein food produced on-site from plant waste that would normally be composted anyway. Selling the idea should include your thoughts on how you could incorporate mushrooms into the existing processes and infrastructure that help the current garden function.

Although it's not exactly community gardening, you can also make use of alleyways, entryways, and other vacant spaces to grow mushrooms. Of course, you should first inquire about whether a permit is required, and you should consult with area shop owners, city officials, and neighborhood residents to solicit their approval and cooperation. Where space is extremely limited, such as at the entrance to an apartment or store, you can hang columns of oyster mushrooms or stack buckets filled with fruiting biomass directly up against the building wall. Like adobe structures in the Southwest, the building functions as a heat sink, absorbing heat during the day, releasing it at night, and helping to keep temperatures stable for the fungi.

Alleyways can make ideal mushroom cultivation spaces. They are often shaded, and if they're not, you can easily shade them by constructing simple lean-tos or using shade cloth. Old shipping containers can be nestled into the dead ends of alleys and other spaces, in storage lots, or anywhere you have empty space. At Mepkin Abbey, a Trappist monastery in South Carolina, I helped develop a mushroom production system using old refrigerated shipping trucks, which were insulated but had lost their air-conditioning units to vandals. We covered the interiors with plastic and sealed them to make the units watertight, and we added a regular air-conditioning unit on one end. We installed a series of pipes overhead, for hanging oyster mushroom columns. We ran automatic misting tubes down the length of the pipes, fitted with micronizing misting heads and timed to gently water the mushrooms for thirty seconds every hour. When we were done, each 40-foot unit could produce an average of 65 pounds of mushrooms per week. This system could be used in any setting and would be especially applicable to urban environments, where real estate is limited. The modular units could even be stacked to minimize their footprint.

It is, of course, possible to cultivate mushrooms on logs in urban areas, but keep in mind that it takes up a lot of space compared to growing mushrooms in stacked buckets or vertical columns. If you're interested, you might check with your local or municipal landscaping crews to see if they have any logs available from tree removal; you can often procure fresh-cut branches and logs, along with a mountain of leaf litter, at local mulching sites. You can use these materials to make raised beds, inoculating them with species that benefit from ground contact, such as shiitakes or reishi. My advice is to avoid mushrooms that have long incubation times or are difficult to fruit. The best strategy for urban farming, for vegetables and mushrooms alike, is a rapid turnover, given the limited amount of space. Think of it as speed dating for food production—fast rotation and optimum return.

GUERRILLA MUSHROOMING

Some of my favorite mushroom-hunting spots are in urban parks, where the green, carefully tended, and typically irrigated open spaces are attractive to mushrooms. (Be sure to ask the workers if they are spraying the beds with any pesticides before consuming these mushrooms, since they may catch chemical drift on their caps.) Sometimes I carry king stropharia and blewit spawn to parks and college campuses and broadcast it into the beds in the hope of establishing a wild patch. In the two years since I inoculated a certain area campus with king stropharia spawn, several patches have caught on and spread, continuing to sporulate and spread into the fresh mulch that grounds workers add twice a year.

I have also located and spawned the "mother pile" of woody debris at my city's recycling center and inoculated logs waiting to be shredded. This kind of guerrilla inoculation magnifies the mycelium downstream, with particles sticking to the tub grinders and shredders, enhancing the spread of edible mushroom species. This is a fun (and easy) way to cultivate mushrooms—leaving the mycelia alone to fend for themselves through periods of rain and drought, so that they develop unique resiliencies, given the particular growing substrate and environmental conditions.

INDOOR URBAN FOOD FACTORIES AND VERTICAL FARMING

Over the past few years I have seen many fabulous projects—both in the planning stages and operational—that involve the multilevel repurposing of old buildings and abandoned factories for closed-loop urban food production. For example, the Plant (www.plantchicago .com) in downtown Chicago operates three demonstration farms at an old meatpacking plant: an outdoor vegetable farm, an indoor aquaponics operation, and an indoor mushroom cultivation operation. At an operation like this, there are many ways to create more sustainable urban food production, such as using the waste from the aquaponics system and spent mushroom substrate to fertilize the vegetable garden. Mushrooms, by their very nature, incorporate beautifully into systems like this, helping to create and perpetuate a circular food system.

There are many creative ways—both simple and sophisticated—to develop this kind of closed-loop system. For example, not only can spent mushroom substrate be used to build soil, but mycelia can function as biological filters to supplement the biofiltration units in an aquaponics operation. In fact, there are many ways to sequence the biomass or by-products (especially gases) to help keep the system closed. While you might grow vegetables in rooftop greenhouses and mushrooms in the low-light indoor spaces, you could design a system to circulate the carbon dioxide from the fungi up to the rooftop to benefit the plants, helping to maintain temperatures and manage the gas exchange in the mushroom fruiting room. Or you could force the mushroom fruiting room's exhaust air up to tanks full of high-fat algae for making biodiesel from the pressed algal biomass. Another possibility is to use water hyacinth for biofiltration in an aquaponics operation. You can continuously harvest that water hyacinth for use as a growing substrate for oyster and paddy straw mushrooms. Or you can compost the water hyacinth to produce a substrate for cultivating almond portabellas and eventually vermicomposting elsewhere in the facility. (See page 71 for more on water hyacinth.)

CHAPTER 12

Shroomin' Off the Grid

Maybe you live off the grid, or maybe you simply want your mushroom operation to use as little energy as possible. Whatever your situation, if you want to align your cultivation skills with planetary stewardship and reource conservation, there are a lot of ways you can develop or tweak your operation to make it ultra low-tech or off-the-grid.

At our farm we employ several methods of cultivation. We practice some large-scale industrial cultivation to make ends meet, and we try to use those profits to develop methods and strategies that are more sustainable and aligned with our ecological values. My goal, for every aspect of the business, is to find the lowest common denominator between accomplishing a goal and minimizing impact on the planet. That can mean weighing the costs and benefits of, for example, cultivating a mushroom on a seasonal basis rather than paying for fuel to heat a fruiting room. Everyone's solutions will be different.

Preparing the Growing Media

You can grow mushrooms on a lot of substrates—paper, cardboard, dried vegetable waste, cotton clothing, et cetera—so finding growing media will probably be one of your less difficult challenges. Preparing those media in a way that minimizes energy and resource consumption, however, poses more difficulty. However, there are options worth considering in a pinch.

Even without electricity, you can pasteurize or sterilize with heat, whether over a fire or with solar pasteurization; see chapter 7 for more details. In very cold environments, consider the opposite: freezing. Soak shredded straw in water for several days, freeze it for a week, and then inoculate it with spawn. I have tested this method several times with my home freezer at 0°F (–18°C) and found that the majority of contamination was destroyed by the cold treatment. So if your climate is this cold, or becomes this cold seasonally, you can cold-treat growing media outdoors before bringing it indoors for mushroom cultivation.

Fruiting Infrastructure

If you are off the grid or simply trying to make do with whatever resources you have available, it's unlikely you'll be constructing a fruiting room with elaborate systems for controlling temperature, humidity, and light. The challenge—and the fun—is that instead of thinking about what you want to grow and then building the infrastructure and buying the materials to match it, you have to think about the resources you have available and then figure out what to grow and how to grow it.

One of the best things you can do is choose a mushroom strain appropriate for and adapted to your climate. (See part 4 for information on temperature fruiting windows for various mushrooms; this will give you a place to start in your decision making.) If you do this, you'll need less in the way of artificial means to meet your mushrooms' needs for heating, cooling, gas exchange, water, and light. Consult chapter 11's discussion of cultivation on balconies and rooftops, since many of those ideas can apply to outdoor off-the-grid cultivation as well. If you decide to grow indoors, underground spaces such as bunkers, caves, wine cellars, and basements can naturally provide relatively even temperature and humidity levels, but you'll need to provide a means of gas exchange so that carbon dioxide doesn't build up, such as using low-light plants that will thrive in the light conditions provided by solar-powered LED lights or exhaust fans to expel the air that sits at the lowest point of the structure.

CARDBOARD CULTURES: LOW-TECH SPAWN CREATION

Although spawn isolation and expansion in the lab will be covered in detail in chapter 18, many wood-loving saprophytes can also be propagated using a primitive technique called cardboard inoculation. This may be a good option if you are off the grid, on a budget, or simply interested in low-tech methods. The easiest mushrooms to expand with this method are oysters, blewits, and king stropharia. It can be used with other species, but it will not be as successful with compost and secondary decomposers such as almond portabella, paddy straw, and shaggy mane, and it definitely is not for use with mycorrhizal species such as truffles, chanterelles, and porcini, since they must have a living host to bond with.

You can use extremely fresh cultivated mushrooms, but mushrooms picked fresh from the wild are best since they'll come along with beneficial microbes from the surrounding soil, which some species require for fruiting. Remove the base of the mushroom (where it was attached to its substrate), and chop it into small pieces. (You can save the cap for eating.) Soak a piece of cardboard in water until it is thoroughly saturated. Scatter the the bits of chopped base over the wet

cardboard, roll it up like a burrito, and seal it in a plastic bag. Let sit for a few weeks, during which time the bits of stem will revert to mycelium. If you are storing the cardboard culture at temperatures above 60°F (16°C) you will need to poke a few airholes to allow the developing culture to breathe. I like to incubate my cardboard cultures under refrigeration (or the coolest place possible), which minimizes molds and bacterial rot.

After a few weeks, remove the cardboard from the bag and unroll it to see if the mycelium has started to spread. If you see any signs of contamination—such as patches of green, yellow, or even black mold—remove the good-looking portions of the spreading mycelium, transplant it onto fresh cardboard, and repeat the process. Once you have a fully colonized your first sheets of cardboard using stem bases, you can easily expand the mycelium by layering the colonized sheets with fresh cardboard, layering them in a plastic tub, bin, or bucket. Each expansion will take from just a few days to a week. Repeat the process to build up a volume of "sheet spawn," which you can then use to inoculate growing media. Mist as needed to maintain moisture for the colonizing mycelium. This cardboard spawn can now be used to inoculate logs, stumps, wood chips, and pasteurized media or heat-treated compost in the place of plug, sawdust, or grain spawn created in a laboratory. The spawn, although not sterile in the sense of pure, is quite clean and maintains its vigor when transferred to different media types.

The cardboard culturing process takes practice and patience, and it may take a few tries to get right, but once you do get it right, the expansion of your cultured mycelium usually proceeds quickly, with little risk of contamination. But it's important that you can tell the difference between mushroom and mold mycelium. Mold mycelium can resemble mushroom hyphae, and you don't want to transfer that mold to your entire culture—or to your entire cultivation operation, if you're not being careful about sanitation measures. So before embarking on the culturing journey, familiarize yourself with the common contaminants discussed in chapter 9, such as *Trichoderma* spp., so that you don't accidentally perpetuate them. If you see any powdery growth on your cultures, toss them immediately, and

Wild-harvested blewits (*Clitocybe nuda*) showing stem bases loaded with fuzzy mycelium bound to decomposing leaf particles. The stem bases harbor beneficial microbes and bacteria that the fruitbody has selected as collaborative partners.

Remove the stem bases of the blewits (and save the caps for eating).

Chop the stem bases into smaller pieces and spread them across wet cardboard. Roll up the cardboard into a "blewit burrito."

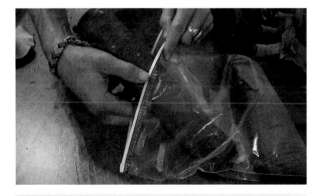

Place the burrito in a plastic bag and store it in a cool place for several weeks.

Once the cardboard is colonized with mycelium, you can layer it with fresh sheets of cardboard to further expand it, whether for spawn or fruiting.

CULTURING ON THE GO

There have been times when I've been traveling without my lab supplies and have stumbled on a great species that I wanted to culture and take home to my lab. In that case, I roll the stem bases in the cardboard. It is often enough to regenerate the stem base into mycelium even if the species doesn't particularly like the cardboard as a food source, buying me enough time to make an additional transfer several weeks later to a sterile agar plate to isolate the species.

sanitize the area, your hands, and any tools or equipment that touched those cultures.

Maintaining the culture is actually simple enough. You just expand the mycelium on different substrates, rather than propagating it over several generations on the same growing medium, which decreases vigor and fruiting capacity.

CARDBOARD CULTURES AS SPAWN: EXPANSION AND FRUITING

Once your cardboard cultures have expanded and you are actively culturing the mycelium you wish to grow, you can allow the mycelium on the cardboard to proceed to the point of fruiting, but it will ultimately deliver poor yields compared to cultivating on other substrates, such as pasteurized plant debris, wood chips, or logs. A better option is to treat the cardboard culture as spawn, using it to inoculate various substrates. There are a couple of different methods for doing this.

The Log Disk Method

This technique is mainly for growers who have access to a chain saw for cutting 6- to 12-inch-diameter logs into smooth rounds approximately 6 to 8 inches thick.

(If you do not have a chain saw, refer to the wafering method below, which calls for a machete or hatchet.) For this example I'll start with two rounds and one colonized sheet of spawn (I suggest oyster or shiitake), but you can do this with as many rounds as you want, and you can use sawdust starter culture instead of a cardboard culture if that's what you have.

Place your colonized sheet of cardboard between two stacked disks, watering generously to wet both the cardboard and disks. Cover the stack loosely with plastic or leaves to preserve humidity and moisture. After two weeks inspect the disks on the face touching the cardboard for visible signs of mycelium. Look for an even white layer across the entire face of each disk. If there are gaps, replace the rounds and wait another week or two. Once there is an even white layer on the cut face that is in contact with the cardboard, flip the disks so the cardboard is sandwiched between the uncolonized sides.

To expand the mycelium at this point, you can add a fresh sheet of wet, uncolonized cardboard at each end of the stack, sandwiched by two more disks. The mycelium from the colonized sides will quickly run through the wet cardboard and infect the new disks, giving you a total of four disks and four sheets of cardboard that are inoculated, in just under a month. Once the new rounds have the even white layer of mycelium, you can repeat the flipping and addition of fresh rounds and cardboard.

As you proceed, you can begin to move colonized rounds from the expansion group to a separate fruiting area, which is best in stacked totems so they can fuse and combine water and energy resources to create more prolific fruitings. Locating these rounds with the lowest round either flush or partially submerged in the ground can also provide access to extra ground moisture. To keep the cultures strong, try not to fruit the original four disks. Use them instead to start new rounds. This will keep the downstream fruiting disks more vigorous, since the cells have not replicated so many times. You can use this method to expand rounds into perpetuity if you keep the culture strong and the rounds shuffled every few weeks. Allowing the mycelium to grow into different wood types or

You can use cultured log disks much like icing on the layers of a cake, expanding rounds into perpetuity if you keep the culture strong and the rounds shuffled every few weeks.

media can extend the strength of the culture by challenging its enzymes to adapt rather than keep feeding on the same food source.

The Wafering Method

Using the wafer method is a long-term project. You use a chainsaw, machete, or hatchet to slice into the bark of a log, making grooves or flaps. You then insert wafers of colonized cardboard into the flaps and let the mycelium get to work. It will be several years before the mushrooms fruit, but the log will produce for up to a decade.

Slice the flaps at regular intervals along the length of the logs in a diamond pattern, much like the drilling

you do for plugging with dowel or sawdust spawn (see chapter 5). Insert small pieces of colonized cardboard into the flaps; if you can find any bits of moss, use it to cover the cardboard to trap and preserve moisture. These wafers will dry out quickly if it is not raining, so commit to a daily watering for two to three weeks. After a few weeks, inspect a few of the inoculated sites to see if the mycelium is infecting the wood. Once a white mat has formed on the wood tissue, you can reduce your watering to once a week for a month and eventually just let Mother Nature take over. You can remove the pieces of cardboard and reuse them on other logs and trees after a month or two, making the best use out of your mycelium.

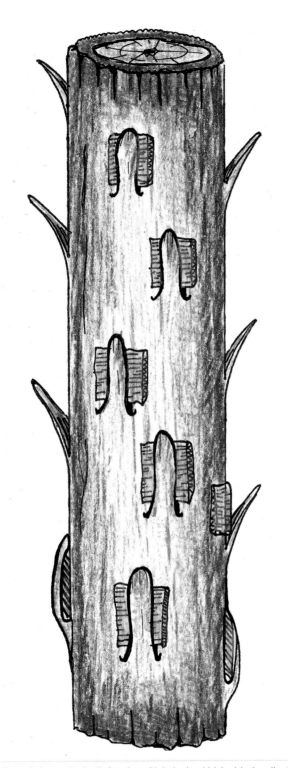

The wafering method calls for a log with its bark, which is nicked or sliced with a chain saw, hatchet, or machete to make grooves or thin flaps for holding wafers of cardboard culture.

STORING CULTURES AND SPAWN WITHOUT ELECTRICITY

In growing situations that are truly off the grid, it can be a challenge to store the mushroom cultures you have isolated using the cardboard technique and any spawn you have created through the off-season, when temperatures or other conditions are not conducive to mushroom cultivation. To begin, you have to convince the mycelium to slow down and enter a sort of stasis. Two methods that can trigger the mycelium to suspend operations without damaging it are dehydration and freezing. Whichever method you choose, store a lot more than you think you'll need, since some of your cultures and spawn may not survive. This gives you the option of selecting the healthiest-looking specimens from among those that revive and begin expansion or colonization when the growing season resumes.

Dehydration

Dehydration is the best option for off-the-grid storage in warmer climates. Mycelia in growing media can survive a long time without water if you do not expose them to extreme heat and keep them relatively dry. Allowing the substrate of a colonized column, bag, or pot to dry completely can provide a means of storing the mycelium until you are ready to use it.

You can also dry mushroom stems (with their bases) and store them in bundles in a cool, dry location. When you're ready to revive them, wrap them in wet cardboard and proceed following the instructions for making a cardboard culture (see page 98). I have experimented with cloning mushrooms, both in the lab and with cardboard expansion techniques, using the dried stem bases of species such as oyster mushrooms, and they have remained viable for nine to twelve months as long as I don't expose them to extreme heat or cold.

Freezing

In freezing environments, you can store spawn or even just mushrooms (whole or just the stem bases) outdoors in insulated containers, such as sealed thermoses, for several months. Drying the mushrooms first helps prevent crystallization of the cells and minimizes damage to the mycelium. Bundle the stems tightly in clean, dry cardboard to help insulate them and protect them from sudden temperature fluctuations. You can also freeze a cardboard culture: Submerge it in water, freeze it for up to several months, thaw it slowly, and then place the culture on fresh wet cardboard to expand it.

Disaster Relief with Mushroom Rescue Modules

Disasters can strike anywhere at any time. In a moment, hurricanes, tornadoes, tsunamis, earthquakes, and other disasters can turn life upside down. In remote or impoverished regions of the world, where resources are already stretched thin, food and water can be quickly depleted. Lack of electricity and sanitation can lead to contaminated drinking water and outbreaks of disease. When the news is fresh and the international community is watching, help is usually quick to come, but once the news crews pull out for their next crisis, the victims are left behind to fend for themselves. And often they simply don't have the resources to do so. Two years after the 2010 earthquake in Haiti, over half a million people were still living in camps with deplorable, life-threatening sanitation conditions. Sadly, the $2.4 billion in international aid that had made it into the country was unable to provide long-term solutions for food and clean water, making people even more vulnerable once the aid shipments started decreasing and then ceased altogether.

We need lasting solutions to meet human needs in the face of crises and disasters, especially in the more vulnerable areas of the world. What if there were a low-tech way to provide a consistent source of highly nutritious food in a matter of weeks? Something that not only "buys time and saves lives" while recovery efforts are under way, but could have a lasting effect and help rebuild communities? Something that is inexpensive and transportable, able to withstand the heat of tropical climates, and easy to learn, even to the point of being diagrammed in pictures for people who cannot read? Something that could help bring about a new wave of humanity, a paradigm shift for food security, a boost to worldwide health and prosperity?

What are we waiting for? Mushroom rescue modules—as I call them—are inexpensive, easy to use, and highly productive protein generators. They can generate golden oyster mushrooms in as little as three weeks and are also capable of filtering contaminated water and reducing mosquito populations (and therefore the incidence of malaria). I have tested these modules at our farm. They are an idea that I think needs to be shared and tried multiple times in the field to improve their design and effectiveness given the ease with which they can provide food from inedible debris.

For an estimated cost of just $50, you can assemble a mushroom rescue module containing everything someone would need to successfully cultivate nutritious edible mushrooms. The basic kit should contain the following:

Fruiting containers. Twenty plastic 5-gallon pots. These can be collected from plant nurseries for free, or purchased for approximately $1 each. After two weeks, additional fruiting containers, eventually over a hundred, will be needed to expand the process, which can presumable be found on-site, since mushrooms can fruit in just about any partially sealed container. If no containers are available, it will be necessary to provide as many as needed to sustain the harvest.

Growing medium. Fifteen pounds of dried, chopped straw. This amount of growing medium will ensure that there is enough spawned biomass to create many more modules downstream of the initial expansion, so that the mycelium can adapt to the climate and kick-start the operation until more substrate can be collected on-site.

Oyster mushroom sawdust spawn. Five pounds. It's important to use species and strains that are tolerant of the climates where they will be cultivated. For tropical destinations pink, golden, and phoenix oysters are good choices. For temperate and colder

Training students at an agricultural trade school located in Mirebalais, Haiti, on how to expand mushroom cultures using the waste and debris commonly found on hillsides, streets, and ditches. This is how we will start the mushroom revolution—forming a grassroots community of knowledgeable students who can share their experience with others.

climates, the kit should include spawn for cold blue, elm, or brown oyster. One 5-pound bag is enough to inoculate the first twenty 5-gallon buckets.

Fuel stick. A fuel stick, or compressed fuel log, capable of burning for four to six hours can be used to heat-treat the dried straw for one to two hours, as well as to pasteurize drinking water.

Waterproof matches. Preferably the self-striking type. These will be needed to ignite the fuel stick or to use for starting fires to pasteurize growing media, if necessary, and for boiling drinking water.

Clear plastic sheeting (12 feet square). This will be used to cover the colonizing and fruiting pots to keep the substrate humid and prevent developing mushrooms from drying out.

Rope. Fifty feet to suspend the plastic sheeting between two posts or trees

Five-gallon bucket. Punched with side holes. This bucket is optional, but useful to trap mosquitoes in areas where malaria, yellow fever, and other diseases they carry are of concern. The bucket is partially filled with water, dosed with Bti, and placed near the mushroom cultivation area to attract feeding female mosquitoes.

***Bacillus thuringiensis* var. *israelensis* (Bti) culture.** Dried. The bacteria is a larvicide for fungus gnat and mosquito larvae and can be used in 5-gallon mosquito trapping bucket or diluted and mixed directly into the growing media.

Upon delivery, recipients should find in the module a detailed illustrated diagram, labeled in the native languages or dialects, explaining the contents, expansion and cultivation process, harvesting, and perpetuation of the system. The only missing component is a large cooking pot, which would need to be procured on-site (but which can be shared between individuals).

The Mushroom Rescue Module

Component	COST (USD)
Twenty 5-gallon fruiting containers (recycled)	$0
Growing medium	$3
Oyster mushroom sawdust spawn	$25
Fuel stick	$5
Waterproof matches	$2
Clear plastic sheeting and rope	$5
Five-gallon bucket	$5
Bti culture	$5
Total cost (before shipping)	**$50**

Using the Module: Week by Week

Week 1. Inoculate a small amount of the substrate—about enough to fill a 5-gallon pot, using about 1 pound of mushroom spawn for every 5 gallons of growing media to create your first mother culture. (I'll refer to pots here, but you can use any available containers.) Include in the substrate any growing media available on-site, from small, saturated pieces of cardboard to dried plant material. Store the culture in the coolest place you can find in a tropical environment, or between 65 and 85°F (18–29.5°C) in temperate climates, covered tightly with plastic (provided as part of the module) to help maintain humidity during colonization. Remaining spawn can be kept cool and saved for expansions by rolling it up to keep it closed until needed for future expansions. It does not have to be refrigerated. (If mushrooms have formed on the spawn, you can eat them before using the spawn to inoculate the substrate.)

Week 2. When the substrate is fully colonized, expand the culture tenfold by mixing the colonized substrate with enough soaked fresh growing medium to fill ten additional pots. It is okay to use paper and cardboard or clothing that has been soaked in collected water, but if you are going to use agricultural wastes, you will need to pasterurize the media first (see chapter 6).

Week 3. Once you've colonized ten 5-gallon pots, they will become the "master" spawn pots. If you let the pots sit, they will fruit, so if you need mushrooms immediately, allow a few to produce, but keep the others expanding to magnify the mycelium as quickly as possible. Make ten more containers from each master spawn pot. Theoretically you could produce 100 containers, but since there will be some losses due to contamination by mold or bacteria, overproducing is the key. Make a lot more than you think you will need.

Week 4. By week four you will ideally have 100 "expansion" pots, each filled with colonized growing medium. Dispose of any that are contaminated with visible molds or undesirable growth. Save the best pot to make ten more pots, which will become your new master spawn. Use all of the rest to inoculate fruiting containers. Each 5-gallon expansion pot can be used to make thirty 5-gallon fruiting containers filled with a pasteurized growing medium like dried water hyacinth, dried banana fronds, dried vegetable waste, prairie grasses, shreds of cotton clothing, cardboard, or whatever is available.

Weeks 8–11. Mushrooms will emerge from the fruiting containers in about one month and will continue flushing for several weeks, yielding an average of 3 pounds per fruiting pot. Assuming 20 percent losses due to contamination, that's eight master spawn pots (80 percent of the ten that you attempt to make), each producing eight expansion pots (again, 80 percent of the ten that you attempt). You hold back one pot as the new master spawn, leaving sixty-three expansion pots. You use each expansion pot to inoculate thirty fruiting containers, of which twenty-four (80 percent) are successful, each producing 3 pounds of mushrooms: $63 \times 24 \times 5 = 7{,}560$ pounds of edible mushrooms within eleven weeks, from one mushroom rescue module. Since you held back that pot of master spawn, you're able to perpetuate this production cycle indefinitely, producing thousands of pounds of mushrooms weekly. And you can scale the process upward or downward as needed.

Using the Mushrooms: Resiliency and Adaptability

Mushroom rescue modules produce a food that is 20 to 30 percent protein (dry weight). In fact, over a fourteen-week period, oyster mushrooms produce almost double the amount of protein as free-range chickens (grown to fourteen weeks and weighing 3.5 pounds).

Although it's best (and simplest) to eat mushrooms, overproduction, gaps in production, or climate fluctuations can leave you with too many at certain times and not enough at others. In this case, mushrooms can be dried and stored indefinitely in airtight containers such as plastic bags or jars; they will not lose their nutritional value. As long as the mushrooms are dried to the point of crumbling, they are safe from deterioration. Powdering itself can be desirable; it renders the mushrooms into protein-rich flour that can be used to make pastries and breads. Sun-drying in tropical climates where the humidity is extremely high is difficult, so in that case I recommend building a solar dryer or dehydrator for the rapid and efficient processing of fresh mushrooms. This can, in turn,

distill safe water from the fruitbodies. Whatever you do, just do not let the mushrooms rot and go to waste.

As part of the production cycle, you can also use or sell the valuable mushroom by-products. You can use them as livestock feed, or you can use the spent substrate as mulch to improve soil fertility and counter erosion and runoff. You can compost mushrooms and their growing media, producing a fertile and microbiotically active soil able to grow nutrient-rich sprouts such as sunflowers in as little as one week. Or you can convert mushrooms and spent growing media into a vermicomposting operation, producing rich soil and a high-quality source of worms for chicken feed. With organization and creativity, there are many opportunities to continually improve the cycle of resources and production.

By now, I hope it's clear that mushroom rescue modules have many potential applications and benefits. In the same way that mushroom production can expand to huge volumes, just one module in the hands of an experienced person who can train others can have a ripple effect. There is also a sort of security with this kind of food cultivation. If one person's crop fails, adjacent mushroom farmers can provide immediate assistance and replacement spawn. One of the wonderful things about mushroom rescue modules is that they can provide the very tangible benefit of producing large volumes of nutritious food and valuable by-products, and they can also foster the less tangible but equally important benefits of building community and trust.

Using the Module for Mosquito Control

Any mushroom cultivation operation will attract a variety of insects. At my own operation, I have noticed that mosquitoes are attracted to and feed on the oyster mushrooms fruiting in our greenhouses. At first I didn't understand why they were voraciously feeding on fungi. I thought that perhaps they were mistaking the mushrooms for an animal. (Remember that fungi create heat, carbon dioxide, and water, just as mammals do.) I discovered that many commercially available mosquito killers use an "octenol lure" as their attractant—and their octenol is actually

---

MAKING CARDBOARD TOWERS

In an urban area, cardboard may be more readily available than plant material. If that's the case, you can use your initial pot of colonized growing medium to make cardboard towers. Just scatter the colonized medium between layers of wet cardboard until you've made a tower approximately 4 to 5 feet tall. Cover the cardboard tower with a sheet of plastic, and keep it wet. The mycelium will colonize and fruit in the cardboard in much the same way it does any other growing media. Position the tower near a waterway and in the shade if possible. For expansion, separate colonized sheets of cardboard and shuffle them between sheets of fresh wet cardboard to make new towers.

a mushroom extract that mimics the chemical composition of human breath and sweat. Light bulb! I set to work creating "mushroom mosquito traps," and for three years now we have been focusing on mosquito control as an added benefit to mushroom rescue modules. This technology has potential for applications in places where malaria, dengue fever, and yellow fever are a continuous, emerging, or reemerging problem. It is simple and effective, and it can easily be adapted and taught in local communities. Here's how it works.

To make a mosquito trap, you will need a bucket or container with a removable lid. Cut or drill small holes in the container about 4 inches from the base around the entire container spaced 4 to 6 inches apart, large enough for mosquitos to enter, which is typically around 1 inch. The hole height will be the level of the water inside where mosquitoes can lay their eggs. Add Bti (*Bacillus thuringiensis* var. *israelensis*), a biological larvicide, commonly sold and marketed under names such as Thuricide or Gnatrol. Place small chunks of mycelium in the bucket inside a small cup or container, either floating or weighted down to keep it from overturning, that fits inside the bucket to keep it from sitting in the Bti liquid; these will offgas carbon dioxide. Then cover the bucket and drill a few large holes just above the waterline. The holes allow the carbon dioxide to "ooze" out, which will attract mosquitoes. Once the mosquitoes are drawn in to feed, the females will look for stagnant water in which to lay eggs—which is conveniently right below them. Larvae that grow from eggs laid in the solution ingest the larvicide, which destroys organisms in their guts critical to their survival. This makes them unable to feed, killing them quickly and soon reducing the local population of mosquitoes.

The Bti bacteria remain viable for weeks in dark locations, such as in the closed bucket of water. And perpetuating the bacteria is simple. Bti is a gram-negative, soil-dwelling, spore-forming aerobe that grows easily on straw or shredded agricultural waste—the same sort of stuff you use as a substrate for growing mushrooms. You can culture the bacteria simply by soaking unpasteurized agricultural waste by-products, such as cereal straws or dried plant debris, with

Compost from the spent growing media can be used to produce protein-rich sprouts, such as these sunflower seeds, using little soil and with an extremely rapid production cycle that requires little space.

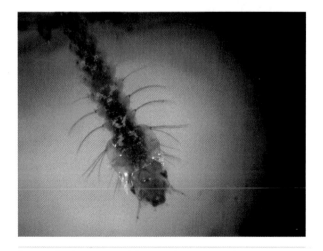

A mosquito larva captured in a mushroom mosquito trap near our composting operation. The mycelium creates heat and outgases carbon dioxide and octenol, a chemical similar to what is found in human and animal sweat.

the bacteria culture and allowing the cultures to perpetuate over several days to allow the bacteria to multiply. Harvest small quantities of this inoculated vegetation when you need it, or to create additional buckets, and add a small amount, a handful or so, to seed the traps with Bti to use for mosquito control.

Instead of the Bti culture, you can also use diluted apple cider vinegar mixed with a small amount of liquid soap (1 teaspoon of soap per cup of vinegar) for mosquito control. The scent of the vinegar attracts the insects, and when they touch the surface to feed, the surface tension and adhesive effect of the soap pulls them under, rapidly drowning them. Carefully swirl the soap into the vinegar without creating bubbles. If soap and Bti are not available, a sticky mat or any type of gooey adhesive can be placed in or near a mushroom trap to capture the insects (which also makes them easier to count).

Using the Module for Water Purification

In addition to providing a reliable source of nutritious food, mushroom rescue modules can be used to filter water of biological and chemical pollutants. Although I'm still in the process of researching this technology, it shows promise. The hope is that water from contaminated water supplies can be percolated through the same biomass used for mushroom production as a means of eliminating pathogens—for example, for filtering out *E. coli* and other coliforms endemic to livestock farming, as well as the bacteria that cause cholera and other serious diseases.

Chemical pollutants would need more contact time; several passes through the mushroom growing media are generally required. As an alternative, spent mushroom substrate can be floated in the water, releasing enzymes that degrade toxic compounds. For a complete understanding of how mycofiltration works and diagrams on how to construct and operate mycofiltration modules, see chapter 22. At this point, anyone who is considering using mycofiltration to produce drinking water should have their operation thoroughly tested by a qualified microbiologist or research lab to verify the efficiency of the filters and removal of contaminants.

CHAPTER 13

Mushroom Products and Cutting-Edge Applications

You can make many products directly or in part from mushrooms. Some of these applications have been around for a long time; others are on the cutting edge, and in the years to come we'll see more and more innovative entrepreneurs developing fungus- and mushroom-based products. (You could be one of them!) This chapter explores the many exciting possibilities, both existing and in theory, of using mushrooms to improve our health, homes, and consumer products. Because not all of these applications have yet been fully developed, there aren't detailed do-it-yourself instructions for everything. But it should be enough to get you thinking, experimenting, and imagining a future where our evolving understanding of the fungi kingdom can be used to make our lives better and solve some of our most pressing problems.

EDIBLE MUSHROOM POWDERS

One of the best ways to extend your use (and sales) of mushrooms is to create value-added products, especially those that use the so-called waste from your operation, such as perfectly fine but unsightly mushrooms culled from your production line. Powdering dried mushrooms not only is a good option for long-term storage, but also yields a complementary product to fresh and dried mushrooms. By drying and powdering, you are increasing the surface area of the mushroom, allowing for more intense flavor and a different texture from that of fresh mushrooms. The most frequent complaint I hear about mushrooms from people who don't like them is the soft, sometimes slippery texture. Powdering can help someone finally find pleasure in the flavor and health benefits of wild

You can powder dried mushrooms and stem bits using a grinder or blender. Mushroom powder can be incorporated into dishes on its own and also makes a nice flour for breading fish, chicken, and vegetables.

and cultivated mushrooms. You can mix powdered mushrooms into many recipes where solid mushrooms just don't fit. One of my favorite uses is to dust scallops or fish with mushroom powder, sear them in hot oil, and then bake them gently to produce a juicy mushroom-encrusted dish.

Step-by-Step Mushroom Powder

Step 1. Dry your mushrooms in a dehydrator or solar dryer until they're brittle (see chapter 2 for more discussion of drying).

Step 2. Break up the mushrooms as best you can with your hands. Then use a blender or mortar and pestle to reduce them to a fine powder.

Step 3. Store the powder in an airtight container, such as a mason jar. Add small bag of rice to keep moisture at a minimum (optional). The powder will keep for years if kept dry.

Mushroom Powder Honey

One clever mushroom storage and delivery system is mixing your powdered mushrooms into honey. Honey is, of course, naturally antibiotic and resists spoilage, meaning that it will keep your mushroom powder viable virtually indefinitely. (Ancient Egyptians stored mushroom extracts in honey to preserve the potency; analysis has revealed that the extracts retained close to their original potency for more than four thousand years.) And mushroom honey has a wonderful flavor; our favorite way to use it is as a sweetener for teas.

Honey on its own has proven antimicrobial properties. When you combine it with mushrooms that have antibacterial and antiviral properties, you have a potent medicinal remedy. Good candidates include turkey tail, shiitake, maitake, reishi, and almond portabella; using these species in combination has been shown to be even more effective than any one species alone. Mushroom honey is a wonderful aid during cold and flu season. It not only soothes the symptoms but helps prevent illness. When I'm sick I dip a spoon in the honey and let it slowly dissolve in my mouth, so that it coats the back of my throat; the effect lasts for hours, easing a cough and soothing a sore throat. And I give mushroom honey to my friends, family, and coworkers to help them resist illness, creating what I hope is a sort of herd immunity. Mushroom honey also works well on minor cuts and scrapes in place of antibiotic ointments.

To make mushroom honey, just mix 3 tablespoons of mushroom powder into 1 pint of raw honey. Store in an airtight container. You can use it daily as a sweetener; to treat a cold or the flu, take 1 teaspoon for every 50 pounds of body weight, daily. (Mushroom honey is perfectly safe for children over the age of two.)

Mushroom Powder in Pasta and Breads

You can mix mushroom powder directly into your favorite homemade pasta or bread recipe to give it a unique flavor and nutritional boost. You can replace up to 25 percent of the flour in your recipe with mushroom powder. One challenge with substituting larger quantities of powder is that mushrooms don't bind as well as wheat flour; if your breads or pastas are not holding together, increase the binding agents, such as egg whites, or simply experiment with using smaller quantities of mushroom powder.

The more pungent mushrooms, such as shiitake, maitake, almond portabella, and morels, work especially well with pastas. My favorite mushroom pasta is ravioli. I supplement the pasta dough with powdered shiitakes, and I fill the raviolis with grated Parmesan and sautéed morels. To serve, I smother the ravioli with a vanilla, red wine, and Parmesan cream sauce. Using different mushrooms for the pasta and the filling makes for a truly special dish. Share these ideas with your local chefs (and encourage them to buy your mushrooms).

MUSHROOM EXTRACTS

People sometimes buy mushroom extracts (also known as tinctures) for medicinal use, but they tend to be expensive and it's hard to gauge their quality, so you might consider making them yourself at home. You can prepare mushroom extracts using a variety of methods depending on the resources you have available and end products you desire. Since the structure and properties of mushrooms can vary, you may have to experiment with different methods to achieve the result you desire. The simplest way to make mushroom extracts is from fresh or dried mushrooms. Another method, which is more complicated, is to make extracts from grain spawn. I'll discuss both methods briefly below.

Making Extracts from Fresh or Dried Mushrooms

There are two widely accepted forms of extraction: hot water bath and alcohol extraction. I recommend the alcohol extraction. Making extracts from hot water alone can be risky, since pathogenic bacteria can flourish on bits of mushroom tissue, won't be affected by the hot water bath, and can pass through filters and into the extract. They will, on the other hand, be killed by the alcohol treatment. If you want to be sure to capture not just the alcohol-soluble but also the water-soluble constituents, after the alcohol extraction you can treat the leftover mushroom tissue to a hot water bath, and use that water to dilute the alcohol extract.

For the most rapid extraction, you'll want a high-proof alcohol, such as Everclear, which is 95 percent alcohol by volume. The ratio of mushrooms to alcohol will depend on what type of mushroom you are extracting and whether it is fresh or dried. For therapeutic use, your extract will require 1 gram of dried mushroom (or the equivalent fresh) per teaspoon of alcohol. An even 1-liter bottle of alcohol equates to approximately 200 teaspoons. So you need about 200 grams of dried mushroom. With oyster mushrooms, as an example, which are 90 percent water when fresh, that equals ½ cup of dried or powdered mushroom or 4¼ cups of fresh mushrooms.

Put your fresh or dried mushrooms, after they have been cleaned and chopped or blended, in a glass or ceramic container with a lid. Add the alcohol, put on the lid, and let sit for at least fourteen days at room temperature (65–80°F/18–27°C). Then strain out

Pure ethanol extracts of fresh turkey tail mushrooms (*Trametes versicolor*) using a blender to maximize surface area and the extraction process.

the mushrooms. You'll now need to dilute the extract down to about 20 to 23 percent alcohol by volume, to keep the alcohol from degrading the active ingredients you've extracted from the mushrooms. With 95 percent alcohol, such as Everclear, that means dilution at a ratio of 1 part extract concentrate to 4 parts distilled water.

If you like, you can perform a hot water extraction of the spent mushrooms for use in diluting the extract. Collect the spent mushrooms in a glass or ceramic container. Bring a kettle of water to a boil, then pour the boiling water over the mushrooms, adding just enough to cover them by an inch or so. Cover the container and let sit for several hours, up to overnight. Then strain out the mushrooms, let the liquid cool, and use it to dilute your alcohol extract. Store your extract in a dark container in the refrigerator. It will keep for several years if the alcohol content is maintained by keeping the container sealed.

If you can't find Everclear or just would like to use something different, you can also use almost any type of drinkable alcohol. However, the extraction will take longer at lower proofs—the alcohol will need more time to break down the mycelium's molecular bonds and release the active constituents. Use the formula below to recalculate the extraction time and dilution ratio, where 95% is Everclear's alcohol by volume (ABV), and ABV ALT is the percentage of alcohol by volume of the alternative alcohol you intend to use.

95% – ABV ALT = X (written as a decimal)

X + 1 × 14 days = total extraction time

ABV ALT ÷ 20% (the desired alcohol percentage) – 1 = ratio of water to use in dilution

For example, if you're using gin that is 35 percent ABV to prepare an extract:

95% – 35% = 60%, or 0.6

0.6 + 1 × 14 days = 22.4 days to soak the mushrooms in the gin

35% ÷ 20% – 1 = 0.75

In other words, you'll soak the mushrooms in gin for 22.4 days, and then you'll mix that alcohol with distilled water in a 1-to-0.75 ratio (1 part alcohol concentrate to 0.75 part water).

Making Extracts from Grain Spawn

If you plan to use grain spawn for making extracts, you will need to be well trained and experienced in the techniques of sterile tissue culture, or "cloning" (see chapter 18), and you will need to have a lab where you can screen your cultures for contaminants; extracts of unknown bacterial or fungal culture can contain toxins and/or make you sick. If you do not possess these tools and skills, make your extracts with fresh or dried mushrooms, as described above, until you have developed expert techniques. If you have mastered the techniques described in chapter 18 and feel confident in your culturing tools and skills, and specifically in expanding cultures onto sterilized grain, you may want to consider using grain spawn to make extracts. The big advantage to this method is that expansion is easy when the cultures are pure and you can produce a large volume of mycelium (and hence extracts) in a relatively short time period. This is especially attractive if the mushrooms are difficult to fruit or extremely slow to form, such as agarikon (*Laricifomes officinalis*) and other large polypores. Plus, as with alcohol extracts, doing this yourself means you know the quality of your product—an important consideration since there are many inferior products with low (or absent!) levels of active compounds.

To begin, allow your mycelium to "supercolonize" the grain, which magnifies the volume and surface area of the mycelium, converting the grain into as much fungal biomass as possible. Give it adequate time. Making grain extracts after only one to two weeks of colonization will give you an extract of primarily grain, whereas after three to four weeks the mycelium will have produced a wave of metabolites that can build up in the grain, visible as a clear or colored liquid. These fluids are highly desirable and can be further increased by subjecting the grain spawn to fluorescent or natural light.

MUSHROOM EXTRACT DOSAGES AND CONSTITUENTS

The daily dosage for your mushroom extract will vary depending on which mushrooms you have used, the targeted pathogen or therapy, and body weight. A good source for dosage information is the *International Journal of Medicinal Mushrooms*.

If you develop your skill and interest in mushroom extracts, you may wish to experiment with growing mushrooms in a way that maximizes the production of particular vitamins, antibiotics, amino acids, proteins, and other beneficial compounds. The wavelengths of the light under which they grow would be an interesting place to start. Like chlorophyll, mushroom pigments serve a specific purpose, allowing mushrooms to respond to temperature and UV radiation in different ways and with different biochemical compounds—that is, active constituents, the same compounds that make mushroom extracts medicinal. For instance, a golden oyster mushroom (*Pleurotus citrinopileatus*) may use light in the spectrum of 570 to 590 nanometers to manufacture a carotenoid entirely different than, say, what a pink oyster (*Pleurotus djamor*) would produce.

An interesting experiment would be to cultivate mushrooms using different wavelengths of light to determine the morphological and biochemical differences that result. This would be one more way in which we can attempt to better understand mushrooms' unique capabilities and the potential for variability in their genetic expression to improve our breeding and cultivation of potent medicinal strains.

Supercolonized reishi (*Ganoderma* spp.) grain spawn, exposed to light and extracted in ethanol. It can develop the varnished tone similar to that of an extract taken entirely from mature fruiting bodies.

Place the grain spawn in a high-speed blender and top it with enough alcohol to keep the slurry moving. A watery paste is ideal. Transfer it to sterile jars and seal them. Place the jars in a dark location and let sit for fourteen days if you used 95 percent ethanol (Everclear), and longer if the spirit is a lower proof (see the calculations on page 126 to determine the steeping time).

When you're ready to use the extract, do not strain the sediment that separates from the clear extract above. Instead, shake the jar just before dosing, and consume the particulates as part of the therapy.

TEMPEH: CULTURING *RHIZOPUS OLIGOSPORUS*

Traditional tempeh is made by fermenting soybeans with *Rhizopus oligosporus*, a sporulating mold (commonly referred to bread pin mold) that appears like a furry growth on decomposing fruits, breads, and vegetables. A great overview of tempeh culture can be found in *The Art of Fermentation* (Katz, 2012), which describes the process of culturing tempeh from purchased strains, as well as perpetuating and expanding tempeh culture. Although commercial tempeh starter cultures are inexpensive, if you master the techniques in part 3, you can keep your cultures alive for years, if not indefinitely, by subculturing the mold just like you would mushroom mycelium. I maintain these fungi in the lab on agar media, which allows me to make tempeh year-round without investing in new cultures.

You can also isolate and purify a tempeh culture of your choice and transfer the culture to sterilized grain, such as cooked organic rice. Remove the grain from the jar or bag once it is fully colonized and has formed black-spored tips. Spread the colonized grain in a shallow pan in a layer about ½ inch thick and cover with a clear pan inverted on top, offset slightly to create a dome that breathes, to allow for gas exchange and a space for the sporulating heads to form. To increase spore production, place the pan under a grow lamp. Once the top of the grain is covered with thick black "fur," transfer the grain into a food dehydrator at low heat and dry it completely. Then place the grains into a blender to make a powder. This powder contains billions of spores and bits of dehydrated mycelium ready to make tempeh in a moment's notice. You can store the powder in jars in the refrigerator for months.

MUSHROOMS AND FUNGI IN LIVESTOCK FEED

Medicinal mushrooms, delivered as colonized grains or into the feed itself or as powdered coatings or additives, offer tremendous potential as food supplements for pets and livestock. There are already several patents in place for myceliated livestock feed as veterinary and medicinal supplements, but none have yet become mainstream. The first to use these supplements was the equine industry, which uses them to improve joint function and available oxygen, for increased speed and endurance, in racehorses. These same powdered mushroom formulas are now spilling into the pet industry, specifically for dogs and cats, as adjunct anticancer therapies and a natural means of improving overall health. Some companies have blended mushroom powders into pet treats—which can also be done at home, of course.

Supplementing livestock feed with dried, colonized grain—such as wheat, rye, whole corn, or other cereals—could delay and inhibit infection and transmission rates for various pathogens, boosting immune levels to help prevent spread and mutation. That sort of medicinal feed is not yet available on the marketplace, but any small to midsize farm could, with a laboratory and appropriate training in sterile tissue culture and spawn expansion (see chapter 18), produce enough for its own needs. Another option would be to supplement livestock feed with spent mushroom growing substrate from primary decomposers, such as oyster mushrooms. This can provide a wealth of nutrition and immune stimulation for ruminants such as cattle, goats, and sheep. This practice can be especially valuable in arid climates, where 30 to 40 percent of the water in a fungal colony resides in the growing substrate. Feeding that "waste" substrate to livestock not only conserves that water, converting it to meat and milk, but also endows the herd with disease resistance (Oei, 2006).

MUSHROOM PAPER AND INK

Many mushrooms can be used to make paper, especially the tougher inedible species that are often overlooked. Polypores are often used since they have a high fiber content, persist a long time in the environment, and are usually available year-round. Infusing mushroom particles into a recycled cardboard slurry can also create a living packaging system, an upgrade to traditional packaging systems.

To make paper you will need a few supplies:

Mushrooms. You can use dried or old mushrooms, but extremely fresh polypores are best.

A screen and deckle. This is basic papermaking equipment. The screen and deckle are two wood frames, hinged together on one side. The deckle is open, while the screen has a wire mesh stapled to it.

A high-speed, large-volume grinder or blender. Since the mushrooms may be leathery and hard, make sure it is not expensive or one that you will be using for anything else.

A tub for floating the slurry. The size of the tub depends on the size of paper you want to make. It must be larger than your framed screen. A shallow plastic tub, about 2 feet square, is plenty large enough for a deckle that is 1 foot square.

Towels and a stack of newspapers. These are used for drainage, for wicking extra moisture away from the screens, and for stacking with the paper to help it dry.

Sponge. You'll use a sponge to wipe the back of the screen, helping to remove the newly formed sheet of paper.

Weights. These are used to press the paper between layers of towels or newspapers, speeding the drying process. You can use anything heavy.

Clothesline or string. This is used to hang the paper for drying. Don't forget the clothespins!

Step-by-Step Mushroom Paper

Step 1. Process the polypores in the blender with enough water to make a homogeneous slurry. If they are large, break them or hammer them into pieces before adding them.

Step 2. Fill the shallow tub with a few inches of water. Lay a towel down on a table or other hard surface, and stack about ¼ inch of dry newspaper on it.

Step 3. Slowly pour the slurry into the shallow tub. The slurry should float on the water. Add enough slurry to make an even layer; if it seems thin or if you can see through it in spots, add more.

Step 4. Hold the screen and deckle with the screened frame on the bottom and the empty deckle on top. Dip them in at an angle to slide them underneath the floating slurry, then lift up and gently shake to smooth the mixture over the screen as the water drains through back into the tub.

Step 5. Carefully flip the settled slurry onto the prepared layers of newspaper, facedown. With a wet sponge gently wipe the back of the screen as you lift it slowly. This will separate the paper layer from the screen. Top the newly formed sheet of paper with additional dry newspaper and another towel.

Step 6. Repeat the papermaking process until you have used up all your mushrooms. Then cover the stack of paper with a final layer of newspaper, and stack weights on top to press out any moisture.

Step 7. After six to twelve hours, remove the weights and separate the layers, leaving the mushroom paper sheets on their base of newspaper but with their tops exposed to the air. After about one hour of air exposure, they can be removed and dried.

Step 8. Peel the mushroom papers off their newspaper bases, slowly and carefully to prevent tearing, and hang them on a clothesline or drying rack to dry completely.

Mushroom Inks and Dyes

We have been experimenting with using spores as ink and infusing ink with viable spores to create a biologically active printing product. The ink has viable spores that not only create the images we see and read, but are also capable of germinating and colonizing the product once they are discarded and wet, speeding composting and decomposition. We have had success using colored spores of various species fermented in culture to produce a solution suitable for use as dyes

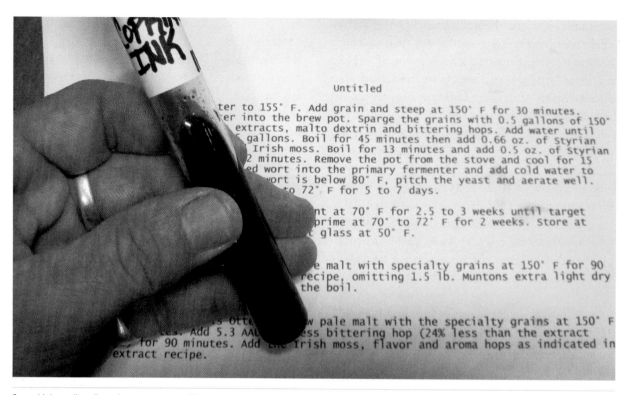

Spored ink can literally make paper come to life.

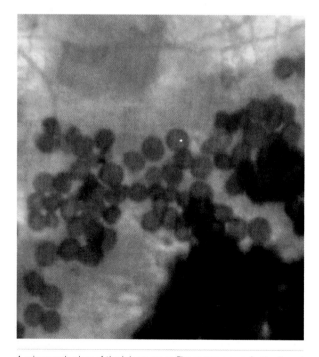

A microscopic view of the ink on paper. The spores are perfectly aligned and adhering to the plant fibers.

and inks. We have also had success adding dried spores to commercial inks and dyes; since chemical solvents can damage these spores, we add them just prior to printing to keep them viable. It is unclear whether or not spore-based inks will develop into a marketable product, but this feels like a major accomplishment. I encourage readers who are skilled in chemical engineering or product design to help develop mushroom-based inks that can be used to create biologically active packaging—and thus promote a hastier decomposition of consumer product packaging.

ANTIMICROBIAL CUTTING BOARDS

Surfaces that tend to spread infection, whether via food or by touch, are often treated with synthetic antimicrobial compounds or are coated with antimicrobial nanoparticles such as silver. Products treated with nanoparticles for antimicrobial purposes include computer keyboards, handrails, elevator buttons, hospital catheters, toothbrushes, cutting boards—the list is enormous, and increasing at a rate of 500 percent per year. Since no regulation is currently in place, manufacturers of these products are not required to disclose them. There is a need, and an opportunity, for alternatives to these products, whose effects on human and environmental health are not well understood. The microbial cutting board I'm presenting here is currently under investigation (as are many of the concepts in this chapter), and although it's not yet ready for prime time (and you should experiment with it at your own risk), I wanted to include it as an opportunity to call for additional research that can further expand the field of mycological innovations.

Old wooden cutting boards are easily recycled into antimicrobial mushroom-infused cutting boards. My wife Olga and I are always on the lookout for hardwood cutting boards at thrift stores and recycling centers. The wood in these boards is still biologically active and will support fungi, it will not rapidly degrade because it is dry most of the time, and fruiting will not occur since the fungus does not have a chance to feed on and store enough energy to fruit.

You can easily make a mushroom-infused cutting board at home. You will need pure mushroom spawn (sawdust, a colonized layer of cardboard, or colonized coffee grounds), a hardwood cutting board, a plastic bag, and a pot of boiling water that your board can fit in. If you do not have a pot large enough, you'll need a heat-proof tray or tub large enough to accommodate the board; you'll place the board in the tray, weigh down the board so it will not float, and then pour boiling water over the board until it's covered by about 1 inch of water.

Step-by-Step Antimicrobial Cutting Boards

Step 1. Boil the cutting board, or soak it in enough boiling water to saturate the board. It usually sinks or barely floats, which typically takes about an hour.

Step 2. Open a large plastic bag and lay it flat. Spread a ¼-inch layer of spawn in the shape of the cutting board inside the bag.

Step 3. Place the cutting board on the spawn layer, then cover the upper surface with another ¼-inch layer of spawn. If you are making more than one board, you can stack them, being sure to use a layer of spawn above and below each board.

Step 4. Trying not to disrupt the spawn layers, carefully tie the bag shut. Poke a few small holes into the sides of the bag, around the perimeter of the cutting board, to provide aeration.

Step 5. After three to four days, check the progress by opening the bag and carefully tugging at the layers of spawn to make sure they are sticking, allowing the mycelium to drill its way into the wood from above and below. Press the layers back together and allow them to colonize for another few weeks.

Step 6. When the spawn has finished colonizing the cutting board, remove the layers of spawn by scraping with a metal or wooden scraper, wash, and it is ready for use. If you wish to expand your cutting board to make more, consider sandwiching the new board with two newly soaked boards, then separating them and flipping so all sides become inoculated, giving you three cutting boards to use or expand with. Once the cutting board is clean, it can live for years if washed with water occasionally, since the mycelium can dehydrate for up to a year.

You can use the spawn or cardboard you removed to make additional cutting boards or to inoculate logs, stumps, or pasteurized media so you keep the mycelium moving and productive. Never waste your spawn! Keep it moving as long as it has gas in the tank to push mycelium into new territory.

PACKAGING, INSULATION, BUILDING MATERIALS, AND BIOTEXTILES

There is a ton of product development under way with the goal of looking at how fungi can be used to create more environmentally responsible consumer goods. Companies such as Ecovative Design, which

A hardwood cutting board layered with sawdust spawn. The spawn completely infuses the surface of the wood in less than a week. This shiitake strain is identifiable by the dark brown zones it creates as it ages.

Once it's colonized, remove the board and inspect it for contamination.

Using a putty knife or other scraper, remove the spawn layer from the entire surface and scrub it clean.

Thanks to the mycelium infused through the wood, the finished cutting board retains a natural microbial resistance for years.

pioneered the use of myceliated biomass for packaging, have opened the door to much broader experimentation, by both experts and home growers. There is plenty of room for innovations in this field, and we can expect to see some exciting advances in the years to come. For anyone interested in entering this wave of product development, I recommend bioprospecting among the thousands of fungi for traits such as high tensile strength or natural antibiotic properties.

Much of this work can be done on a small scale at home or at the end of a production line in a fruiting facility. Spent oyster columns or sawdust-based fruiting substrate can be shredded, mixed with waste paper, and hydropulped into new cardboard or paper products. The mycelium also helps glue paper particles together and may be able to replace the binders or chemicals used by the paper industry to lock these fibers together. One advantage of using myceliated biomass for paper replacement is that initially these fibers are hydrophobic, meaning they repel water. It is only when the paper is soaked that it breaks down, giving it a well-timed shelf life and the ability to be converted to compost at the end of its useful life. Regular paper bags, by comparison, absorb water easily and disintegrate quickly, losing their strength in seconds.

Mushroom cultivators may wish to recycle a percentage of their spent growing media by channeling the waste to specialized machinery that could cut and shape the biodegradable containers used in their retail packaging! These hydropulpers can be purchased used, or you can collaborate with an existing company specializing in box production to mix the mycelium with paper fibers to create your own living packaging.

You could also dry spent growing media for use as insulation. Mycelium, with its nonconductive nature and web-like architecture, resembling that of fiberglass, has a comparable R value (2.9) at a 2-inch thickness to commercially available insulation. What is also surprising is that it's fire-retardant and hydrophobic, since dried mycelium (that is, skeletonized chitin) beads water away unless it is completely saturated. The mycelia of many mushrooms also possess antifungal and insect-repelling properties, offering a huge advantage over traditional insulation. Columns

of spent oyster mushroom substrate, for instance, can be dried and cut with a band saw to shape into myceliated and insulative boards of any thickness. The columns themselves can also be shredded to make a fluffy insulation that can be blown into structures. However, if you try this, be sure to apply an exterior layer of vapor barrier to keep the fungal sheeting or loose shredded mycelial mass from getting wet and converting back to an active vegetative state; this can trigger the fungus to attack the very structure it was protecting in the first place (but only if there is an overwhelming amount of continuous moisture, which would cause molds and rot anyway). Of course, if this interests you, experiment with it and build prototypes before you fill up your home with mycelium.

When air- or freeze-dried, colonized growing medium also has the characteristics of a durable building composite in that it can be formed or cut into shapes (such as blocks) suitable for building a structure. These blocks will need a light moisture application between adjoining layers, rather than chemical adhesives or mortar, to affix the pieces. Once the blocks are naturally fused together (a matter of days), they can have the same tensile strength as concrete and steel, with less of a tendency for fissures due to the slight flexibility of the mycelium, even when it's completely dried. If kept dry, blocks of mycelium can last indefinitely. Mushroom species of interest for this application are oyster and reishi, the latter forming a dense, rubber-like material that holds its shape and repels water.

I have isolated a few strains of fungi that would make great rubber replacements, since the actual fruitbodies of these fungi are difficult to tear, resist moisture, and possess antibiotic properties. Once I isolate the cultures, I expand them into a biomass on sawdust and form them into molds or sheets that can be cut to shape. We have made several small prototypes using this material, including flip-flops.

These and other mycelium-based products could replace consumer goods that have a short life and build up in landfills, such as plastic bottles, disposable cups and food packaging, dog chew toys, and more. The possibilities are endless.

Growing Mushrooms in Space

On your way to the moon or Mars? Consider cultivating some mushrooms on the way! Okay, most of us reading this don't have rocket ships yet, but who knows what the future holds? We do know that anyone traveling in space needs a good source of protein, which mushrooms provide, and that mushrooms can be grown with relatively little light and are easily grown on different waste by-products. Creating a perpetual food cultivation system within a small, enclosed (and hopefully healthy) living space can be thought of as extreme permaculture, since everything would need to be recycled and fed back into the system. So let's just imagine what this could look like.

One challenge to cultivating mushrooms in space is gas exchange. Half of Earth's oxygen comes from the phytoplankton and algae in oceans and surface waters, not land plants. Using phytoplankton and algae, it is possible to create flat-panel cultivation units for oxygen production in space. In fact, this is being researched at the Institute of Space Systems in Stuttgart, Germany, with the goal of creating systems that could eventually support human life—and also scrub the carbon dioxide from mushroom cultivation chambers.

Algae and mushrooms are also an excellent combination due to their complementary nutritional benefits. One cup of dried algae contains about 375 calories, 57 grams of protein, and 19 to 20 percent of the recommended daily allowance for all nutrients and minerals combined; combine that with vitamin D supplements and protein-rich mushrooms, you have practially a complete diet. Although aquaculture is currently being explored for its potential to provide protein during space travel, it would require copious amounts of water, and fish are difficult to breed and raise from eggs to adults. Mushrooms, on the other hand, propagate by division and are capable of expanding significantly in just a few weeks.

Some species of algae, when cultivated in anaerobic reactors, produce hydrogen, which can conceivably be separated to produce hydrogen fuel. Mushrooms have similar applications. One mushroom group in particular, false morels (*Gyromitra* spp.), produce gyromitrin, a water-soluble hydrazine that hydrolyzes into monomethylhydrazine, a component of some rocket fuels. Coincidentally, the high-protein mushroom is edible if the hydrazine is removed. Thus it doubles as a food and fuel source in one organism, just like the algae.

Excess algal biomass could be used as a growing substrate for mushrooms. Preparing the growing medium can be accomplished by exposing it briefly to the external space environment, since the extreme heat or cold would kill molds and bacteria in an instant, just as if you were pasteurizing or sterilizing it in autoclaves on Earth. The biomass would then be returned to the inner, pressurized cabin to inoculate with spawn and use for mushroom cultivation.

Mushrooms only need a few pulses of light every few minutes, not continuous light exposure, to maintain the growth and development of the fruitbodies. Compared to continuous light, pulse timers can save over 90 percent of the energy needed to cultivate full-size mushrooms, and LED lights in the blue-green spectrum require much less energy and can be powered by solar arrays. In fact, savvy engineers should be focusing on the use of LED lights to cultivate mushrooms under different wavelengths with an eye toward mediating the development of various nutritional and bioactive constituents to benefit nutrient deficiencies and maintain the health of astronauts in space. These experiments can also help mushroom farmers here on Earth reduce their energy costs, since flooding our fruiting rooms day in and out with artificial light is expensive and wasteful.

It is important to test these parameters for each mushroom species and strain being considered for space trials. Plants and many fungi produce the gas by-product ethylene, a plant hormone that is known for its effect in ripening fruit. Accumulations of this trace gas can become a problem unless it is removed or filtered from the air. Filters of potassium permanganate—which is nonflammable and safe to use with general precautions like wearing gloves and safety glasses to avoid skin and eye contact, and so ideal for use in space—can be used to capture the ethylene, oxidizing it into water and carbon dioxide. Of course this additional carbon dioxide must then be offset, likely by more plant or algal culture.

Allowing mushrooms to fully mature into a sporulating state in space is probably not ideal. In fact, I would venture to say it should be completely avoided, since the spores, as airborne particles, can clog filters and lower air quality—another element in the system that needs to be dealt with. Instead you could cultivate the mushrooms in a growing chamber that has lower light and higher carbon dioxide levels than is typically desirable to induce a long, stringy mushroom with little or no cap capable of producing spores. Harvesting the mushrooms young, as buttons rather than full-size fruitbodies, would also help keep the threat of airborne spores to a minimum. Working together with breeders or discovering a strain of mushroom with a very low spore load, but that still produces great yields, will be a challenge, but a part of the overall solution to help engineers design mushroom cultivation systems that intersect seamlessly with other life support and filtration systems on aircraft and base modules.

What about cultivating mushrooms on Mars? The atmosphere of Mars is primarily composed of carbon dioxide, so in order to cultivate mushrooms there, the initial priority would be to establish enough algal or plant growth well ahead of initiating mushroom cultivation to generate sufficient oxygen for the mushrooms to develop correctly. Ethylene would also be an issue, though likely it could be managed through the use of potassium permanganate. Every effort should be made to recycle the growing substrate and create soil in which the next generation of seeds from vegetative modules could be planted. It would also be important to design an extremely well-insulated growing chamber that would protect the fungi from radiation while still allowing diffuse light to enter, since the high levels of radiation would damage dividing cells, causing mutations that would likely kill or prevent mycelia from mating or forming fruitbodies. Temperature differentials would also play a role in siting operations; Mars at noon near equatorial

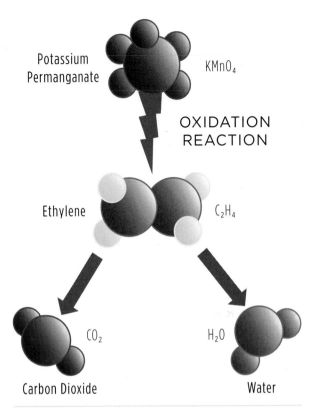

One obstacle to producing vegetables and mushrooms in space is the production of ethylene. Ethylene can be oxidized using potassium permanganate to produce water and carbon dioxide.

regions can reach temperatures around 70°F (21°C) but drops to nearly –225°F (–143°C) near the poles.

Whether or not it will be possible to cultivate mushrooms on Mars remains to be seen, but I believe regardless that there is value in exploring the methods and engineering that can lead to successful space cultivation of plants and fungi. I encourage people to experiment with these ideas. At the very least, this kind of experimentation reinforces environmental responsibility, since one of the inherent challenges of growing mushrooms in space (or at least in a space vessel) would be maintaining balance in a closed system—exactly like we are challenged with doing here on Earth.

Mushroom Space Cakes

Do we actually need to *fruit* mushrooms to benefit from the protein and nutrition we associate with them? Why not eat the myceliated substrate? After all, cultivating actual fruitbodies, which eventually produce spores, is not the best thing for air quality in an enclosed environment. And producing mycelium, rather than actual mushrooms, would take up less space, consume less oxygen, take less time (mycelium is ready to eat in about a week), and require no light. Sure, in eating colonized mycelium we would also be eating the substrate, but by the time the mycelium colonizes the growing medium and thickens, it will have converted much of the biomass to mushroom protein while also breaking down the lignin and other structural components of the plant cell walls, so that it will be much more digestible. Mycelial "space cakes" may be the superfood of the future even right here on Earth. Using fungi to create a new food product can be done and enjoyed anywhere in the universe, starting in your kitchen.

To make a "space cake," you need to be able to create liquid mushroom spawn using the lab skills described in chapter 18. You'll create liquid agar spawn or grain slurries and pour them directly into a prepared growing substrate, such as beans, grains, or rice. In space, the growing substrate may be dried bean leaves or other by-products of a hydroponic plant system. Once the substrate has been supercolonized by the mycelium, it can be be cooked fresh or dried and compressed for long-term storage. Fresh cakes can be sliced and baked to reproduce the flavor and texture of the original mushroom species. Dehydrated cakes simply need a little warm water to spring back to life within munutes of rehydration.

Think of it like making tempeh (see page 128), except that instead of using tempeh starter culture (which sporulates everywhere; not good in space), you'll use nonsporulating cultures that will still bind the food product without the airbone contamination. Choose mushrooms that have a high mycelial binding strength and rate of growth, such as shiitake, oyster, and reishi, to hold the cake together and outcompete other organisms.

Growing mushrooms under specific light wavelengths could improve the nutritional biochemistry of the myceliated cakes—most mushrooms are capable of synthesizing additional vitamin D. The cakes can also be manufactured on Earth and dehydrated for use on spacebound missions until food production systems are well established. They can also be used as starter cultures, since the cakes themselves are compressed mycelium and can be pulverized in water to make slurries capable of substantial expansion.

CHAPTER 14

Mushroom-Infused Beer, Wine, and Spirits

What is the perfect mushroom delivery system? Combine the health benefits of mushroom extracts and the most consumed cold beverage in the world to create a mushroom-infused beer! In this chapter, you'll learn how to introduce into healthy and medicinal brews different fresh or dried mushrooms, as well as extracts that you can prepare in advance of brewing. (See chapter 13 for more information on preparing extracts.) To brew mushroom-infused beer, you need to understand some brewing basics, so we'll cover that, along with considerations to weigh when you're choosing a mushroom to add. We'll finish up with a few recipes and by taking a look at mushroom-infused wine and spirits.

Humans have been brewing alcoholic beverages for centuries, even since the time of the ancient Egyptians. But only recently have people started to experiment with adding mushrooms and mushroom extracts. Granted, mushrooms don't immediately come to mind as a source of fermentable sugars, and indeed, you can't brew beer from fungi alone; you need the malted grains to produce the classic flavors (and alcohol content) of modern beer.

Although this is not a detailed chapter on homebrewing, I've included information below on sourcing quality ingredients and considerations to bear in mind

if you plan to add mushrooms or mushroom extracts. If you're not yet an experienced brewer, I recommend that you invest in a comprehensive text dedicated to brewing beer. In many areas, there are local brewing

At the international debut of medicinal mushroom beer at the 2012 Telluride Mushroom Festival (a.k.a. "Shroomfest") in Colorado, 240 gallons sold out in under forty-eight hours and tasted great! In 2013, we quadrupled production to handle the demand.

THE GROWTH OF A PASSION

In 1992, at the age of twenty, I brewed my first beer. I didn't follow the directions and ended up brewing a beer with twice the amount of malt and half the amount of water I should have. The beer was too strong to enjoy casually, over 12 percent alcohol. I immediately knew that I needed to better understand the process before brewing any more. After a few more trials using basic beer recipes and following the exact methodology, I successfully brewed a delicious beer. I soon graduated from using malt syrups to all-grain formulas. In 1995, when I started to identify and cultivate wild mushrooms, I decided to combine two of my favorite hobbies! I brewed my first mushroom beer by adding dried and powdered shiitakes.

Sourcing Yeast, Grains, and Hops

Fermentation of beer begins with yeast. Start with a culture from a reliable company or brewer. Pure cultures of beer yeast strains are available worldwide and are easy to ship. If your yeast culture can be delivered in one to three days, choose liquid; it will be more active and invade the brewed medium quickly. But dry yeast is better if you live someplace remote and it might take a long time to arrive. If you know a local microbrewer who keeps his or her fermentation discharge to perpetuate the yeast strain with minimal cost, you might be able to get a culture from him or her. Look for someone who keeps the yeast strains in cold storage, maintains cleanliness, and alternates growing media to keep the strains strong and viable. Expanding yeasts indefinitely on the same formula or media will result in strain weakening and performance, called strain senescence—just like with mushroom cultures. So if you are going to maintain the cultures in peak performance, you will need to keep masters and backups, just as you would with your cultivated mushroom cultures. See chapters 17 and 18 for discussion of basic home laboratory equipment and spawn generation techniques. You can apply that information to perpetuating yeast cultures as well, so that you don't have to purchase yeast starter cultures every time you brew.

Organic malted grains are increasingly available, and it's worth seeking them out. Although you might pay double in price for organic, these grains are worth it. (Mushroom-infused beer is about quality, not the quantity consumed.) I am trying to encourage my local brewing companies to grow their own grains organically, so that I can source grains for my lab's spawn production from them and use the the cereal straw left in the fields for cultivating oyster mushrooms.

Organic hops are expensive, since they are vulnerable to powdery mildew in most growing areas. I grow my own and freeze the excess for brewing in the off-season. Hops grow from rhizomes that

clubs that are open to new members and creative ideas. Their experienced brewers can help you fine-tune your brewing techniques and formulas (and one day you can return the favor). Often these groups schedule gatherings where they brew from start to finish. Attending a meeting and teaming up with a few local enthusiasts is a cheap way of getting way ahead of the learning curve. Regardless of how you develop your expertise, you should have at least a few successful batches of good-tasting regular brew under your belt before you start adding mushrooms or extracts.

As with mushroom cultivation, cleanliness is critical when you're brewing beer. Although brewing is more forgiving than sterile tissue culture, in order to maintain the integrity of your beer, you should strive for complete cleanliness or suffer the consequences. Contamination may not be evident at a certain batch level, but it can accumulate and alter the dynamics of a fermented beverage if the expansions become increasingly open to competitor microbes.

Fresh hop cones fruiting in our garden. I purchased a few small rhizomes and now have a thriving, organic hops garden for my personal use.

you can order in early spring. Plan on searching for availability in January and February for a March purchase and delivery. If you experience disease problems, release beneficial insects such as ladybugs, predatory mites, aphid predators, and assassin bugs. Use only organic fungicides to suppress molds and mildews. Liquid copper is a potent antifungal compound that can be applied as a foliar spray, attacking the fungi by breaking down the cell walls, and killing little if anything else. It is a good idea to plant hops in more than one location and to have a backup source available. Once you have a hops garden, you should be able to perpetuate the rhizomes from cuttings if you wish to expand the production or share some cuttings with friends. Dig in early March and cut sections that are branching near the surface and showing small swollen buds. If in the future you have problems with your patch, perhaps the friend you gave cuttings to can give you cuttings back or share the harvest.

Incorporating Mushrooms

For homebrewing, you can use either fresh or dried mushrooms. If you use fresh mushrooms, the damp tissues can be covered with active yeasts and bacteria, which could affect the quality of your beer, since the wort temperature (160°F/71°C) is not high enough to sterilize the fruitbodies. One way to get around this, if you have the equipment, is to autoclave the fresh mushrooms for an hour before adding them to the beer. In fact, even when using dried mushrooms, in which most of the potential contaminants will have gone dormant, I still recommend autoclaving for an hour, at 250°F (121°C), to eliminate the possibility of contamination and bad (or failed) beer. In my own brewing, I mostly use powdered dried mushrooms to capture the essence of the mushrooms and add intense flavoring. (You can even gently roast the powder, browning it just before you add it, to impart a unique flavor.) The dry powder has a lot of surface area, facilitating exposure to hot water and alcohols, which improves the extraction of flavor and medicinal compounds.

You can add fresh mushrooms or powders at a few different stages. If you add them to the grain and mash, you'll extract flavors and bitterness (not a bad thing in beer), and you'll also unlock any water-soluble medicinal compounds during the cook. In "dry shrooming," the equivalent to dry hopping, you add the powder (or even whole dried fruitbodies) into the fermentation container along with your hops, so they ferment together. This can impart unique flavors, complex aromas, and different depths of bitterness, which will vary depending on the mushroom species you choose. Be sure to use a hops bag to make it easier to remove the bulk of the mushroom pieces and powder before kegging or bottling, since this debris may clog filling and dispensing lines.

You can also use mushroom extracts in your homebrew (see chapter 13). Alcohol extracts are preferred over hot water extracts because they better release the beta-glucans and other molecules that are trapped in the chitin matrix, which is what you want when creating medicinal brews. You can add extract at any stage of the brewing process, but I like to add it while kegging or just before bottling, shaking the extract well to be sure all of the active ingredients are added to the beer.

Choosing a Mushroom

There are a few factors to consider before choosing the species—aroma, medicinal properties, nutritional properties, bitterness values, and cost. Modifying any of these factors could impact the others, so experiment on a small scale before producing large batches. Availability is an important factor and for wild mushrooms will vary by season. In fact, for some mushrooms that are rare or endangered, like the agarikon (*Laricifomes*

Hemlock reishi (*Ganoderma tsugae*) is a commonly used medicinal mushroom typically harvested from higher elevations. For your beer production, you can either hunt for these mushrooms or collect, clone, and cultivate them on conifer stumps outdoors or on sawdust formulas indoors.

officinalis), cultivation may be your only option for procuring sufficient quantities.

Flavor

Much like hops, mushrooms can provide preliminary and lingering flavors when they hit your palate. Start off by adding small amounts of mushrooms or extracts to small batches of beer, taking into account that some types of beer can take several weeks to months to mature and fully develop. You can then make a judgment call as to how much mushroom to use with a larger volume of beer. Experiment with the kinds of mushroom you add, the form you use (fresh, dried, or powdered), the quantity, and at what stage you add it to determine the flavor you like for the kind of beer you are brewing.

Aroma

Pouring your beer into a glass is the only way to really appreciate the aroma, since your nose is, shall we say, strategically positioned, and when you drink, you taste the beer and smell its aromas simultaneously. This synching magnifies the experience.

Many mushrooms, such as birch polypore, contain vibrantly aromatic essential oils that mix and bind with other oily molecules present in hops. When you're considering which mushrooms to use with your brew, think of the impact aroma will have on your drinking experience. You may wish to choose mushrooms that have a pungency and sweetness for brown ales and lagers, or savory and hearty species for darker brews such as stouts.

Medicinal Properties

The reason I started brewing mushrooms and beer together was to infuse the medicinal properties of mushrooms such as reishi and shiitake into a beverage that tastes good, has high market demand, and stores the medicinal compounds safely for years. I also later found beer to be an easy delivery system for extracts—and a little more interesting than just taking the extracts themselves.

I do not condone massive consumption of mushroom-infused beer to "get your medicine." Use mod-eration, not only because you're drinking an alcoholic beverage, but also because taking any medicinal mushroom continuously for too long, just like with any food or medicinal product, can have negative effects. Although the effect will depend on how much or how little mushroom powder or extract you add to your beer, I recommend drinking a bottle of mycobrew on a weekly, not daily, basis, as well as alternating recipes so that you are exposed to different properties from different mushrooms. If you are looking for a particular medicinal effect, such as anti-inflammatory actions to relieve the pain associated with arthritis, observe how your body responds to the different formulas, noting any positive or negative reactions. And try the mushrooms or extracts themselves, without the beer, to see if they work better for you on their own.

Bitterness Values

As you likely know, hops are added to beer to provide a bitterness that counterbalances the sweetness of malt. Mushrooms also vary in their bitterness; some mushrooms, such as many polypores, are very "hoppy" and are typically used not only for their medicinal compounds but also to impart a controlled bitterness to the beer. Wild mushrooms such as *Tylopilus* spp. are extremely bitter and should be used and experimented with sparingly.

The best way to test an edible mushroom's bitterness is to chew on a fresh one. Assuming that you have correctly identified the mushroom in question as safe, bite or chip off a small chunk and start chewing. Allow a good sixty seconds before calculating the bitterness. Keep chewing, occasionally swallowing the juices, but do not swallow or spit out the piece. Note the initial and lingering qualities of bitterness, or lack of, with each species you try.

Hops are experiencing a spike in disease problems in commercial production fields, encouraging growers to spray pesticides and eventually likely impacting the price of beer. Breeding new hops varieties that are resistant to pests and diseases, along with using beneficial insects, is improving yields at organic farms, but the exploding worldwide demand, due to the proliferation of microbreweries and popularity of homebrewing, has hops growers fighting to keep up. This being the case, we can potentially develop a strategy to substitute some hops with specific mushrooms, partially or entirely.

Cost

The key to a great beer recipe is that it is not only simple but affordable too. The mushrooms you add are a key part of this, so use them efficiently and effectively. Don't just pour a slurry of freshly blended shiitake mushrooms directly into your wort or finished beer and expect an award-winning brew. To minimize costs, use powdered mushrooms or an extract. You will use substantially less to achieve the same end product than you would with larger pieces in bulk form.

MUSHROOM-INFUSED BEER RECIPES

Below you'll find some basic beer recipes that incorporate mushrooms. Each recipe is sized to make 5 gallons of beer. However, given the broad range of ingredients available to homebrewers, you don't have to limit yourself to the ingredients you see here. You can modify any existing recipe. Just be sure to formulate your beer to balance sweetness and bitterness. And consider making it a specialty regional beverage that incorporates local and organically grown ingredients. Have fun making your first mushroom beer, and "hopfully" many more!

T. J. Daly, the head brewer at Smuggler Joe's in Telluride, Colorado, strains a batch of medicinal mushroom extracts into a sanitized bucket to remove the grain solids before adding them to the kegs. The director of the festival, Scott Koch, helped design the formula and assisted with the brewing.

T. J. doses each keg with a precalculated amount of extract to ensure that every pint has a certain dose of targeted medicinal compounds. This blend was from reishi (*Ganoderma* spp.), birch polypore (*Piptoporus betulinus*), and turkey tail (*Trametes versicolor*).

Turkey Tail Ale

A golden pale ale with turkey tail (*Trametes versicolor*)

6 pounds extra-light malted grain
4 ounces fresh or 2 ounces dried turkey tail mushrooms
1 ounce hops
Ale yeast

Iceman Amber

An amber ale with birch polypore (*Piptoporus betulinus*) and amadou (*Fomes fomentarius*)

6 pounds extra-light malted grain
1 pound amber malt
12 ounces carrot juice (preferably organic)
4 ounces fresh or 2 ounces dried birch
    polypore mushrooms
4 ounces fresh or 2 ounces dried amadou mushrooms
1 ounce hops
Ale yeast

Reishi Red

A natural red lager with reishi (*Ganoderma* spp.)

6 pounds extra light malted grain
1 pound amber malt
12 ounces red beet juice (preferably organic)
4 ounces fresh or 2 ounces dried reishi mushrooms
1 ounce hops
Ale yeast

Agarikon Stout

A dark, earthy brew with agarikon (*Laricifomes officinalis, Fomitopsis officinalis*)

1 pound chocolate malted grain
1 pound crystal malted grain
5 pounds pale malted grain
2 ounces liquid agarikon mushroom extract
    (from colonized grain)
1 ounce hops
Trappist yeast

Note: Agarikon is endangered and should not be harvested in the wild, but it is a powerful medicinal mushroom worthy of a stout beer. Since this mushroom cannot be cultivated in large quantities and the mycelium is slow to grow, the best option for this recipe is to allow myceliated cakes to supercolonize over the course of several months, until they are thick with biomass, and then prepare an extract from them (see chapter 13 for information on making "space cakes" and extracts).

MUSHROOM-INFUSED WINE AND SPIRITS

Much like mushroom beers, wines can also be infused with mushrooms or extracts. Dried mushroom powder will impart mushroom flavors to the wine, but be sure to sterilize the powder before adding it, since dried mushrooms can contain wild yeasts and bacteria, such as *Acetobacter* spp., that may be undesirable or spoil the solution. To sterilize the powder, autoclave it for one hour at 250°F (121°C). Then portion the powder into the wine fermentation tanks or aging barrels. For every liter of wine, calculate a starting dosage of 5 grams of mushroom powder to achieve a basic medicinal dose in an average-size wineglass. Increase the concentration as you like to achieve a higher degree of flavor or therapeutic dose.

Olga, my wife, loves to make chanterelle vodka, and we serve it often at our workshops and family gatherings. It is easy to make infused spirits, and since the alcohol concentration in spirits is so high, there is no risk of spoiling the beverage by adding unsterilized mushrooms, whether fresh or dried. We usually infuse our spirits in large mason jars. We add cultivated or wild mushrooms, whole or chopped (chopping increases the surface area of the mushrooms and improves color and flavor extraction), and fill the jar with spirit, such as vodka. Olga sometimes adds spices such as lemon peel, hot peppers, and even cinnamon sticks to give it a kick. We let the infusing spirit steep for about a month, shaking occasionally. At first the mushrooms tend to float, but eventually they become saturated and fall to the bottom. Then we strain out the mushrooms. You can eat those infused mushrooms as a "mushroom shot," or you can cook them to remove the alcohol by searing them in a pan over high heat for just a few minutes. Spirits infusion is a great way to store fresh mushrooms long-term, allowing you to enjoy a shot of flavor every now and then, as you daydream about the next flush of mushrooms in the coming fruiting season.

CHAPTER 15

Mushroom Marketing

Imagine that you have mastered cultivation of the mushrooms you want to grow, whether they're oysters or shiitakes (or morels!). You have a bumper crop of a consistent product, more than you and your family and friends can eat. So you want to start selling them. It is one skill to grow consistently good-quality mushrooms. It is another to locate buyers and to market and sell them. Although not everyone is born with a salesperson's intuition, there are some simple skills and strategies you can employ to help you market your product. Although I'll focus mostly on small local markets (such as farmer's markets and restaurants), you can employ some of these ideas if you decide to scale up to a larger market as well.

There is a popular saying in the agricultural market: "Don't sell what you can grow, grow what you can sell." By understanding demand, or by creating a demand by educating your consumers, you can steer your production toward what sells, as you simultaneously develop your cultivation skills. Depending on where you live, a variety of mushrooms may be more or less recognizable to your local consumers. In many European and Asian countries it is not uncommon to find over a hundred species of mushrooms sold in public markets and herb shops. Selling your mushrooms in this sort of environment can be difficult, since so many mushroom vendors have already entered, or even inundated, the market, making it difficult to compete. In this way, the lack of a large existing local market can be an advantage, even though you may have to work harder to find and educate your customers. Understand and consider the challenges and opportunity of your particular market. You might, for example, consider developing a steady supply of two easier-to-grow-mushrooms such as oysters and shiitakes. Alongside this production, you then might grow smaller batches of more difficult or unusual species, which may vary with the season. This allows consumers to rely on your consistent stable of products (and you to rely on those sales), while also giving them (and you) something new to get excited about, and it helps you stand out from other producers.

Attracting Customers

One thing that always attracts a crowd at a farmer's market is the smell of fresh mushrooms, garlic, and onions caramelizing in a pan on a small portable burner—it's a wonderful attraction that draws in passersby without making them feel pressured to buy something. When I'm introducing a new mushroom to the market, I cut up a few, cook them briefly, and place samples in small paper cups. I often use only the minimum preparation, such as oil and a sprinkle of salt or pepper, to allow the mushroom's flavors to shine. Explain the value of this simple preparation to your potential customers, who may ask how to cook the mushrooms before deciding to buy any. It's a pity to ruin a mushroom the first time you cook

Olga has finished setting up our booth while I unload the car.

it, so give them some information to help them succeed with it. And, of course, be sure to check your market's regulations on cooking and handing out prepared samples.

When you design your signs, bear in mind that the average consumer has an attention span of less than five seconds. Keep your signs simple and quick to read. Offer a brief description of your product—just enough to catch someone's interest. If customers want to know more, they can ask you about your goods. Mushrooms that aren't commonplace or easily recognizable may require a thirty-second tutorial on flavor and preparation, possible recipes, and any

other information you have that will persuade customers to take a chance. Vegetarians are often on the lookout for foods that offer comparable protein to meats, so you might add a few basic nutritional comparisons. Consider adding descriptions that reflect your region, customs, and anything else that makes your product unique. Think of your mushrooms like wine: Winemakers document and display the country and region where their wine is made—its terroir—noting key flavors, tones, and notes about each batch.

My wife Olga and I frequent farmer's markets with a variety of fresh and dried mushrooms as well

At larger venues we use a flat-screen TV with a looping series of basic spawning directions and beautiful pictures of mushrooms to grab the attention of people passing by.

as spawn and kits for growers. I wonder sometimes why consumers are quick to snatch up the shiitakes right next to a beautiful display of rainbow-colored oyster mushrooms. What is the problem with oysters? I think some people are turned off by the name "oyster," which implies a food that traditionally is salty, slimy, and not well received by finicky eaters of any age. I sometimes recommend that growers consider the various common names of a given mushroom and highlight the one that is most appealing on their signs and other marketing material. For example, "trumpet" in Europe describes a vase-shaped gilled mushroom. Instead of "lion's mane" for *Hericium erinaceus*,

which has an amazing flavor, you might consider using "pom-pom" or "waterfall"—names that European chefs have capitalized on for many years.

Price your mushrooms. Way too many vendors fail to display a price. It is like going to a restaurant where the menu is unpriced; the bill is a mystery and ordering is stressful for anyone on a budget. For the most part you will be selling mushrooms to everyday consumers, not the fraction of us that are so wealthy we don't need to worry about cost, so make your prices clear.

Above all, make your signs clear and easy to read. For example:

GOLDEN OYSTER CLUSTERS

Organic—$10/pound

Nutty cashew flavor when sautéed

Excellent when fried like calamari and served with a spicy dipping sauce

High in protein: 20 g/100 g (eggs average 13 g/100 g)

POM-POM or WATERFALL MUSHROOM

Organic—$12/pound

Flavor and texture of crabmeat and lobster

Sauté in butter or sear and bake

Although you may want to supply some nutritional information to your customers, never exaggerate nutritional data, and do not make medical claims. Do not, for example, label mushrooms in a way that suggests they can "cure," "treat," or "eliminate" diseases or disorders. The U.S. Food and Drug Administration (FDA) enforces advertisements on labels, boxes, packaging, websites, and any other form of promotional material that involves herbal supplements and neutraceuticals. Fresh and dried mushrooms, powders, extracts, and all other fungi products described as having health-promoting properties fall under this regulation.

Language that makes no concrete claims yet recognizes that some mushrooms *are* valuable in the health and healing process can include phrases such as "supports" or "enhances," as in "supports healthy immune function," "enhances healthy liver function," or "supports T-cell function." If you make statements such as these, include the common FDA disclaimer: "This statement has not been evaluated by the FDA. This product is not intended to diagnose, treat, cure, or prevent any disease." (For additional information regarding dietary supplement regulations, visit the FDA website at http://www.fda.gov/food/dietary supplements.) And when you're selling mushrooms with such claims, keep with you a few photocopies of legitimate scientific research in a file or binder; if someone inquires, you can hand them a copy. Post these references on your website, if you have one, and offer direct links to them when possible, regularly updating and searching for the most current research available.

Packaging

There are two ways to sell mushrooms at an open market: bulk or packaged. Mushrooms sold in bulk are more frequently and easily damaged, since consumers will sort through the mushrooms until they find what they want. Unfortunately many mushrooms are extremely delicate, to the point that no one will pay them much attention after a few people mishandle them. There is also something to be said for cleanliness; many

Packing I helped design for Mepkin Abbey is biodegradable, includes recipes, and has a viewing window to display the fresh mushrooms.

people do not wash their mushrooms before cooking with them, and probably don't want strangers fondling their produce before they chop it up and throw it in the pan. Consider a different strategy—weighing out portions of mushrooms and packaging them in clear biodegradable containers or bags so that customers can see the quality and freshness, while eliminating contact. With no need to weigh every purchase, this also makes pricing your display and the speed of transactions extremely efficient! If you do decide to sell in bulk, consider handling your booth like a butcher or deli counter, where you keep the mushrooms behind a table or counter and take orders from your customers. Though they won't be able to reach the mushrooms themselves, customers can still see every mushroom and cluster and point to or pick out the ones they prefer.

Consider your packaging. You don't have to use plastic. You can offer your product in an inexpensive unbleached paper bag or a deep cardboard vegetable tray. In this sort of natural-fiber packaging, the mushrooms will release spores, impregnating the packaging with millions of inoculum that will assist in the composting of the packaging when it gets wet.

Using biodegradable packaging has one potential, though solvable, drawback: We once assisted in the design of earth-friendly, biodegradable mushroom packaging for Mepkin Abbey in South Carolina. After about a month of use we received a phone call from a grocery chain. The manager said that the mushrooms looked strange, and he was worried that they had become moldy. He sent me pictures. They showed that the oyster mushrooms, fused to the base of the box, had begun to feed on the box, causing the fruitbodies to completely fuzz over! Although there was nothing wrong with the mushrooms, this was clearly a problem, since buyers were passing them by, thinking they were past their prime (when in actuality they were just trying to stay alive longer!). The simple solution was to place a small piece of waxed paper on the bottom of the boxes to keep the mushrooms from finding the food source, keeping them fresh for at least a week on a refrigerated shelf or storage unit.

So keeping the mushrooms from "eating" their packaging over the course of their shelf life is one consideration. The breathability of the package is also important. If you scale up to the point that you are shipping mushrooms, your packaging should have ventilation holes to allow for breathability and slight drying under refrigeration. Federal regulations also require that storage requirements and contents—for example, "Fresh shiitake mushrooms/keep refrigerated at 38–45°F(3.3–7.2°C)"—be clearly marked on the outside of the packaging on all sides. If you are designing or running a commercial operation, consult the Mushroom Good Agricultural Practices guidelines, developed by Penn State University and the American Mushroom Institute (May 2010; http://www.mgap.org). These guidelines lay out a set of standards for producing, harvesting, and shipping mushrooms that conform to the current FDA food safety regulations. And arrange for an inspection from your local health inspectors if you are seeking certification of food safety compliance.

BRANDING

When you cultivate mushrooms at a commercial volume, selling them all can be a challenge, so I always recommend starting small and increasing your volume as you develop your market and add accounts. I spend a lot of time talking to local chefs and giving presentations to regional cooking schools and food writers, so they not only are familiar with our mushrooms but also trust my experience and expertise—this helps sell the mushrooms.

As you develop your markets and accounts, branding becomes important. You want your clientele to recognize *your* mushrooms as something special, and *your* operation as one they trust and want to do business with. You want name recognition. Have a recognizable logo that suits and best describes you and your growing operation, so when people see it on a sign or banner at the market, on a chef's menu, or on your packaging, they will associate it with the quality of your product. Conversely, bad product will equate to logo repulsion, so be sure you are always selling the best-quality mushrooms you are capable of cultivating, regardless of the size or savvy of your market.

CHAPTER 16

Fungi in the Classroom

There are many ways in which educators and naturalists can incorporate fungi into learning opportunities and lesson plans. Though reading and watching videos about fungi in class are always options, hands-on activities offer greater impact, allowing sensory and tactile exploration and the opportunity to follow a living organism through its stages of development. This helps students of all ages develop valuable skills in observation and scientific methodology in a way that is both memorable and enjoyable.

I've divided this chapter into sections based on grade level, as well as by five "tracks": fungal ecology, cultivation, recycling and composting, mycoremediation, and research and science projects. These lessons include both indoor and outdoor activities and can be tailored for a wide range of situations and educational goals. Most lessons are based on information covered in part 1 and the earlier chapters of part 2, although in a few cases, especially for the college-level lessons, you'll need to refer ahead and master some of the material in part 3.

There are no limits to the creativity and experimentation you can incorporate into these lesson plans if you are capable of preparing a growing substrate and introducing a fungi to colonize it. Experiment on a small scale and achieve a level of success before introducing the units to the classroom. If the simplest lessons, such as growing oyster mushrooms in paper cups, fail to produce any fruitbodies, students can easily become disinterested in fungi, so my advice is to plan ahead, fruit several batches of mushrooms a few different ways on different growing media on your own, and show them to the class (when they are fruiting perfectly) in advance of the lesson to pique the students' interest. This will also help you decide which parts of these activities you want to do as prep work, versus what you want to have students do as part of the lesson, resulting in better use of your class time once you are ready to move forward. Since it can take several weeks for mushrooms to fruit from spawned material, you can use the waiting period to cover supporting material and exploration, such as a quick stroll around the building looking for native fungi and exploring their role in the ecology of your region.

GRADES K–4 (AGES 5–10)

Fungal Ecology Track

Mushroom Art

Use store-bought mushrooms with gills or other patterns to dip into dyes and make prints on paper. Draw outlines around mushrooms on paper, and use the outlines to make collages. Draw mushrooms and label them with the basic parts such as base, stem, and cap.

Making Spore Prints

Collect fresh mushrooms and place them on different-colored sheets of construction paper to collect spores of different colors. (See chapter 2 for more information

Fungi in the Classroom

Using this chart, teachers and other educators can pick age- and skill-appropriate content to develop into classroom lessons and activities.

Project type	Grades K–4	Grades 5–8	Grades 9–12	College
Fungal ecology	Mushroom art Making spore prints Basic life cycle	Insect-killing fungi, or "zombie ants" Mushroom dissection	Identifying mushrooms Mycorrhizal relationships Isolating and cloning fungi	Entomopathogenic fungi Mycorrhizal fungi
Cultivation	Fruiting cups	Gas exchange basics Expanding biomass	Space lab: Build an enclosed ecosystem Basic mushroom cultivation	Space lab: Build an enclosed ecosystem Region-specific mushroom cultivation technology Maintaining and storing fungi Bacterial-fungal interactions
Recycling and composting	Paper and cardboard Coffee cultivator Mushroom paper	Waste reduction using mushrooms	Waste reduction using mushrooms	Waste reduction using mushrooms
Mycoremediation	How mushrooms "eat" pollution	Plate contamination observations	Using fungi to control soil erosion and filter contaminants Remediating an oil spill	Screening fungi for use in mycoremediation Engineering mycoremediation prototypes
Research and science projects	Observation and communication	Mushrooms in the news Gravitropism	Report: mycoremediation projects Testing mycofiltration of water	Packaging, insulation, building materials, and biotextiles Isolating and extracting novel fungal metabolites Fungal foods of the future Biofuel production

on making spore prints.) Allowing the kids to explore outside and look for mushrooms is a fun activity; teachers should warn students that mushrooms are safe to touch, but not to put mushrooms in their mouth.

Basic Life Cycle

Help students learn and understand the difference between mushroom spores and plant seeds. Use graphics to show how spores float around and land on their food source, germinate somewhat like a seed, and colonize the food source. When mushrooms run out of food, to keep their species alive they fruit mushrooms, which produce spores, and the cycle begins again. (See chapter 1 for more information on the mushroom life cycle.) Plants, in contrast, do not

colonize and consume their substrate but take root in it, drawing nutrients from the soil and sun and producing seeds at the end of their growth cycle.

Cultivation Track

Fruiting Cups

I have taught many basic classes using plastic or paper cups filled with pasteurized, inoculated growing medium. The class can work individually or in groups depending on materials and budgets. Teachers should pasteurize the growing medium on a stovetop the night before and allow it to cool. In class, students can mix the substrate with oyster mushroom spawn and pack it into cups. They should label their cup with their name, the type of mushroom being grown, and the date and place it in an old aquarium (or terrarium) with a lid to preserve humidity. Place a small amount of water in the bottom of the aquarium to maintain moisture; don't let the cups dry out. Students can take turns misting the cups daily to maintain a moist surface; the cups should not be submerged or excessively wet. After a few days the spreading mycelium will become visible. Allow students to examine their cups and describe what the mycelium is doing (colonizing the food source). In just over two weeks the cups will capable of fruiting mushrooms; at this point, they should be misted several times daily. Once mushrooms start growing, students can measure their rate of growth. (The mushrooms double in size daily; students will be amazed!) When the mushrooms are fully mature (they'll stop growing), students can harvest them and make spore prints to help them understand how the cycle begins anew. (See chapter 1 for more information on mushroom ecology and life cycle.)

Recycling and Composting Track
Paper and Cardboard
Show how mushrooms can eat cardboard and paper by soaking paper sheets in water and layering them with spawn in a container. Within a week the mycelium will spread, completely covering the sheets, and then it will begin decomposing the paper and recycling it back into soil. (See chapter 10 for more information on cardboard cultivation and recycling with mushrooms.)

Coffee Cultivator
Collect spent coffee grounds from the teachers' lounge, students' homes, or an area coffee shop. Place them in a bucket, and inoculate the grounds with an oyster mushroom starter culture. Leave the lid cracked open, or drill a few holes in the lid to allow for gas exchange. (Explain that mushrooms are like people and need to breathe.) Observe the bucket every day. (You can treat it like a pet and name the bucket!) Keep adding coffee until the bucket is two-thirds full. Then stop adding coffee and begin misting the bucket every day. Mushrooms will form within a few weeks. When the mushrooms are fully mature (they'll stop growing), students can harvest them and make spore prints to help them understand

how the cycle begins anew. (See chapter 10 for more information on cultivating mushrooms on coffee grounds.)

Mushroom Paper
Mushrooms have been used to make clothing, fiber, and paper for centuries in different parts of the world. Many inedible mushrooms, such as turkey tails and other polypores that typically cover fallen logs and branches, are suitable for use as paper fiber. Collect enough fungi for a good-size papermaking project. Work with students in class to blend the mushrooms with water into a slurry and to form the slurry into sheets, which can then be left to dry overnight. (See chapter 13 for more information on making mushroom paper.)

Mycoremediation Track
How Mushrooms "Eat" Pollution
Use examples such as your paper and cardboard or coffee cultivation project (see above) to describe how mushrooms can "eat" or break down waste so it doesn't go to landfills. Engage students in brainstorming what things mushrooms are capable of breaking down. What would the world look like if there were no mushrooms? (See chapter 10 for more information on recycling and composting with mushrooms.)

Research and Science Projects
Observation and Communication
Following a short class discussion on fungal ecology or a walk outside to observe mushrooms feeding on their favorite foods, have students communicate their new understanding by drawing or painting images of mushrooms as important recyclers in the environment, in relationship to trees, composting, insects, and pollution. Ask students to describe their drawing or write a poem about it. (Send me the drawings!)

GRADES 5–8 (AGES 10–14)

Fungal Ecology Track
Insect-Killing Fungi, or "Zombie Ants"
I challenge you to find a child in this age group who will not love the subject of fungi attacking and taking over

the brains of insects! Entomopathogenic fungi, such as cordyceps (*Cordyceps* spp.), colonize not plant debris but insects, and they make for fascinating discussion of the fungal life cycle. You can find photos of cordyceps colonization online using the keywords "cordyceps mushrooms insects." Class favorites will likely be photos of ants with mushrooms popping out of the back of their heads and giant tarantulas covered with fruitbodies. Many of these fungi thread their way into their host's brain and steer them around like a remote control car while they are alive! (See chapter 1 for more information on the ecology and life cycle of cultivated mushrooms.)

Mushroom Dissection

Dissecting a mushroom can help teach students the basics of fungal anatomy. Whether you're using wild mushrooms or store-bought varieties, use specimens that have the basic features such as cap, stem, and gills. Have the students cut the mushrooms from top to bottom, and ask them to draw what they see and label the parts. After the dissections, use the cut-up parts of the mushrooms, such as the separated caps and stems, as paint stamps, dipping them in paint or dyes to make prints on paper.

Cultivation Track
Gas Exchange Basics

Build on the "fruiting cups" lesson in the cultivation track for grades K–4 (see above). The cups will produce a large amount of carbon dioxide, which will pool in a covered aquarium like invisible water. To illustrate the importance of gas exchange for fungi (and the symbiotic relationship between plants and fungi), at the start of the "fruiting cups" lesson, place half of the cups in an empty aquarium, and place the other half in an aquarium that also contains some low-growing plants. Cover both aquariums with plastic or a solid lid, and set them side by side in a spot where they will have equal lighting. As the mushrooms form, those in the aquarium without plants will feature elongated stems, as they search for fresh air. The mushrooms in the aquarium with plants should look considerably different, with much shorter stems, as the plants do the work of "scrubbing" carbon dioxide and exchanging it

for oxygen. Students can measure and graph the stem length and cap width of the mushrooms, noting the differences between the aquariums. They can come up with a hypothesis about why they are seeing differences between the two aquariums and develop a new experiment to test their hypothesis.

Expanding Biomass

After a mushroom has fruited several times, it will have exhausted its substrate, but the mycelium is still alive and active. Students can recycle and reinvigorate the mycelium by expanding it on paper or cardboard. Have students spread the mycelium on wet cardboard and roll it into "burritos" (yes, use terms they like to hear!). Store the burritos in a lidded plastic container with a few holes drilled in it to allow gas exchange and limit dehydration. In just a few days the burritos will be fully colonized and can be used as spawn for starting new experiments. Liken this process to cloning or "copying" a fungus. (See chapter 12 for more information about expanding spawn on cardboard.)

Recycling and Composting Track
Waste Reduction Using Mushrooms

Encourage students to bring recyclable items from home or observe ways in which their school handles waste. Students can interview personnel such as cleaning crew, cafeteria staff, and administrators to gather facts about the disposal of all waste on the school grounds and brainstorm ways to redirect biodegradable waste to an on-site mushroom composting or vermicomposting operation. Gain permission from school officials to designate a mushroom growing and recycling area where students can collect and sort the appropriate waste by-products that can be used as mushroom growing substrate; they need only soak the materials, add mushroom cultures, and create compost. (See chapter 10 for more information on recycling and composting with mushrooms.)

Mycoremediation Track
Plate Contamination Observations

Prepare a series of agar plates sealed in ziplock bags to maintain their purity. The plates can contain bacteria alone, fungi and bacteria on opposite sides, and fungi plated onto agar with herbicides or fungicides in it. (Some fungi, such as oyster mushrooms, are excellent at degrading chemical herbicides but may not grow well on plates treated with fungicides.) Have students monitor the plates, noting their observations once a week. You can also give students uncultured agar plates and have them expose the plates briefly in a few different areas, such as the classroom, the school bathroom, and other locations. Students can then observe any mold and bacteria growth that results, counting, describing, and drawing the different types that form. (See chapter 17 for more information on preparing agar plates.)

Research and Science Projects
Mushrooms in the News
Ask students to research a particular topic related to innovative uses of mushrooms, such as mycoremediation, medicinal mushrooms, or mushroom-based packaging materials, and ask them to prepare a short essay or presentation about their research.

Gravitropism
This lesson builds on the K–4 "fruiting cup" lesson and evaluates how mushrooms are influenced by gravity. Most plants exhibit gravitropism: Their roots grow in the direction of gravitational pull (positive gravitropism), while their stems grow in the opposite direction (negative gravitropism). To find out how gravity affects the growth of mushrooms, cultivate a group of mushrooms in cups. I suggest using reishi mushrooms (*Ganoderma lucidum*) for these trials, since they form antlers, long thin pins that are sturdy and slow growing, making observations very easy. Reishi mushrooms also require very little water and can tolerate high carbon dioxide, which means you can add just a little water to the cups to keep them moist and covered tight with a domed lid with just a small airhole in the top to prevent spillage. Set aside half of the cups as your control group, in a plastic bin that's left with its lid cracked open or has holes drilled through the lid. Set up an old record player inside a second, identical plastic bin. Attach the cups of the other half of the colonized cups to the turntable's platter (the spinning disk). To make it easy, you can affix containers of the same size to a thin piece of plywood or any other building material and screw the containers directly into the wood that is shaped the same size as the spinning disk, then you can simply drop in the fruiting cups, which can be added and removed easily. When primordia appear in the cups, turn on the turntable and leave it running. The spinning platter creates a sort of sideways gravitational pull by virtue of its centrifugal force. Allow the mushrooms to mature over one to two weeks, observing and making note of their growth pattern, versus that of the control group.

GRADES 9–12 (AGES 14–18)

Fungal Ecology Track
Identifying Mushrooms
All students need a break from the classroom, so take them out once in a while for a stroll to see what kinds of fungi are growing. Keep your plans flexible, so you can factor in the weather and scout for fungi in advance. And encourage students to bring in mushrooms that they find on their own. Using basic keys, which can be found online or in field guides, examine the mushrooms and try to ascertain their genus. Teachers should familiarize students with dichotomous keys here, perhaps using basic gilled fungi (Agaricales) as a starting point, and also using spore prints and general dissections to assist with making the distinctions.

Mycorrhizal Relationships
Perform small experiments with vegetable seeds known to form mycorrhizal relationships with fungi (fast-growing plants such as beans are a good option). Sterilize the surface of the seeds by dropping them into ethanol, stirring or shaking the solution for thirty seconds, and then straining out the seeds and rinsing them twice with distilled water. (If no ethanol is available you can use 50 to 70 percent isopropyl alcohol.) Using distilled water, moisten enough potting soil to germinate all the seeds you plan to use in a disinfected container (that has been wiped down with diluted

bleach [10 percent] and dried) until you can squeeze a few drops from a fistful. Then sterilize in an autoclave or cook it in an autoclavable bag or oven bag for two hours in a large covered pot with just a little water in it, adding water to maintain high steam.

Clean the number of small pots you will need with 10:1 diluted household bleach or rubbing alcohol (isopropyl, 70 percent) and, using gloves that have been disinfected, portion the sterilized soil into the containers that are going to be seeded. Using groups of five seeds for each experiment, plant seeds in plain, sterilized potting soil as a control. Plant the remaining seeds in conjunction with a treatment of your choice, such as a mycorrhizae inoculation or a dose of nitrogen fertilizer (with and without accompanying mycorrhizae). Keep the control group and the treatment groups in the same light and water conditions. Observe and measure their rates of growth, such as overall height and the distance between leaf nodes (if they are elongated, that can be a symptom of lack of light or excess nitrogen). Consider adding soil pathogens as a variable to determine the degree of disease resistance imparted by the mycorrhizal partners. You can also stain and view mycorrhizae under a microscope to detect and confirm the bonding and presence of the fungi inside the roots.

Isolating and Cloning Fungi

Have students make oyster mushroom spore prints on aluminum foil (see chapter 2 for information on making spore prints). You can order prepoured agar plates or you can pour them the morning of the experiment, adding penicillin to the formula if it is available to minimize bacterial contamination, making a selective medium for fungi. Have students transfer the spores to the agar plates. All transfers should be performed either at a laminar flow hood or near a Bunsen burner flame, with sterilized instruments. Alternatively, oyster mushrooms clone easily from cap tissue; students can take small fragments of the interior, staying away from the gills or outside surfaces, and transfer them to the agar plates.

Place the cultured plates into individual ziplock bags, labeled with the date and other culture information. Store these plates at room temperature or in an incubator set at around 75°F (24°C). Mycelium should begin to radiate out on the plate in a few days. Oyster mushroom mycelium is white. If there are any other colonies of molds or bacteria on the plate that do not appear white and strand-like, have students draw what they see and offer an explanation as to how they got there. (For more information about culturing agar plates and cloning mushrooms, see chapters 17 and 18.)

Cultivation Track

Space Lab: Build an Enclosed Ecosystem

Many have tried (and failed) to create replicable long-term enclosed functioning ecosystems that could produce food and handle water filtration during space travel. Students can create a small-scale model with plants and fungi; with the right balance, they provide a suitable gas exchange for each other, while also producing food: fruits, vegetables, and greens from the plants, and protein-rich mushrooms from the fungi. Students will need to consider how to proportion the biomass to balance the system, since too much or too little of any one component will throw it off. Mushrooms grown in such a system are bioindicators of gas levels. If oxygen levels are high enough, the mushrooms will have short, thick stems. If oxygen levels are too low, the stems will look like spaghetti noodles—long and thin, and somewhat deformed.

The plants and fungi will also produce other by-products, including ethylene, a natural plant hormone that causes fruit to ripen quickly and then encourages molds, which can manufacture even more ethylene, which can be toxic to developing mushrooms (and people) in an enclosed environment; limiting plant decomposition will help keep it in check. Students can also use a small duct fan to pull ethylene out of the enclosure, filtering it through potassium permanganate crystals, which can be purchased from hardware stores or pet stores.

The system will be enclosed for the duration of the experiment, so clear containers are the only option. If items need to be added to or removed from the chamber, students can install a small access panel, but even with these precautions, gases will ultimately enter and exit, which can alter the results. Alternatively, students can make a "glove box," a set of sterile gloves sealed inside the box and accessible from outside. In any case, I recommend placing all of the materials you think you will need inside the chamber at the beginning of the experiment, whether you need them right away or not.

The projects can be part of a competition to see which model performs the best, by comparing favorable growth of both plants and mushrooms. Students can work in teams, and they should document the entire process, including daily notes on changes, challenges, and successes. Who knows? Perhaps one of your students will be responsible for cultivating mushrooms in space! (See chapter 13 for more discussion of the potential for growing mushrooms in space.)

Basic Mushroom Cultivation

Students with access to indoor or outdoor space such as fruiting rooms or greenhouses can create small-scale production trials where they can experiment with cultivating mushrooms on different substrate formulas. They should record yields, calculate biological efficiency, and experiment with methods of natural pest control.

Recycling and Composting Track
Waste Reduction Using Mushrooms

In addition to scaling up the waste reduction project described for the recycling and composting track for grades 5 through 8 (which is a project that can be done at any institution by anyone of any age), high school students can evaluate the efficiency and performance of composting operations. They can begin a basic mushroom-based recycling process by inoculating paper and cardboard with spawn, allowing it to colonize and fruit, and then adding red wigglers to create a rich compost that can be used by the campus's horticultural program or in other gardening activities. Students can test the compost for fertility, pH, and nutrient content using basic kits that test for nitrogen, phosphates, and potash—all compounds that affect plant growth. (See chapter 10 for more information on composting with mushrooms.)

Mycoremediation Track
Using Fungi to Control Soil Erosion and Filter Contaminants

Students can construct small biological soil erosion and pollution control units in trays or other containers by planting them with varying ground covers, plants, and fungi in an appropriate substrate. Students can mix and match the components as they like, so long as they keep track of what they use and use the same soil base for all units. Allow the plants to take root and the fungi to colonize for a couple of weeks before testing the units. For a control, set up a few identical containers with the same soil base that is used in the planted containers.

To simulate site runoff or water flow, set up a "water run," a long trough or channel angled at the desired slope to fill with myceliated mushroom biomass, where water runoff can be simulated by adding a known volume to the upper part of the channel and collected at the base for testing. Units should be measured for rate of flow using clean water before filling and running the experiment. Since water filration and site runoff is based on retention and contact time, especially for mycoremediation experiments, rate of flow can tell you how much time the water is interacting with the mycelium in your mushroom treatment. After construction, use your testing volume of water and pour it into the channel with a collection tub waiting. Time the runoff, as well as measuring the amount of water collected, which should be near identical to the original volume. Units can be filled with different erosion treatments such as none (bare soil), soil covered with erosion fabric, shredded wheat straw, and finally a mushroom substrate. Interview your local landscaping companies or department of transportation for advice on what they would use to reinforce your experimental design. For the mushroom substrate experiment allow the mycelium to recover and thread mycelium into the soil, which will help bind the soil before testing the runoff. Levels of silt or soil erosion can be measured by dry weight after each run using the option of treatments, where the ideal treatment would minimize to eliminate silt or soil runoff. Be sure to run the trial in replicates, recording the data and averaging it to compare the efficiency of the treatments. Students can also become engaged in determining the cost of the erosion prevention. Give them the cost of the materials, and they can determine the best erosion control for cost.

For testing contaminated water, such as diluted herbicides, fertilizers, and other possible environmental pollutants, the water should be collected and examined for each run, noting retention time (the amount of time it takes the water to filter through each unit). Be sure to exercise safety and use protective gear such as gloves and safety glasses when using chemicals. You can test the level of contaminants remaining in the water using various test kits, but for a low-tech method you could also use the dilution "bean test" or "worm assay" described in chapter 22. The results from all modules can be graphed and analyzed to determine the best treatments and the variables that contribute to the erosion prevention and remediation of polluted water.

Remediating an Oil Spill

Review the span of oil spills with your students, from the larger historic oceanic spills down to homeowners spilling small amounts during an oil change for their car. However it gets there, all that oil adds up—and ends up in the environment, interacting with plants, animals, waterways, and geologic structures. Fungi, however, can be used to remediate an oil spill, using their extracellular enzymes to break apart and cleave hydrocarbon molecules into smaller units that are more easily biodegraded by other fungi and bacteria in a sequence. The mushrooms do this by feeding on the oil after running out of their natural food source, which can be a volume of grown or prepared substrate, such as spent oyster mushroom media. Hydrocarbons are chemically similar in architecture to what fungi decompose naturally, which is lignin, or the woody component of wood, so it doesn't take them long to adapt and begin to decompose the oils, typically just a few weeks.

You can set up a small mock oil spill in a classroom using vegetable oil (motor oil, though more appropriate, is considered a hazardous waste, and using it would require you to secure a special permit from local environmental regulators). All you need is a small aquarium, a few plastic bins, shredded dried grass (or hay or straw), some type of vegetable oil, and of course mushroom biomass. The biomass can be sheets of soaked cardboard colonized with oyster mushroom mycelium two weeks ahead of the "spill."

Fill the aquarium about half full with water, and pour a measured amount of the oil onto the surface. (How much you use is not important; it's only important that you measure it.) The oil will float. Have students add small amounts of the shredded dried grass, which will also float, and stir it with a large metal spoon or stick. The oil will stick to the plant matter. When the plant matter appears to have taken up all the oil, remove it from the water and layer it between the colonized sheets of cardboard in small bins, packing it all in tight. Each bin should receive the same amount of oiled plant matter, so you may want to make the bins one at a time, stirring up your plant matter with the measured amount of oil one batch at a time. Cover each bin with a lid that has had a few holes drilled through it to allow for gas exchange.

Once a week, students should inspect the bins by gently peeling back the layers to photograph and observe the mycelium in action. After a few weeks, large yellow droplets or metabolites will form on the myceliated surface, which is the mycelium exuding the enzymes it needs to break down the oil. Once the bins have been sitting for approximately one month, the students can then add a number of red wiggler worms to see if they survive. (Make enough small bins to allow for adding worms at weeks four, six, eight, ten, and twelve, if you like.) Noting the duration of this experiment, it would be wise to start early in the year, since it can take up to three months for results.

Research and Science Projects
Report: Ongoing Mycoremediation Projects

Have students research and report on mycoremediation projects that are ongoing and worldwide.

Students should outline the problem being addressed and the importance of solving it, while also explaining why researchers chose to use mushrooms, rather than any other existing strategy, for the remediation.

College (Ages 18+)

Fungal Ecology Track
Entomopathogenic Fungi

Entomopathogenic fungi—fungi that attack and kill insects, primarily by parasitizing them—can be found almost anywhere, though it can take a bit of work to find them. The fungi colonize their host insect, threading their way inside the insect's body and sometimes even brainstem, to the extent that they can take control of the insect's activities and movement, like steering via remote control. It takes a special eye to locate and find these infected insects. Encourage students to browse and sift through gardens, horticultural plots, and organic farming enterprises nearby, collecting insects that appear sluggish or, even better, to be "mummified" on a plant, having died while gripping plant tissue. Not all fungi produce elaborate and noticeable fruiting structures; any insect that appears to be engulfed by or swollen with mycelium, powdery growth, or tiny bumps that are not part of its normal exoskeleton may have been parasitized by a fungus.

Every attempt to identify the host insect and isolate and identify the fungal parasite should be made, especially if the insect is a pest that frequents and damages local agricultural crops. DNA analysis can provide identification clues, and the fungus can be cultured, purified to eliminate contaminants, and reintroduced to behavioral or mortality galleries, where the pure culture is added to a living collection of insects (for example) to determine its singular pathogenicity, excluding all other variables.

Mycorrhizal Fungi

The scientific community is in need of in-depth studies that examine the relationships between plants and mycorrhizal fungi. In recent years I have had many requests from gardeners, landscapers, nursery owners, and native plant experts to seek out and isolate strains of fungi that have mycorrhizal relationships with not just agricultural crops but also prairie grasses, orchids, and endangered plants whose populations are dwindling due to habitat destruction or illegal harvesting. These efforts deserve grant funding, since many of these plants cannot survive without specific strains of fungi—fungi that need to be collected, characterized through DNA analysis, cultured, and potentially developed into commercial products that can help plants become more resilient and able to thrive. The fact that every wild isolation, or ecotype, of a fungal species can be very different from other ecotypes in its abilities to perform specific metabolic functions means that there will always be a better strain out there waiting for you to find and test it! I recommend that students consider this work to be a promising, and understudied, field of research that can have valuable applications for increasing crop yields while reducing dependence on chemical fertilizers.

Cultivation Track
Space Lab: Build an Enclosed Ecosystem

Building on the space lab project described in the cultivation track for high school students, college students can construct larger chambers to increase biodiversity and the range of variables being examined, all while trying to maintain stability in the system. There are many possibilities for students interested in building this sort of enclosed ecosystem. They may be motivated to explore the possibility of mycoremediation in space, micropropagation—cultivating in very small vessels, with an eye toward automating the process—in space, or providing for all the system's energy needs with solar panels or arrays. They may explore the possibility of fungi as a primary food source, conducting nutrient analysis of various species and surveying their biochemical responses to being grown under various wavelengths of light in an attempt to determine optimum outputs with minimum inputs. Students can design entire processes for media treatment, culture expansions, spawn, and create cultivation chambers with zero gravity in mind. For competitive group experiments, try giving each group an identical set of materials to see what they can make from the items, creating an opportunity for collaboration and creative problem solving. (See chapter 13 for more discussion of the potential for growing mushrooms in space.)

Region-Specific Mushroom Cultivation Technology

Students can focus on indigenous, locally available forms of agricultural and industrial waste on which to grow mushrooms native to their region. Students can isolate and acclimate (or "train") native fungi to feed and fruit on the growing media, measuring biological efficiencies to fine-tune formulas and identify high-yielding strains. They can incorporate lessons on strain isolation and expansion, graduating up to trials of specific cultivation methods in order to master their skills and then design and implement improvements. Mixing students who are skilled in biology, microbiology, agriculture, and engineering can be beneficial, bringing insight from multiple academic disciplines and encouraging cross-communication and collaboration.

Maintaining and Storing Fungi

Most microbiology students learn to culture bacteria but not fungi. Teaching students the techniques of spore germination, cloning, media preparation, and storage can eventually develop into new lab exercises and experiments related to metabolic testing, biofuel production, mycoremediation, and other projects. Distribute fresh agar plates (with antibiotics) to students and provide them with both a fresh mushroom and a spore print. Teach methods for both spore germination and expanding live tissue. Distribute stock plates with cultured mycelium and have students create test tube slants, which can be refrigerated for several weeks or months and then plated to test for viability. Using strains such as pink oyster (cold-sensitive) and cold blue oyster (cold-tolerant) will demonstrate the relative differences in storing fungi long-term. (Covering slants with mineral oil before refrigeration offers better long-term protection for cold-sensitive species.)

Bacterial-Fungal Interactions

Contact with bacteria can stimulate fungi to release various metabolites to encourage or limit bacterial growth, based on sequences of need in various fungal signaling pathways, so bacterial partners can be identified as necessary and indispensable players in triggering fungi to manufacture unique chemical compounds. Students can create galleries, or behavioral plates, which are simply petri plates that have been subplated with two or more species with space between them to study basic interactions between bacteria and fungi—for example, between penicillin and certain species of gram-positive bacteria such as *Streptococcus* or *Staphylococcus*. They can also identify isolates produced by various fungi in response to interactions with bacteria and then analyze those isolates for their specific functions and mechanisms.

Recycling and Composting Track
Waste Reduction Using Mushrooms

The waste reduction projects in the recycling and composting track for grades 5 through 8 and for high school can be scaled up for college-level applications as well. Students can organize a campus-wide project using mushrooms to effectively divert biodegradable material from the waste stream and explore other ways of recycling, such using mycelium to convert waste paper into packaging material (see chapter 13). The potential applications are limited only by the students' imaginations. Students can gain support for their project by discussing waste management with campus managers and by offering a short introductory program on composting with fungi so interested students, faculty, and staff can see how it works. Networking with different departments can build healthy bridges, adding cross-curriculum interest to mushroom research, which helps advance the study of mycology.

Mycoremediation Track
Screening Fungi for Use in Mycoremediation

The basic life cycle of fungi is to take hold, eat, stay alive, and develop a biomass capable of producing spores. Any fungi that can do this in the presence of a substrate contaminant show promise for remediation of that contaminant. Students can prepare "traps"— enriched substrate samples containing baseline levels of a particular toxic soil contaminant. They can expose the traps to the air (perhaps even in known sites of contamination) and then monitor the samples to see which naturally occurring fungi develop into viable mycelium. The colonizing mycelium can be subcultured and soil levels of the contaminant increased to gradually "train" the fungi (genetically speaking) to express those traits that allow them to break down the contaminant and thus thrive in the contaminated substrate. Expect many molds here, which is appropriate; they are all fungi and are all welcome to the dinner party. You want to find the organisms that devour the main course and ask for dessert!

Students can also collect fungi at known contaminated sites and clone them (see chapter 18) to reproduce strains that have already acclimated to the contaminant, site conditions, and surrounding ecology, and they can compare those strains' mycoremediation effectiveness against that of the lab-trained samples. They can also screen and "improve" these native strains to select those favorable mutations that improve their mycoremediation potential and then reintroduce the improved fungi at the original contaminated site for in-the-field mycoremediation studies.

Engineering Mycoremediation Prototypes

Interdepartmental collaborations can create strong alliances within a college or university, and amazing results. For mycoremediation, I can think of no better combination than engineering and microbiology. Engineering students are being trained to put data to work in the real world, designing methods, systems, and equipment to produce particular results or solve particular problems. Young scientists are being trained to explore data, to ask questions, and to test hypotheses; for aspiring microbiolgists, all this happens in the fields of biology and cell function. To facilitate such collaboration, identify a site of pollution that could benefit from remediation. Working together, engineering and microbiology students can design and test methods for mycoremediating contaminated soil and water, moving from data collection to prototype testing to the eventual construction and implementation of a filtration system.

Research and Science Projects

Packaging, Insulation, Building Materials, and Biotextiles

Fungi are extremely diverse, and although the cultivation of mushrooms is the main focus of this book, don't forget about their ability to form the kinds of dense mycelial mats that can be used to manufacture various consumer products, from textiles and packaging to building materials and insulation. Students can test different kinds of fungal mycelium for qualities such as tensile strength, antibiotic properties, and insect-repelling or water-repelling properties, with an eye toward using fungi to replace an existing product with a known negative environmental impact. And students can create and test prototypes of biodegradable packaging, insulation, building materials, and biotextiles by cultivating mycelia and molding them into the shapes needed for a given application. (See chapter 13 for more discussion of these and other cutting-edge applications of fungi.)

Isolating and Extracting Novel Fungal Metabolites

Fungi are like assembly-line factories, manufacturing metabolites in response to their environment. Students can dose fungal cultures with particular chemicals or biological organisms and then harvest and analyze the resulting metabolites, comparing them against control cultures to identify which metabolites are produced in response to which trigger. Testing the metabolites back against the biological or chemical trigger will provide clues as to whether the fungal culture wishes to either eradicate or promote the trigger. An example of this process would be stimulating a mycelial biomass with a particular bacteria, harvesting the resulting metabolites, adding the metabolites to an agar formula, and then culturing the bacteria on the agar to observe and record inhibition rates. Testing with different dilutions of the metabolite can help students determine the concentrations at which the metabolite is effective.

Fungal Foods of the Future

Many fungi, including molds and yeasts, make food that humans can use as an energy source, such as bread, beer, tempeh, and mushrooms. Students can practice with known cultures to reproduce existing foods, and then they can research and experiment with using fungi to develop a new food product, with an eye toward its market potential. For example, students can work with polypores, which are highly desirable as medicinal mushrooms but generally considered to be too tough to consume, and culture the mycelium to create space cakes (see chapter 13). The potential for experimentation is limitless, and students can seek out combinations that satisfy dietary and medicinal needs, potentially developing food that future generations will enjoy.

Biofuel Production

Fungi are well known as decomposers, but they are also molecule builders. Their enzymes perform molecular disassembly of complex substrates and compounds and then reorganize these elements as new molecules. Small-scale assays and experiments can provide clues to a fungus's metabolic pathways and substrate preferences based on the by-products it produces, especially gases. Continuously sampling levels of carbon dioxide generation, for example, can reveal the progression of lignin degradation in the substrate, helping students understand the cycle of bioavailability of by-products during the degradation process.

Matching and isolating fungal strains native to the carbon sources, while also isolating bacteria that perform collaborative, metabolic processes, can enable the production of alcohols and biofuels, which is an excellent area of focus; achieving any level of the sequence is a victory. Additional experiments can encourage the expression of fungal enzymes up to a specific point, such as for primary lignin degradation and rendering nutrient availability, and then "kill" the fungal biomass to end the primary metabolic sequence. The mass (and its component enzymes) can then be added to an anaerobic digester or bioreactor utilizing bacteria that are known to produce biofuel compounds of interest, such as gut bacteria from termites or mammals known to produce gases such as methane, which can be converted to methanol. Similar sequences can be created to produce alcohols such as ethanol and butanol using a limitless combination of fungi and bacteria in sequence systems.

PART III

Advanced Techniques and Research

CHAPTER 17

Basic Laboratory Construction, Equipment, and Procedures

A milestone in anyone's development as a mushroom grower is the construction and maintenance of a sterile culture laboratory capable of producing a large volume of spawn of many different species for personal use or small-scale commercial production. Although hobby growers may not see a need for this investment in time and money and might prefer the ease of ordering quality spawn from a reputable supply company, if you are already accomplished at consistently cultivating several species of mushrooms using purchased spawn, you may be ready to upgrade to a home laboratory.

A major benefit of making your own spawn is cost savings. Another is having a consistent and fresh supply of spawn. To ensure that they have enough spawn on hand to meet demand, some commercial producers allow their spawn to overincubate or let it sit in cold storage for months. This is not the best strategy, however. Spawn is better off made in small batches as needed. Making your own allows you to take control of every step of your entire operation.

A warning, however, is in order: It can take years to develop the skills and processes needed to become independent of commercial spawn producers. The best strategy is to start small, attempting trials of spawn generation to make sure your technique is at a high standard before you embark on a full-scale operation with its concomitant major financial risk. Although it's ultimately your responsibility to make sound decisions about investments and scaling up, I've tried to present this material in a way that will help you minimize the risks, beginning with basic laboratory construction, followed by simple tissue culture and spawn production techniques. Let's get started with the basic tools and supplies you will need to construct and maintain a small home laboratory.

SETTING UP A LABORATORY

You can turn just about any space into a basic lab as long as it is enclosed. A small closet or extra room is perfect to begin with. In fact, you could even use just part of a room, partitioning off a section for use as a lab. I have seen partitions constructed with metal or PVC frames and enclosed with clear plastic, like greenhouse walls. The point is that there are many options, so think creatively about how to use the space you have available.

The entry should be well sealed to help keep the interior air clean. In a plastic partition wall, a zippered entry works well. The floors should never be carpeted or textured; instead you'll want smooth concrete,

BASIC HOME LABORATORY EQUIPMENT AND SUPPLY CHECKLIST

Here are the basic equipment and materials you'll need to germinate spores, culture fungal tissues, and create grain spawn. Once you invest in the materials below, they can last for years. Then, as you hone your skills, you can expand your culturing or spawn production by installing a larger sterilizer (autoclave) and increasing your laminar space. For five years I was able to generate over one hundred 5-pound bags of spawn a week to supply my spawn business using two tabletop sterilizers and two laminar hoods placed side by side. We grew our product lines and business, and now our new autoclave has the capacity to create over one thousand bags a week, which I use for both spawn and fruiting block production. The point is that if you start small, develop a demand, and are passionate about your work, it can develop into supplemental income and even a dream job.

Basic Equipment

1 laminar flow hood with a HEPA filter (a 2×3-foot filter is best for economical reasons, and the size creates a nice space for transfers)

1 pressure cooker or sterilizer

1 electric impulse sealer for sealing autoclavable bags (12- to 16-inch seal length)

12 narrrow-mouthed mason jars with metal lids (or as many as you need for grain expansions)

1 glass Erlenmeyer flask (1,000 ml), or any wide-bottomed glass container (like a growler) that can hold 3 to 4 cups of liquid

1 alcohol lamp, filled with denatured alcohol

Scalpels or X-Acto knives

Scissors

Wooden clothespins (for clamping autoclavable bags while sterilizing)

Equipment for Advanced Techniques

Blender with all-glass container—no plastic base parts that can melt (for liquid cultures)

Laboratory refrigerator (for storing petri plates, cultures, and spawn masters)

Microscope (for inspecting cultures for contamination, spore observations, and mycelial morphology)

Plate stirrer (for liquid cultures)

Autoclavable mason jar lids with micron filter disks

Supplies

Petri plates (either reusable glass or sterile plastic)

Autoclavable bags with micron filter patches

Gloves (nitrile or reusable rubber dishwashing)

Isopropyl alcohol (70 percent) or half-strength bleach for disinfecting surfaces

Paper towels

Aluminum foil

Surgical or dust masks

Tape or Parafilm to seal petri plates

Ziplock bag (quart or sandwich size)

Colored permanent markers (to label petri plates, jars, and spawn bags)

Laboratory notebook (to keep records, inventory, observations, and notes)

---

tile, linoleum, or any other hard surface that you can sweep and mop daily. The walls of a lab will also need to be kept clean, so they should be made of smooth materials that can handle a weekly wipe-down. If you have no other space available, you can use a spare bathroom: Thoroughly sanitize it first, then run a small portable air filter or ionizer in it for an hour, and when you reenter the room, do so slowly, and allow the air to settle before you start work.

If you're not using a laminar flow hood (see the discussion below), I recommend wearing a simple and inexpensive surgical or dust mask, to keep you from breathing on the cultures, and using a generous amount of 70 percent isopropyl alcohol to sanitize the counter where you will be working. Perform your work next to an alcohol lamp, which keeps the air in the immediate working area cleaner.

Positive Pressure and Gas Exchange

Although not entirely necessary at first, consider installing a positive-pressure ventilation unit. This is a small fan fitted with a HEPA filter that pulls clean fresh air into the lab space. The incoming air "inflates" the room (positive pressure), so that if there is a leak of any sort, clean air pushes out, rather than contaminated air coming in. The fresh air is also important for the developing spawn, which generate carbon dioxide, and also for your own safety, since you would never want to seal yourself into a room saturated with nothing but carbon dioxide!

Commercial spawn laboratories often have positive-pressure air lock entries, to ensure that the laboratory space is never open to the outside. A length of ribbon hanging from the door can serve as a visual cue that airflow is indeed directed out from the lab, and never in toward the heart of the lab.

Constructing a Laminar Flow Hood

A laminar flow hood is a "bench" or small work surface surrounded by a usually open cabinet, or hood, with its own HEPA-filtered air supply. The unit pulls in ambient air at its side or base, filters it through a

My old lab had two laminar hoods set side by side: one for cooling down sterilized substrate and the other to use as a work space and for transfers.

THE LEARNING CURVE TO
PROFESSIONAL MUSHROOM CULTIVATION

The most rewarding gift I ever received was my first laminar flow hood, or sterile tissue culture station, custom-made for me by an orchid culturing couple, Dan and Madeline Nelson from May River Orchids in Bluffton, South Carolina. Dan was a builder, and he designed the hood and installed the motor assembly for my brand-new HEPA filter, allowing me to begin the journey into years of satisfying research and projects. Having this equipment allowed me to pursue my fascination and to pay close attention to the individual needs of every mushroom I tried to isolate.

When I started out with lab work, my efforts were inconsistent; cloning wild mushrooms and germinating spores proved to be frustrating and sometimes costly, especially for a young college student with limited funds. Persistence paid off, as they say, and twenty years later I maintain well over a hundred edible mushroom strains and have mastered spawn generation techniques using equipment that can be bought for a few hundred dollars. I quickly learned that it's not necessarily the size or sophistication of your equipment that leads to success, but rather developing mastery with what you have and can afford.

By starting off with basic tools I was forced to be creative and extremely careful with my techniques, which honed my skills. Specialized scalpel sterilizers, stainless steel blenders, and professional incubation chambers would all come later; it was the Styrofoam coolers, alcohol lamps, and a glove box (a still-air transfer box with hand holes fixed with gloves to handle objects and make transfers with very little air movement) made from cardboard and set up in my parents' attic that made it all happen. Having transitioned from that flimsy cardboard glove box to a dedicated closet, then a dedicated room in the house, and now an 8,000-square-foot building, I can look back and can respect the challenges, losses, and improvements I have made in developing my own processes and procedures, fusing my microbiological education with my personal preferences to optimize my culturing experience. What I've learned—and hope you'll learn, too—is that as long as you don't compromise the integrity of your cultures, there's no one right way to culture fungi. So learn what you can from the processes I recommend, but feel free to modify them to suit your own setup and skills as your techniques improve and expand.

HEPA filter, and gently blows that clean air across the work surface. The constant outward-moving filtered airflow prevents ambient air from entering the work area, thereby preventing contamination from airborne microorganisms such as bacteria and fungal spores. In most home and small-scale commercial spawn laboratories, laminar flow hoods handle the round-the-clock air purification that maintains the integrity of the space.

You can purchase a laminar flow hood as a kit (for a premium), or you can build one inexpensively if you have basic building and electrical skills. If you are purchasing it, expect to pay over $1,000 for a small

prefabricated kit that you screw together and plug in; if you're building it yourself, plan to spend around $400 for a high-quality HEPA filter and a fan or blower unit, and more for any other components (wood, work surface, fasteners) you can't scavenge and are forced to buy.

To begin building your own laminar flow hood, look for a gasketed HEPA filter with an efficiency rating of 99.997 percent at 0.3 micron. The gaskets allow for a perfect seal in the box enclosure, eliminating the leakage of unfiltered air into the work area. For a full-size laminar hood, I prefer a HEPA filter that is about 2 feet high and 3 feet wide. But it can be helpful to have a separate smaller laminar flow hood, with a smaller HEPA filter, for isolating wild specimens or molds, bacteria, or sporulating cultures that could threaten the cleanliness of the main filter laminar flow hood.

The blower or fan is typically a furnace-type, "squirrel-cage" blower that you can purchase at your local hardware store. The blower must have the proper electrical rating for your work space and a force, rated at cubic feet per minute (CFM), that is appropriate for your filter. If the blower doesn't have enough force, the laminar will lack the airflow it needs to maintain an even stream of sterile air across the work space. If the blower has too much force, it can damage the tissues you're working on or create a whirlwind in the lab, blowing contamination all over the place.

Now that you have a filter and a blower, you will need to build a box to house them. Measure the exact dimensions of the filter, and construct a basic box-like structure around it, extending roughly 8 to 10 inches from the back of the filter and approximately 18 to 24 inches from the front. The deeper front space will be your actual work space. The shallower enclosure in the back will be used for the blower housing, where a small hole the size of the blower output is mounted, usually on the top of the unit, which can also be housed inside its own separate box with an open top where a coarse prefilter can be placed to catch most of the larger particulates before they are passed on to the HEPA filter.

For the actual work surface in front of the filter, use a stainless steel sheet or comparable surface that can withstand disinfection without staining or rusting for the work space directly in front of the filter. You can sometimes find and convert old tabletops or commercial kitchen components for a bargain, so take some measurements and start looking around.

To fasten the box together, use small bolts with wing nuts on compression washers so that you can easily open the box to access the filter when it's time to replace it. My friend who built mine (see "The Learning Curve" sidebar) designed the filter housing with a removable lid so I could slip the filter out and slide a new one in without disassembling the entire unit.

You'll typically need to replace the HEPA filter once a year if it's used heavily, but if you install a cheaper prefilter on the air intake to capture the bulk of larger particles before they enter the unit, the HEPA filter can last two or possibly three years.

Although we made our unit out of galvanized steel that was bent at a machine shop, you can easily build the box for your laminar flow hood out of any sturdy material, such as wood or scrap metal, as long as the interior of the work space is impervious to water. If you build the box of wood, you can seal it with epoxy or laminate to give you the wipe-down protection you need.

Once you have built the box for your laminar flow hood, you're ready to seal all joints and seams to create an airtight enclosure, ensuring the optimum push of air from the blower and eliminating the accidental pull of unfiltered air through any tiny gaps or cracks. Some suppliers suggest silicone caulk for sealing, but I prefer aluminum tape, used in the HVAC trade for sealing ductwork. Use it to line all the joints of the box as well as the junctions between the box and the HEPA filter. This not only prevents air leakage but also makes the laminar easier to clean, since you can extend the tape up and over the edges of the filter frame and along the entire inside perimeter of the filter housing where it connects to the work area to keep contaminated air from being sucked into the work space from behind the unit.

Once you have installed the HEPA filter and blower and sealed all the joints and seams, turn on the blower. Check to make sure that you have adequate airflow inside the laminar hood, and then start checking all the seams and junctions for air leaks. A smoldering stick of incense works well; hold the stick close to the seam you're checking, and if the smoldering end brightens, it signals that the air around it is being pulled into the box, and the seam needs to be sealed. Check the entire construction to make sure it is airtight.

If you've built a laminar flow hood in which the blower pushes a solid wall of sterile air across the work surface at 100 to 600 CFM, and the work space enclosure is waterproof, that is quite an accomplishment. (The ideal airflow rate for a commercial laminar flow hood is approximately 550 CFM.) Anyone can build this. Professional units may advertise fan control switches and other gizmos, and they may look nice, but if you just want to practice or work at a small scale, I would keep it basic but functional. It doesn't have to look pretty; it has to work. One of the obstacles and misconceptions in the mushroom industry is that you need to buy an expensive, fancy prebuilt laminar hood in order to become a tissue culturist. It's just not necessary. I still use a laminar hood my friend built for me fifteen years ago. If you construct your hood simply and with easy-to-remove components, you should be able to use it for years of tissue culture experimentation.

MAKING AGAR PLATES

Perhaps the most profound moment in my life was watching my first spores germinate on an agar plate. This is a magical moment for every cultivator, witnessing the first step of a much larger cycle and the foundation of more complex cultivation efforts. At the

time I wasn't sure if I had cultured mold or mycelium, so I might have even thrown the plates out. But it didn't matter. I was off and running.

An agar plate is basically a petri plate (also known as a petri dish) with agar, a gelatinous extract of seaweed that is commonly used as a substrate for tissue cultures. You can buy sterile petri plates; if you buy glass ones, you can wash and reuse them, sterilizing them between uses. (You can also use small jars, much like the ones used for baby food, in place of petri plates.) You can purchase agar in powdered form from mycological supply companies and at most Asian markets (it's used like gelatin in Asian cooking). As you progress, it becomes more economical to purchase lab-grade agar. Once you have the powder, it's easy to prepare the gel, sterilize it, and pour agar plates.

Step 1: Preparing Agar Gel

Slice a small potato into small chunks or shred it with a grater; you do not need to peel it first. Place the pieces in a small pot along with 3 to 4 cups of water. Boil

Boiling chunks of potatoes in water is the first part of making a simple agar formula at home.

for one hour, being careful not to let it foam or boil over. The water will look like "potato tea"; it will not thicken. If the water level drops significantly, simply add water as needed and keep it boiling.

Strain the potatoes from the liquid. Measure the hot broth, and then pour it into an Erlenmeyer flask (or any wide-bottomed glass container). Add 1 teaspoon of agar powder (approximately 5 grams) for every 2 cups of potato broth. Adding the agar to the broth while it's still hot prevents the powder from clumping and helps it dissolve nicely. Never use cold water; the agar clumps instantly. Professional labs use hot plate stirrers to homogenize their agar before sterilizing. I still just swirl the agar into the hot water to dissolve it.

Step 2: Sterilizing the Agar

In order to sterilize the agar solution you will need a pressure cooker capable of maintaining anywhere from 12 to 15 pounds per square inch (PSI). The lower the PSI, the longer the cook time and the lower the maximum sterilization temperature will be, so try to find one that is rated to operate at 15 PSI. The best pressure cookers have "metal-to-metal" seals, without the rubber gaskets that are used in cheaper units, but if the gasketed model is all you can find or afford, go ahead and use it to develop your skills before you make a financial commitment for a better model, which can cost several hundred dollars. Whatever type of cooker you use, it must be tall enough to accommodate the flask with your agar solution.

Fill the pressure cooker with water according to the manufacturer's directions, and place your flask of agar solution in the inner pan, keeping it upright. If it seems at risk of tipping over—such as might be the case if you have to move the cooker to a different spot once sterilization is complete—you can set the flask in an additional wide-based container, such as a wire basket or old metal bowl, wedging the flask into place with wads of aluminum foil to stabilize it.

Plug the open top of the flask with a wad of cotton or polyester fill (like the kind you'd use to stuff a pillow), cover it with aluminum foil, compress tightly so it is not loose, and begin heating. When the pressure cooker reaches an internal pressure of 15 PSI, start

your timer. Process for twenty minutes, then turn off the heat and allow it to cool.

Once the cooker has dropped to an internal pressure of 2 to 3 PSI, it's safe to move it, if you need to. I use a cart to move the cooker, just to be safe. When the pressure drops to just slightly above 0 PSI, it's time to open the cooker. To avoid contaminating the sterilized agar, open the cooker only in a clean room or in front of your laminar hood, and only when the internal pressure is still above 0 PSI. If the pressure drops below 0 PSI, it will form a vacuum, and when you open the cooker it will suck in contaminated air; if it does, you should resterilize the agar. Additionally, if you open the cooker when the internal pressure is still well above 0 PSI, the sudden change in pressure can make the agar boil up and out of your flask; you will lose much of the gel, making a mess and wasting supplies and time.

Before you open the cooker, thoroughly sanitize your work area; ideally this will be at a laminar flow hood. Open the cooker and remove the hot flask, using a paper towel that has been saturated with 70 percent isopropyl alcohol as a mitt, or a sanitized jar lifter (the kind you use to lift mason jars from a canner; these are available at most hardware stores). Never touch the jars your bare hands, no matter how clean you think they are; the oils on your hands can be easily colonized by bacteria, compromising every transfer from the moment you start. Place the hot, sterilized agar on your sanitized work surface to cool.

Step 3: Pouring the Agar

Monitor the temperature of the agar. It will be extremely hot when you remove it—up to 200°F (93°C)—but it cools quickly. I use an infrared temperature gauge that I purchased for $30 at a hardware store to beam the cooling agar every ten minutes. You can swirl it a bit to mix it while it is cooling. Do not touch the flask with your bare hands; wear sterile gloves. The pouring temperature is around 100°F (38°C); if you miss this window and the agar gels, you can reheat it in your sterilizer to melt it back again. If you are adding any antibiotics (see page 175), plan to do so between 105 and 110°F (41–43°C).

An infrared temperature gauge is one of the best investments you can make. It will help you orchestrate the entire sequence of events for agar preparation.

You'll need sterile petri plates at your laminar flow hood. Each petri plate comes with a base and a lid, and normally you'll buy them stacked in a plastic sleeve. Wipe down the outside of the sleeve with rubbing alcohol, and similarly sanitize a pair of scissors. Use the scissors to trim the bottom off the sleeve. Invert the entire stack of petri plates and set it upright on your work surface, still in the sleeve. Gently tug the plastic sleeve upward to remove it; roll it up and store it in on the laminar hood, in case you need to use it for storing plates later. Divide the stack into four or five groups; smaller stacks will make pouring faster and more efficient.

When the agar solution drops to around 100°F (38°C), which will be cool enough to touch, you're ready to pour. Pull one of the stacks of petri plates in front of you. Hold the flask of agar in one hand. With your other hand, grasp the lid of the petri plate at the bottom of the stack and lift straight up; your aim is to lift the rest of the stack, while leaving the base of

the bottom plate on the tabletop. Slowly pour the agar into the petri base until it is about half full, and then lower the stack back down to replace the lid. Then slide up to the next petri plate and repeat the process, and keep going until you've poured the entire stack. When you move a filled stack aside, gently slide it to prevent the liquid agar from sloshing up into the lid and over the edge; this can become an entry point for contamination. Two cups of agar should be sufficient for a sleeve of twenty to twenty-five plates. With practice you will be able to pour an entire sleeve in a minute or two.

Allow the plates to cool to room temperature. After about an hour, check them for gelling by picking up a few and tilting them to the side to see if the agar slides. Agar that is properly gelled should be firm and elastic. Wet agar, or agar that has not properly gelled, is much more prone to contamination and more difficult to make clean spore streaks on, since the surface will disintegrate when you rake it lightly with a tool.

You should label your agar plates with the date, the recipe, and any additives (such as antibiotics, wood screenings, or supplements) you may have used. That can be a lot of information to put on such a small plate, so you may want to construct a basic code, using, for example, particular colors that correspond to particular recipes and additives that you use on a regular basis. For example, using colored permanent markers in my own lab I use the color green to denote the antibiotic penicillin G, and I'll mark an agar plate's outer edge once if it contains 1 ml penicillin G, twice if it has 2 ml penicillin G, and so on. To mark an entire stack of poured plates, I take the colored permanent marker and drag it up the entire stack of plates once they are gelled, coding them quickly and efficiently.

Step 4: Using and Storing Agar Plates

After the gelled agar plates have cooled, you can use them immediately, streaking spores or adding mycelium (see chapter 18 for details). Seal the seam between the base and the lid of the cultured plates with tape or Parafilm to keep them from separating and to prevent air infiltration. If there are any extra plates that you don't plan to use right away, seal them in new plastic ziplock bags (quart or sandwich size) in packs of three or four plates. (Plastic bags are actually very clean inside since they were heat-injected when they were made, so there is a minimal risk of mold contamination in them.) Label them with a date and agar formula code, and store them in the refrigerator, where they will keep for one to two months, and possibly longer. If you have more than a few extra plates, you could stack them and store them in the plastic sleeve they were originally packaged in, taping it shut.

Having extra plates can be a blessing if you are ever out collecting and unexpectedly find a mushroom you would like to culture (almost every day for me); since spore streaks and clones take only a few plates, these prepoured plates are perfect. Inspect the plates in their plastic bag in front of the laminar before you use them. If you notice any circular green, yellow, or black powdery growth, do not open the bag; discard it and the plates inside.

ALTERNATIVE AGAR BASES

As you expand your knowledge and skills in culturing fungi, you will likely want to substitute other bases for the potato one. Potato agar is difficult for bacteria to grow on, so it makes a great starter medium for beginners, but for picky mushroom species that need a more complete or specific food source, you may want to prepare the agar with a different base. Even different ecotypes of the same species of mushroom can differ in their enzymatic ability to break down and use compounds, so once you begin to observe their likes and dislikes, you can customize your agar formulas to improve the performance and health of your mushroom cultures.

Other ingredients you can use in place of potatoes include powdered brewer's malt, dog and rabbit food, tomato juice, cattle feed, cornmeal, beet pulp, and much more. Keep track of the formulas you use in a journal so you know which mushrooms species do best on which particular agar formula. Try mixing small quantities of supplements into your agar to see which additives give the best mycelial growth for each species of mushroom you are isolating. Of course, the more you supplement the agar, the greater your risk of attracting and cultivating contaminants, so your goal is to find the range of supplementation that yields the best formulas while preventing molds and bacteria from outcompeting your mushroom cultures.

ADDING FUNGAL GROWING MEDIA TO AGAR

Although the goal of lab work is to re-create a perfect growing environment for mushroom spores and mycelium to flourish, it's incredibly difficult to duplicate

Dyes can be used to mark zones, or galleries, on agar plates that allow you to observe mycelial behavior and growth in response to different formulas. I also occasionally use colored agars for a more prominent background when photographing or observing species that produce faint mycelium, such as lion's mane (*Hericium erinaceus*).

the complex chemistries and web of interconnectedness that is present in the wild. But we can try! One way to make your agar attractive to mushrooms that favor a particular food source is to add a little of its preferred fruiting substrate to the agar, allowing it to dissolve and release its chemical signature into the solution. The resulting solution is called enrichment medium or substrate agar, and I use it for hard-to-clone mushrooms or picky fungi that prefer one specific food source, to help them "feel at home."

I dry and powder samples of many growing substrates, and I keep those powdered samples in jars in my lab, so I have them on hand to add to agar plates as needed. Adding a tablespoon of finely ground sawdust or screened wheat straw powder (sieved to the smallest particle you can obtain) from your fruiting formula to your agar formula can help the cultures adapt and express the enzymes they need to get their metabolic factories in gear. Likewise, you can use substrate agar

plates as experimental mini trials to see if a certain strain of mycelium can grow on a substrate that is unusual for it (shiitakes with wheat straw, for example). Alternatively, instead of substrate you could add samples of pesticides or other pollutants to your agar plates in experiments to gauge a fungus's ability to break down that pollutant. Alternating media keeps the mycelium's enzyme production strong through many series of expansions in the lab, so this kind of experimentation can give you valuable information.

I like to pour split plates to create transitional plates or behavioral galleries, where I can see how fungi react to different concentrations of a particular supplement or additive in the agar media. One side has an agar formula that the mushroom has adapted to and likes (such as potato agar), and the other side has a more challenging medium. To make the split gallery plate, I set the base of the petri dish at an angle (by resting one end on a dowel, for example) and pour in enough agar to fill half. When that agar has gelled, I flip the base around and do the same to fill the other side with the second formula. When both sides of the gallery plate are set, I transfer the mycelium to the "safe side" to give it the opportunity to expand on something familiar before reaching the middle of the plate, where it will sense the change in the medium and I can observe how it reacts. If I've used an additive to create the challenging medium, I can increase the concentration with each consecutive transfer to see what levels that particular mycelial strain can handle.

ADDING ANTIBIOTICS TO AGAR

Professional lab agar formulas sometimes require the addition of specific antibiotics to minimize or eliminate contaminants early on in the mushroom cultivation process. Ampicillin and streptomycin, for example, are commonly used to limit or remove a broad spectrum of bacterial contaminants. Powdered antibiotics are the most economical and can be purchased from science supply companies, but you need to rehydrate them, vortex (mix) them, and centrifuge the liquid, and then use a syringe filter to slowly dispense the antibiotic into the cooling media. You can also buy agar formulations premixed with antibiotics that are heat-stable and can be sterilized, but they are expensive.

Maybe the most practical solution is the sterile, injectable penicillin G used for livestock inoculations, which is also perfectly suited for agar. Since the solution is sterile, you can add it as the agar is cooling down, when the temperature drops below 110°F (43°C), but before it gels, swirling the solution to evenly distribute the antibiotic. (Adding the antibiotic above 110°F/43°C or autoclaving it will destroy it.) You can use a sterile syringe to add the antibiotic. In fact, you can use the syringe repeatedly as long as you wipe the antibiotic jar and the syringe with alcohol before using them for the first time; keep both in the same ziplock bag in the refrigerator. Antibiotics can also be used in liquid cultures, especially for longer storage to keep bacteria from flourishing in a nutrient suspension that is being used for repeated transfers.

CHAPTER 18

Starting Cultures and Spawn Generation

Once you have agar plates, there are no limits to what you can do with a fresh mushroom, whether collected from the wild, cultivated, or purchased. I have cultured all kinds of mushrooms from all sorts of places. You may undertake culturing to expand a particular strain of mushroom, to make your own starter culture, or sometimes just because you want to experiment with different techniques to hone your skills.

In general there are two pathways you can take if you want to start a culture: You can either start it from spores or "clone" it. Spores offer genetically variable offspring; the mushrooms you culture from spores can be identical to or remarkably different from the parent mushroom in shape, taste, and fruiting parameters. For this reason, spore germinations are ideal for breeding or creating a strain specific to your region or growing substrate; they allow you to selectively culture and expand those specimens that express the traits you desire. And spores are easy to collect (by taking a spore print; see chapter 1) and can remain viable for more than a year if you keep them cool and dry. When I see an unusual mushroom—perhaps one growing on an unusual substrate—that I want to identify or clone, I always take a spore print to make sure I have more genetic material than just the mycelium.

"Cloning" a mushroom is as easy as cutting or peeling it to reveal the clean inner tissue and removing a small fragment of tissue, which is nothing more than compressed mycelium. Cloning mushrooms is easy, and it creates fungi that are genetically identical to their parent mushroom. I use cloning, for example, when I am isolating the same species of mushroom from different climates, seasons, or growing substrates to build a collection of strains; in this way, I can create for that species a "library" of strains that can grow on just about anything, anywhere. Creating a similar collection of strains from spores is possible after multiple transfers and some tissue culture trickery, but finding these mushrooms in the wild and cloning them means they have already done all the hard work for you. Either way, you should become accustomed to both methods and practice on different species, since each mushroom also has its own unique culturing challenges.

SPORE GERMINATION: CULTURING FROM SPORES

To culture a mushroom from spores, you will need to collect the spores. You can make a spore print (see chapter 1), or you can try to harvest spores by using a

Spore prints are a genetic universe. These samples can create new strains that function differently from their parents.

sterilized tool to lightly rake the gills or pores. Prepare a set of agar plates; you may want to supplement the agar with a little bit of the mushroom's preferred substrate, such as a type of wood (powdered) or soil. This can be important for cloning wild mushrooms that are picky about their growing substrate, such as maitake, which have a strong affinity for oak trees. The profiles in part 4 will help you fine-tune each step of the cultivation process for each species you would like to cultivate, including recommended agar formulas.

Prepare your lab or work space about an hour ahead of time, turning on your laminar flow hood and sanitizing the entire interior of the hood. Using 70 percent isopropyl alcohol, wipe down the area several times, and wipe down the tools you will be using. Bring your prepoured petri plates to the work area and wipe down the bag they're packaged in. Then leave the room and shut the door, giving the HEPA filter time to clean the air in the room several times over. Once the hour is up, return to the room, again shut the door, and make sure you have all the tools you need for a transfer handy. Light your alcohol lamp. It's time to streak some spores.

Streaking keeps the spores in a tight line, which helps them to find a suitable mate nearby. Streaking spores also allows you to see different strains, or ecotypes, of the same mushroom separate and leave the pack, known as "sectoring." These growths can look very different from the rest of the culture on the plate, with the healthiest and most efficient strains surging outward much faster than inferior strains that may have mated but never developed a liking for the agar.

Spore Streaking Step-by-Step

Step 1. Organize your work area. Wearing gloves and wiping them down with alcohol, you will now clean and transfer all of your materials to your culturing space. Place your agar plates at the inside edge of the hood work space, up against the face of the filter, where the air is most sterile. Arrange the rest of your equipment from sterile (closer to the filter) to sanitized with alcohol (in the middle of work space, off to the side) to not sterile (closest to you), being sure that any tools you set in the work space are sanitized. You can set your spore print, which is not sterile, at the front of the work space, but I prefer to keep it just near the laminar hood instead, stepping back to reach it when I'm ready.

Step 2. Hold an inoculation loop, scalpel, wire, or whatever metal hand tool you'll be using in the flame of the alcohol lamp until the tip is glowing red. Remove it from the flame, and with your free hand lift the lid of a petri plate just an inch or so, holding it ajar. Dip the hot tip of the tool into the agar a few times, until it stops popping from the heat and is cool. Remove the tool and close the lid.

Step 3. The tool should now have some wet agar on its tip, which is great for transferring spores. Scrape the spores with the tip of the tool and the spores will stick to the wet tip

Step 4. Lift the lid of a petri plate at an angle again and drag the tool's tip through the agar in a wide Z pattern. Then close the lid. Minimizing the amount of time you have the lid off is critical to avoiding contamination, so make the operation fast and smooth, no longer than three seconds.

Step 5. Using tape or Parafilm, wrap the seam of the petri plate to keep the lid in place and to prevent contaminants from entering.

Step 6. Label the petri plate with the date, agar formula, spore type, and whatever other information you are tracking (see the discussion of record keeping at the end of this chapter). Most fungi prefer to incubate at a temperature of 77°F (25°C), and storing them in a small cabinet or insulated box can help maintain these temperatures. Professional labs use expensive incubators with thermostats that maintain the temperature consistently. I would recommend that beginners and homeowners use a small cabinet with a heating pad set near the bottom and adjusted to maintain the desired temperature. The agar cultures, although not entirely necessary, prefer the dark, and spore germinations are fast for some (morels and paddy straw), where most species can take about 7 to 10 days to notice any fuzzy growth and mycelium forming.

Step 7. Monitor the cultures for mycelial growth. Every mushroom has unique mycelium; most mycelium is white, but it can also be other colors. Molds appear as circular masses that produce powdery rings of spores in almost every color of the rainbow, depending on species. The majority of mushrooms do not produce sporulating mycelium, so if you see colorful circular masses (usually brown, green, or black), you'd probably be correct if you guessed that you are growing mold on your plate. In this case, either you should discard the plate, or you can try to rescue the mycelium growth that you want (see the discussion of rescuing cultures from molds and bacteria later in this chapter).

Step 8. When the mycelial growth on the plate is symmetrical and radiating outward after several transfers from petri to petri, you should make multiple backups for storage and use a few of them to make grain spawn (described later in this chapter) in preparation for cultivation and fruiting.

TISSUE CULTURING: "CLONING" A MUSHROOM

What if you find a mushroom growing on a peculiar substrate, or one that is exceptionally large, ecologically unique, or otherwise warranting your attention and possible cultivation? In this case, learning how to clone mushrooms is essential. As is the case for spore germination, every mushroom has its own particular challenges for cloning. It has taken me many years of trial and error to understand the nuances of tissue culturing, such as which area of the mushroom is best to take tissue from, and which tissue cultures do best in which type of agar formula, depending on the mushroom species. Part 4 describes the details for all the mushrooms profiled; here we'll focus on a general overview of the cloning process.

Taking a Tissue Sample

A mushroom is essentially a dense bundle of mycelial tissue that has taken shape and color according to its

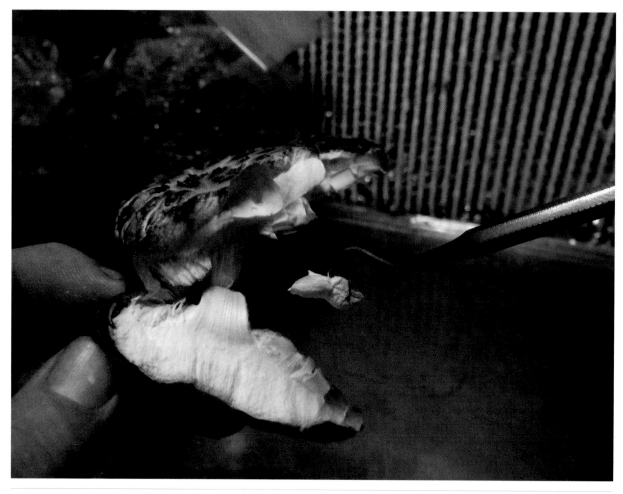

Cracking open the mushroom cap reveals the sterile internal tissue. Culturing this tissue (that is, cloning the mushroom) gives you mycelial tissue that is genetically identical to that of the parent fungus.

genetic instructions, growing into a specific form we recognize as a given species. To clone a mushroom, you need a sample of its tissue. Typically the exterior tissue, which has been exposed to the elements, is not clean enough for culturing; it will carry with it spores, molds, and bacteria that will contaminate your cultures. The best tissue is taken from the interior of the mushroom. What you are looking for is clean, fluffy mycelium, a little on the dry side, that you can remove.

The tissues in a mushroom differ in density from the tip of the cap down to the base where it attaches to the growing substrate. When you crack a mushroom open from top to bottom, you will quickly determine which areas look best—dry and fluffy! Avoid areas of

tissue with holes or discolorations; these might be the result of insect infestation and are likely with a source of contamination. Excise a small sample, using a sterile knife or scalpel, and transfer it to a poured petri plate that was prepared earlier. After removing the tiny fragment from the mushroom's interior, quickly but smoothly crack the lid just open enough to place the tissue on the center of the agar gel and push the tissue down with the tip of the tool to force it into the gel.

If you cut into a mushroom or squeeze a tissue sample and water visibly comes to the surface, it's too wet and will not clone well. Tissue samples must be relatively dry. If you're out in the field and happen upon a mushroom you want to clone, but it's wet, go ahead and bring it home, but let it air-dry for a day or so. Take advantage of that time to make a spore print, just in case the tissue culturing does not go well. Check the progress of the drying period by cracking the cap open slightly. Some mushrooms can be cloned up to a year after harvest by dehydrating the fruitbodies and refrigerating them. This maintains the dryness while preserving the tissue.

If a mushroom you want to clone is extremely small or has little tissue to work with—for example, if you find a rare mushroom that is deteriorating, retaining only a few small fragments of tissue—you'll have to take a risk and proceed with the best tissue sample you can get. You may need to surface-sterilize or sanitize the samples before transferring them to agar. Dirty specimens should be washed first with clean water several times and then soaked in 3 percent hydrogen peroxide for one to four hours in a plastic bag under refrigeration (which keeps it from rotting). Once removed, it can be allowed to air-dry in the lab to reduce the water content of the specimen. When the sample is almost dry, but still fresh, you can wipe the outer surface such as the cap with alcohol, not soaking it again but just to prep the surfaces before cracking it open. I prefer to get rid of as much debris as I can outside the lab and clean and cut the specimen into a fairly clean chunk of material that I can wipe down just a bit before making a transfer in front of my laminar hood.

Some mushrooms, like the polypores, generally do not have good, viable interior tissue that is willing to leap off onto your agar plate in the same way that tissues from some of the fleshy-gilled mushrooms do. In this case, taking a tissue sample from the mushroom's base, where it attaches to its substrate, may be your only option. Just as in the surface-sterilization method above, I prefer to rinse or soak these brackets and conks in hydrogen peroxide for several hours before attempting to crack open the mushroom and look for prime culturing tissue within.

If your mushroom is fresh and possibly actively producing spores, you may not want to bring it into your lab space, where its spores could contaminate your other cultures. Instead, you might excise a tissue sample outside the lab, transfer it to a petri

The mushroom tissue samples are tucked into the agar on a petri plate. I like to place multiple transfers on the same plate to improve my chances and to save plates, although if you are still honing your technique, you increase the risk of contamination with multiple transfers.

A small tissue sample removed from a mushroom. This small amount is all you need to grow millions of pounds of mushrooms!

dish, and bring that tissue into the lab for culturing. Alternatively, you might opt for a "still-air" transfer (described later in this chapter) outside the lab. If you do a still-air transfer, you may run into some peripheral contamination from mold spores, visible along the perimeter of the plate rather than in the middle, but there is less of a risk of contamination than in bringing a potential spore bomb into the lab.

Culturing: Transferring Tissue Samples to Agar Plates

The steps you'll take to transfer a tissue sample to an agar plate are much the same as those of spore streaking (see page 178), with a few exceptions. In this case, the transfer tool will be a scalpel or dental pick. When you've sterilized the tool in the flame of the alcohol lamp and you dip it into the agar to cool it, use the tip of the tool to make a small cut or depression in the agar. Insert the tissue sample into the cut or depression in the agar. Then proceed as described for spore streaking.

Inserting the tissue sample right into the agar means that more of the tissue's surface is in contact with the growing medium, which can stimulate the tissue to react more favorably than when it is laid directly on the surface of the agar. It also keeps the tissue from drying out as it works to revert into its colonizing mycelial state.

Most cultivators are taught to use just one tissue sample per plate, but I prefer setting multiple samples on each plate, spread out, to increase my chances and to budget the plates. Transferring multiple samples onto a single plate can increase the risk of contamination, but I inspect the plates frequently and can sector healthy mycelium to fresh plates once they take, giving them more room to grow.

Special Techniques for Culturing Finicky Mushrooms

Cloning nongilled mushrooms is more difficult than cloning most gilled mushrooms. I have tried for many years to improve on the methods for cloning nongilled mushrooms that are published in widely available cultivation guides. At a point of desperation, I have tried some crazy things, many of which didn't work, but some of which did. The best method I developed is to place the entire mushroom cap and most of its stem underwater, with just the attachment point (most of these mushrooms attach to wood) out of the water. The mushroom will want to float, so submerge it in a glass jar or other container that is small enough that you have to wedge the mushroom in there, which will keep it in its upside-down position. You can also tape the mushroom in place to keep it steady underwater. Place the container and its mushroom in a plastic bag and seal it closed. The lack of oxygen to the fruiting body forces the mycelium up and out of the attachment point, looking for air to transport down to the

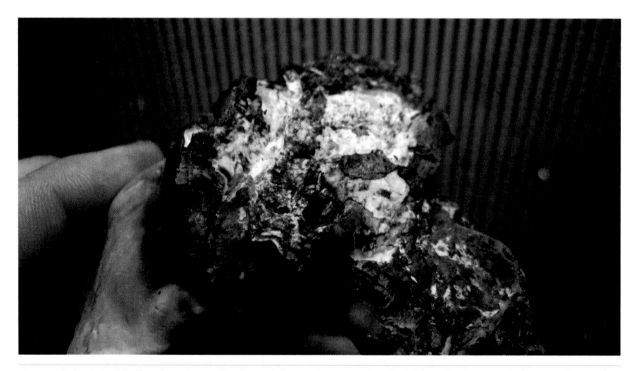

The base of a hemlock reishi (*Ganoderma tsugae*), showing viable tissue inside the attachment point. I have had success culturing tissues from partially colonized wood debris soaked in 3 percent hydrogen peroxide.

oxygen-depleted fruitbody. When it's fuzzed up beautifully, you can snip it and use the tissue for cloning.

If a mushroom is rotten or the tissue is just not suitable, I sometimes cut or hack into the colonized deadwood it grew on (where most white rot fungi are visible as bright white growths), chop it into small bits, and place it in a test tube or plastic ziplock bag. I cover the bits of colonized wood with 3 percent hydrogen peroxide, seal up the container, and refrigerate it overnight. I then transfer the bits of wood to agar. This method can be successful but is at greater risk of contamination. I advise you to watch these samples closely and check daily for signs of mycelium (a fine fuzz covering the wood, visible by holding the petri up to the light) or of molds and bacteria. Bacterial colonies can appear like clear or cloudy smooth zones around the samples, where most molds are colored and powdery colonies. Be prepared to subculture on a moment's notice if you see evidence of bacteria or molds.

RESCUING A CULTURE FROM MOLD OR BACTERIA

At times, especially if you are isolating wild cultures or germinating spores, you will encounter contamination that you will need to deal with. It may take some time to learn what the mycelia of various species look like, but the contaminants are easy to identify in comparison. Molds can be green, black, yellow, or brown circular colonies that keep popping up all over the place. Bacteria and yeasts will appear as irregularly shaped growths that can be cloudy or clear but are almost always smooth and shiny in appearance. Mushroom mycelium in most cases is linear and cottony; it is usually white, though some species show different colors.

If the mycelium you're culturing is growing somewhere on the plate, and the contaminant is some distance away, you might opt to transfer the mycelium to a new plate to try to save it. There have been many occasions when I have found a rare or unusual fungus and have used it to prepare stacks of clones and spore germination plates that result in nothing but mixed

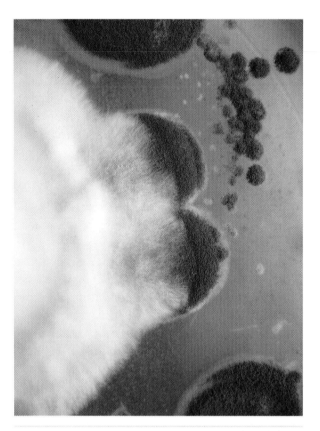

Molds and bacteria can compete aggressively with mycelium. Differentiating what is mycelium and what is mold or bacteria is critical when you are attempting to separate and purify cultures. Here we see *Aspergillus* contamination (green and powdery, circular colonies) competing for agar space with the shiitake mycelium in white.

cultures and no pure isolates. That leaves me no other choice than to take the best-looking of the contaminated plates and perform a transfer to a fresh agar plate, with the intent of leaving the contamination behind.

If you decide to attempt to rescue mycelium from a contaminated plate, you'll need to do a "still-air" transfer, meaning a transfer in a room with no airflow (a glove box would also suffice). Never open a contaminated plate anywhere near your laminar flow hood or in your clean room. Set up your "still-air" workstation, disinfecting the entire work area including the table and walls, cleaning and transferring your tools and supplies to the box: a small alcohol lamp, several fresh petris (enclosed in a zippered plastic bag), the contaminated plate (enclosed in its own zippered plastic bag), and any necessary transfer tools that have been wiped down with alcohol. Slowly remove the petri lid from the contaminated sample and carefully retrieve the healthy-looking mycelial tissue, taking care to avoid any contamination. Then gently close the lid. Transfer the mycelium to the fresh agar plate, seal it, and slip it into a fresh ziplock bag. Slowly place the contaminated culture in a separate ziplock bag and dispose of it. Now you can return to your clean room with the freshly plated mycelium. Monitor it closely as the mycelium grows; you may need to make an additional rescue if any molds came through the transfer.

Bacteria are less of an issue than molds; in fact, you can open a bacteria-contaminated plate inside the lab and make transfers without much risk. There is, however, the possibility that the mycelium is purposefully harboring bacteria, actually cultivating it on its hyphae—what I call "hitchhiking bacteria." In a natural soil or wood ecosystem, fungi and bacteria collaborate, and fungi "collect" favorable strains of bacteria to help metabolize their waste by-products, like a septic tank. You can employ a simple technique to clean hyphae to purify a bacteria-contaminated mycelial culture: the antibiotic "sandwich." Pour antibiotic-supplemented agar plates, place the contaminated mycelium on one side of the plate, and then cut a large wedge of antibiotic agar and flip it over on top of the contaminated culture, sandwiching it between agar layers. The mycelium will now have to drill its way through the agar. The bacteria will not be able to accompany the mycelium through the antibiotic-laced agar and will be left behind. The purified mycelium will fuzz up on the outside of the agar layers, and you can transfer this mycelium to a fresh plate.

I use the following antibiotic agar formula, recommended by Dr. Bob Blanchette from the University of Minnesota, for screening cultures through one or two transfers before switching them back to an antibiotic-free medium:

- 1,000 ml distilled/deionized water
- 15 grams malt extract
- 15 grams agar

- 2 grams yeast extract
- 0.06 gram Benlate (50% wettable powder; a fungicide)

Combine the above ingredients, autoclave for twenty minutes, and let cool to 113 to 122°F (45–50°C). Then add the following:

- 0.01 gram streptomycin sulfate
- 2 ml lactic acid

SECTORING: SEPARATING CULTURED STRAINS

You may sometimes get a "mixed" culture of the same species from a spore streak plate, in the sense that you have produced a few strains with completely different morphology, climate and soil preferences, metabolic requirements, and more. These strains develop as islands of mycelium on an agar plate. Since mycelium will not fuse with mycelium that is not identical to itself, where the islands collide they will begin to build walls and develop thickening lines of division. Once such colonies differentiate, you will notice that some surge ahead and others fall behind, just like watching a race. The colonies that are more aggressive may make better candidates for expansion or further experimentation, and you can "sector" them (select them for growth). To sector a promising mycelium from among a group of strains, excise a small bit of the outermost mycelium, or the hyphal tips, and transfer it to a fresh agar plate. Once the sector is transferred to a new plate, you should see a much more even and symmetrical growth, a textbook sign that you may have just purified a new strain of a mushroom from its original spore group.

MAKING GRAIN SPAWN: EXPANDING YOUR CULTURES

Expanding a culture to grain spawn allows the mycelium a better delivery system into bulk media such as

Jars full of grain spawn, colonized with morel mycelium. The jar lids are autoclavable and have filters to allow for gas exchange.

sterilized sawdust or pasteurized agricultural wastes and composts. Think of grain spawn as bridging the gap between agar cultures and fruiting substrates, acting as the "carrier" of the mycelium. Agar cultures are not suitable for direct planting because the gels dry quickly and offer no springboard for the fungus to outcompete molds, bacteria, and other fungi. Once agar cultures are transferred to grain, they may hesitate briefly to acclimate to a new food source, but then they will quickly begin to run through and colonize the grain. Once fully colonized, each one of these grains is a powerful fuel tank, loaded with the nutrients the mycelium will need to beat out the competition. This first-generation "master spawn" is usually used to inoculate another round of grain that will become the spawn for cultivation and has the ability to expand into thousands of pounds of mushrooms in just a few weeks.

For grain spawn, you can use common seed grains such as corn, millet, milo, wheat, and rye. Our laboratory uses wheat and rye grains almost exclusively for making first-generation master spawn (the leap from petri plate to grain), and then we expand to either additional wheat or soaked millet, a common component of birdseed. You can use other grains, or even beans or seeds, if necessary; they just need to be types that maintain their shape when soaked in hot water and sterilized. Don't use grains that are hulled, become sticky or gummy, or explode easily when they're heated in water or the spawn will be mushy and hard to separate. Choose grains that have a waxy or durable coating so that it will be easy to break them apart and mix them during the colonization process.

You'll need containers for sterilizing grain and making spawn. The most common ones are narrow-mouthed canning jars, but you can use any thick glass jars with metal lids. The jars will need some form of aeration, since the mushroom mycelium needs to breathe. Some suppliers sell filter disks that fit right into the lid of a mason jar. You can also simply drill a hole and stuff it with a wad of cotton or polyester fill, like the kind you use for stuffing pillows. Tyvek, or a similar woven house wrap used to vapor-proof buildings, also works well; you can drill a hole through the lid and secure the house wrap over it.

Bags are typically used for large quantities of grain. Autoclavable bags with aeration filters are available through many suppliers. In a pinch you could use the oven bags commonly sold at grocery stores, or any bag or container that is microwavable or oven-safe. I prefer to go from petri plate to jar to bags for the best results.

Making Master Spawn Step-by-Step

To make master grain spawn, some cultivation manuals advise you to mix dried grain with water in your jars and then sterilize it, so that the grain rehydrates as it is being sterilized. I prefer to precook the grains (by boiling and soaking) before I sterilize them, for two reasons. First, mixing dry grain and water can give you mixed results; since grains vary in their moisture content, your final grain can vary from too dry, which won't let the mycelium spread, to too wet, where bacteria overcome the grain much more quickly than the mycelium can handle. If anything, you want the grain to stay just a little on the dry side, since wet grain is the preferred habitat for bacteria and turns to mush when sterilized, making it difficult to mix when inoculated. The second reason is that by allowing the precooked grain to rest prior to sterilization, the hardy bacteria that survived the boil are "tricked" into letting their defenses down, thinking they have found safety after a life-threatening event. This triggers endospores to germinate, making the bacteria more vulnerable. When you then expose them to the sterilization temperatures, they are more easily destroyed, along with any other contaminants.

Step 1. Since grain will double in size once it is hydrated, measure out half the volume you need into a large pot that is at least three times larger than the amount of grain. Cover it with just over twice that amount of water. Bring to a boil, stirring occasionally to keep the grains on the bottom from scorching.

Step 2. As soon as the water is hot and before it starts boiling—around 160°F (71°C)—remove the pot from the heat and let stand, covered, until the grains are almost completely hydrated, typically about thirty minutes to one hour. Remove a few grains periodically and cut them in half to check their progress. When a grain is ideally hydrated, you'll see that it is mostly moist throughout, with a little bit of a dried center.

Step 3. Once the grains are hydrated, strain them and let them cool. If the grains appear a little overcooked or slimy, rinse them repeatedly with cold water until the grain juice has washed away. As the grains cool, stir them occasionally to keep them from sticking and to evaporate any lingering outer moisture. The grains should be swollen and moist, but never wet on the outside.

Step 4. Fill your containers (jars or bags) half full with the cooled grain. Seal the containers, being sure that each has an aeration filter.

Step 5. Place your jars in a pressure cooker or autoclave and process at 12 to 15 PSI for two hours.

Step 6. Turn off the heat and let cool. Take this time to set up your laminar flow hood: Sanitize the interior of the hood, run the HEPA-filtered ventilation unit for an hour, and sanitize and lay out all the tools you need, along with the cultures you'll be using to inoculate the grain.

Step 7. If you're using a pressure cooker, once it has dropped to an internal pressure of 2 to 3 PSI, it's safe to move it, if you need to. When the pressure drops to just slightly above 0 PSI, remove the jars or bags and place them on your laminar flow hood, as close to the filter as possible to minimize the chances of contamination, and let cool.

I prefer to precook (by boiling and soaking) grain before sterilization because it gives me better control over how much the grains cook. Overcooking can cause the kernels to swell and burst, making grain spawn sticky and more prone to contamination. I monitor the grains frequently during the cook, stopping and draining the grains just before they explode, and rinsing them in cold water to cool them down immediately.

Timing is critical if you are cooking grains before sterilization. Left: Dry grains straight from the bag. Middle: Perfectly cooked grains, boiled for one minute and then soaked for thirty minutes. Right: Overcooked, exploded grains, after boiling for more than ten minutes and soaking longer than thirty minutes.

Once drained and allowed to air-dry briefly, the cooked grain is portioned into jars for sterilizing. Sterilize your grain jars in a home pressure cooker or sterilizer at 12 to 15 PSI for two hours.

Step 8. Once the grain has cooled to less than 90°F (32°C), you're ready to inoculate it. Hold the tip of a scalpel in the flame of an alcohol lamp until it is red-hot, and then use it to cut the most desirable areas of your petri plate cultures into cubes and transfer them to the jars of grain. For a quart-size mason jar half filled with grain, I generally add about ten ½-inch cubes of colonized agar from a petri plate. Liquid culture can be used at a rate of 10 ml per quart-size jar, or the equivalent for larger bags.

Step 9. Shake the jars or bags to distribute the inoculum, label each container, and store at a temperature between 65 and 75°F (18–24°C); temperatures any higher than that will promote bacterial growth. Be sure to label the expansion generation and the origin of the donor culture for your records (see the discussion of record keeping at the end of this chapter).

Step 10. Let the inoculated grain sit, undisturbed, for a few days. If you've used agar-based inoculum, once the agar wedges fuzz up heavily and stick to a few grains (evidence of colonization), shake the jar or bag to distribute the inoculum more evenly. (There's no need for shaking if you've used liquid culture as your inoculum.) If you shake the container right after the transfer, like some manuals suggest, the cut sides of the agar wedges, which are smooth and sticky, will stick to the container in places, wasting valuable inoculation surfaces and slowing colonization. If any agar wedges get stuck to the upper parts of a container after you shake it, you can try to knock them loose by shaking again, or you can simply set the container on its side so that the grain falls over and covers the stuck wedges.

Step 11. Shake the containers every few days, as soon as the agar wedges colonize the surrounding grains. When you notice that the myceliated grains have begun to inoculate adjacent grains on their own, shake the jars gently one last time to check for contamination.

Once you have used an isolated culture (generation 0) to inoculate a master culture of grain spawn (generation 1), you can use the master spawn to inoculate more grain (generation 2), and then use that grain spawn to create more grain spawn or switch to substrate spawn (generation 3), which incorporates elements of the mycelium's preferred growing medium and can be used to "seed" a mushroom cultivation that will proceed all the way to fruiting. Grain spawn can also be used directly for a few species and applications, such as *Agaricus* on compost and *Pleurotus* or *Calocybe* on pasteurized agricultural waste products indoors. In most other cases, you will need to expand the culture through three generations before you're ready to inoculate a substrate with spawn and begin cultivating mushrooms.

Inspecting Grain Spawn for Contamination

You will need to know what the mycelium of your mushroom looks like to follow its progress here and monitor for contamination. (Part 4 gives details on mycelial characteristics.) Daily inspections are a must. Molds can be somewhat granular to extremely wispy in texture. The most common mold contaminants in grain culture are light green, dark green, or mustard yellow. Some molds, such as the common bread mold (*Rhizopus* spp.), look like white or clear fur with little black pinheads on the upper surface where the spores discharge.

Some bacteria are easy to spot, quickly becoming "greasy spots" in the container, and often producing a cloudy, colored fluid around the grains. They commonly appear at the bottom of the containers, where excess moisture collects. (If you are having problems with bacteria, the answer could be to reduce the presoak time of your grains prior to sterilization, thereby reducing the moisture in your spawn.) Scent, too, can be an important indicator. Each mushroom mycelium has its own particular scent (as documented in part 4). If instead your grain spawn has a musty or apple cider smell, bacteria may be present.

If any containers of grain spawn show signs of contamination, get rid of them. Remove the container from your lab or incubation space immediately, discard the grain, and sanitize the container with diluted bleach.

If you are noticing persistent or widespread contamination, you will need to determine the source

MAKING LIQUID CULTURES

Making "dry" transfers, or agar-to-grain and grain-to-grain transfers, is a traditional method of creating grain spawn, but what if you want to create larger volumes of grain spawn for a commercial operation? Liquid cultures may be the answer.

The limitation is that you will need to have a blender jar that can be sterilized in an autoclave. I use a regular Oster-type blender; when I'm ready to autoclave it, I just remove the plastic lid and base, leaving the metal blade and rubber base, since they are heat-tolerant. I cover the top and bottom of the pitcher with aluminum foil and place it in an autoclavable bag. I fill a separate jar with water and autoclave both at the same time for thirty minutes and then let them both cool down to room temperature in front of the laminar hood's sterile airflow. Once cool, I disinfect the plastic base and lid that I removed earlier with alcohol and put them back on the blender jar. I pour a small amount of the sterilized water into the blender and add cut cubes of colonized agar or chunks of myceliated grain spawn. I blend the mixture until it is thick but pourable, and I store this liquid culture in sterilized tubes or jars.

The advantage of using liquid culture over agar-based culture is that it has a much more rapid colonization rate and you'll use much less of your culture. The disadvantage is the possibility of encouraging bacteria, which thrive in high-water-content media, so maintaining sterile conditions is critical. Before expanding liquid cultures, I recommend performing a streak test (see page 191) to screen the liquid for contaminants. This will prevent problems further along in the transfer process.

Blend grain and mycelium to make liquid cultures, which will colonize grain quickly and store well in sterile jars or tubes.

Once you transfer the mycelium to grain, inspect it daily. Greasy spots like these are a sure sign of bacteria. If you see signs of contamination, remove the container of grain spawn from the lab immediately. Discard the contaminated grains, and clean the jar with diluted bleach. Never expand the grain from any jars that show signs of contamination.

of contamination. You can do this via the process of elimination. First test the agar and the grains. When you prepare a batch of grain spawn, leave a few jars of the cooked grain un-inoculated (that is, don't add agar culture to them). If the un-inoculated and unopened jars of grain develop contamination, the problem lies with the grain and may be the result of inadequate sterilization. Make sure that you are autoclaving the grains at the appropriate pressure and for the appropriate length of time.

If the inoculated grains develop contamination but the un-inoculated grains do not, the problem may be in your agar cultures, your air filtration, or insufficient disinfection of the work space. To test the agar you can transfer a wedge to a fresh plate, or use the streak test (described in the sidebar) for liquid culture. To eliminate your filter as the cause you can leave a fresh petri plate open for sixty seconds in your cleaned laminar work space, taping or Parafilming it shut and observing it for contamination over several days. If your

cultures are indeed contaminated, you may be able to salvage them; see the discussion of rescuing cultures from molds and bacteria earlier in this chapter.

Next, test your HEPA filtration to make sure it's working properly. To test the filter, open a few freshly poured agar plates one at a time in different areas in front of the filter. Leave them exposed for about thirty seconds, then close and tape them up, labeling the location of their exposure. If contamination shows up on the plates (which, if it's going to happen, will be in just a few days), the filter is the likely culprit; you may need to replace the filter or reseal the laminar flow hood.

MAKING SUBSTRATE SPAWN STEP-BY-STEP

Larger growers sometimes use grain spawn (generation 2) for spawning fruiting substrates; however, most growers can't afford this option, so my recommendation is to learn how to make good sawdust-based substrate spawn (generation 3) and use that for spawning fruiting substrates. Though the fruiting yields will be slightly lower than if you used grain spawn directly, you'll be able to inoculate more fruiting substrate with your substrate spawn, using much less grain spawn.

The substrate spawn is made similarly to grain spawn—by first soaking and rehydrating a growing medium (primarily sawdust) to create a carrier, and then inoculating that growing medium with mycelium. The lesson here is that mycelium prefers to go from "like to like," especially for wood decomposing species. Mixing the sawdust with some of the growing medium you intend to use for fruiting the mycelium, and perhaps also some of the growing medium you used in the preceding expansion, is a good strategy for acclimating the fungi to its current and future homes in sterile culture before it is forced to deal with competitors. For example, with wood-loving fungi that you have cultured on grain and intend to fruit on wood chips, you may create substrate spawn using sawdust, powdered grain, and powdered wood chips.

Substrate spawn benefits from supplementation, or adding further nutrients to the media to aid the mycelium in a rapid colonization response once it is spawned into a bulk fruiting substrate. These additives can be any number of available grains that retain their shape when cooked and do not become "gummy," such as whole corn, wheat, rye, oat, millet, milo, and similar whole cereal grains. These larger grains are "gas in the tank" for the mycelium when it is running through the fruiting substrate. Grains should be hydrated before sterilizing. Wheat or rice bran is commonly used in small quantities, which is a fluffy, flaky additive that gives mycelium a jump start on colonizing the spawn substrate in the lab. My simple and favorite substrate spawn recipe, per 5-pound bag, is soaked and drained hardwood sawdust (5 pounds or 2.36 kg) and ½ cup of hydrated wheat bran (roughly 200 grams). Adding more supplements can be risky, since the more nutritive the media is, the more prone it can become to contamination. Also, additives can increase the heat generation, or thermogenesis, of the

STREAK TESTS

Streak tests can be used to test liquid cultures (grain spawn prepared as a slurry) for contamination. You simply take a loop, dip it into the slurry, and drag it across a fresh agar plate, making a large Z pattern. Use an agar formula that has no added antibiotics, and don't use a potato-based agar, since many bacteria species do not propagate well on it. I like to use a malt extract agar to be sure every bacteria species that is present becomes visible.

Seal the plate with Parafilm or tape and let it incubate for several days. You should see a solid line of mycelium form along the streak. If you also see bacteria or mold colonies, the culture is contaminated.

If the streak test is positive for contaminants, repeat the entire process using every preceding expansion of the contaminated culture, from the original isolate (tissue culture or spore germination) on up. It can take some time to conduct this series of streaks to establish the site of contamination, but once you have it, you'll know which expansions are potentially contaminated, and you can revert to an uncontaminated expansion generation to maintain purity in your culture lines.

Remove a small sample of the liquid with a sterilized loop to streak across a fresh agar plate. Any contamination will be evident in one to two days.

substrate, which can cook the mycelium if the temperatures rise above 100°F (38°C) internally for a few hours, so err on the lower side of supplementation when creating substrate spawn. I prefer to supplement my fruiting media, which is either pasteurized or sterilized, and not risk the potential contamination at an early stage of pure culture. Some cultivators mix and soak their sawdust, substrate, and supplements together, but I prefer to hydrate each component separately before mixing them together to ensure that each particle is optimum.

Inspect your grain spawn and choose the best-looking specimens to expand into substrate spawn. For sawdust, you'll want a type that is compatible with the mushrooms species you are growing. Aged sawdust works better than extremely fresh sawdust since it holds moisture better, but be sure it is not too old—no more than three months. Once you have the sawdust, it is time for yet another expansion—one step closer to planting and cultivating mushrooms from spawn that you created yourself!

Step 1. Soak the sawdust in water for a few hours, and up to overnight. Do not let it sit longer than twelve hours; you may notice a sour smell (signaling bacterial contamination) if you soak it too long. I prefer to hydrate supplements or soak grains just before adding them to the sawdust and sterilizing them since they spoil much more quickly than sawdust. To soak sawdust, I created a simple two-bin system where the outer bin has a single drain plug and a few bricks on the interior bottom to keep the second, inner tub (which has holes) slightly separated to ease drainage. The inner tub has ½-inch holes drilled over the face of the bottom with a piece of door screening material over it to keep the particles from racing out when it drains. I can plug the lower bin and fill the bins with water, allowing for a long soak time of 12 to 24 hours.

Step 2. Drain the water from the sawdust by pulling the plug. Allow the sawdust to drain for about an hour, until it is moist but not wet. Squeeze a handful. If squeezing produces water, the sawdust is too wet. If the sawdust does not stick to your hands, it is too dry.

Step 3. Bag the sawdust, using heat-tolerant bags with filters, and add any supplements you need for the mushroom species you are cultivating. Wheat bran, moistened with water and added at a rate of ½ cup for every 5 pounds of hydrated sawdust, is a good additive.

Step 4. Fold the tops of the bags over, pressing down gently to force out any air inside. Secure the tops of the bags; wooden clothespins work well.

Step 5. Process the bags in an autoclave or pressure cooker for one and a half to two hours at 15 PSI, or about 250°F (121°C).

Step 6. Turn off the heat and let cool. Take this time to set up your laminar flow hood: Sanitize the interior of the hood, run the HEPA-filtered ventilation unit for an hour, and sanitize and lay out all the tools you need, along with the grain spawn you'll be using to inoculate the substrate.

Step 7. If you're using a pressure cooker, once it has dropped to an internal pressure of 2 to 3 PSI, it's safe to move it, if you need to. When the pressure drops to just slightly above 0 PSI, remove bags of substrate and place them on your laminar flow hood—as close to the filter as possible to minimize the chances of contamination—and let cool.

Step 8. Once the substrate has cooled to less than 90°F (32°C), you're ready to inoculate it. For each 5-pound bag of sterilized substrate, you should use roughly 1 cup of colonized grain spawn, but most cultivators simply pour and do not measure. The more you use, the faster the colonization. The less you use, the higher the risk of the mycelium taking too long to colonize and potential contamination.

Step 9. Shake the bags to distribute the grains, label each one, and store at a temperature between 65 and 75°F (18–24°C); temperatures any higher than that will promote bacterial growth. Be sure to label the expansion generation and the origin of the donor culture for your records (see the discussion of record keeping below).

Record Keeping: Tracking Cultures and Generations

To begin, assign every isolate (tissue culture or spore germination) an origin code. You use a code simply because, for example, "PDJAMOR1" is easier to write on a specimen label than "Tissue culture #1 from *Pleurotus djamor* collected on 7/1/2013 from Red Maple at the southeastern base of Big Mountain." You will, of course, keep a record of all origin codes with their full source information.

Starting with your initial isolates, you will need to label every plate, jar, and bag of culture and spawn with generation codes that will help you keep track of how many expansions the culture you're using has been through. The original pure isolates are considered generation 0. Every subsequent expansion—whether to master grain spawn, liquid culture, or any other form—progresses in number. For example, an isolate (generation 0) can be used to make a liquid culture (generation 1) that you store by freezing; you can later use that culture to create master grain spawn (generation 2), which can be used to inoculate a greater volume of grain (generation 3), which can be used to create substrate spawn (generation 4).

The goal is to maintain the lowest generation numbers to produce the best spawn possible for your production needs. Mushroom cultures that have

been expanded many times (or cultured on identical media many times) become distanced from their wild origins, suffer from weakened genetic expression, and can lose key fruiting traits—all of which can result in

Every form of a culture, from isolate to spawn, should be labeled so that it can be tracked. Tracking is important to maintaining cell lines and allows you to trace all expansions back to the original isolate.

STRAIN SENESCENCE

As fungal cells replicate, their DNA needs to divide and copy over into the new cells. There are proofreading mechanisms to prevent errors, but the more cultures divide and age, the more prone they are to making genetic errors that can alter their original vigor and efficiency. This is why it is so important to track and maintain cell lines closest to their wild isolates: to preserve the unique properties every strain or ecotype possesses.

Most colleges and research universities purchase "zoo animal" fungi strains that have been "in captivity," or kept in culture, for so many years that they sometimes lose their ability to adapt to variables in their environment. A zebra born in captivity and released into the African wilderness may not stand a chance at survival compared to one born into a wild herd that has learned to understand and react quickly to threats and rely on a variety of food sources. Similarly, fungi that have been grown on laboratory media, without a diversity of natural food sources, and without the pressure of competition from other organisms, are progressively weaker as they divide and create long highways of cells far from their original spore germinations, which is why many cultivators vary media between transfers—to keep the mycelium exercising its metabolic "muscles." Use 'em or lose 'em, they say, and for fungi it's true. Subject to monoculture conditions (the same nutrients and culture conditions, over and over), they develop monoculture qualites—that is, they become dependent on those nutrients and conditions, and unable to thrive when those factors change.

lower yields, disease susceptibility, diminished cap color, and the loss of other distinctive features that define a young strain. Cells that divide repeatedly, far from their original spore germinations and mating origin, are subject to mutations that can lower yields and alter their behavior. This is called strain senescence.

You may also wish to give every plate, jar, or bag an individual tracking code. Subtle differences in temperature or minute levels of contamination mean that not every container, even within a batch, is identical. Tracking individual cultures allows you to better respond to contaminations and to better evaluate particular culture lines. And with each culture specimen labeled with its original isolate, generation, and individual tracking code, you can track the history of each culture from its mycelial source to the fruiting room.

I use alternating letters and numbers in my codes, which I find easier to read than a string of numbers. For example, if I transfer first-generation agar culture to a jar of sterilized grain, I would give that jar the code of 2-A (2 = second generation; A = first batch). If I used that same first-generation agar to make more jars that day, the codes for each jar would be 2-B, 2-C, 2-D, and so on. If I then expand jar 2-B to bags of grain or sawdust spawn, the bags would be coded 3-B1, 3-B2, 3-B3, and so on.

Record all of this information in a lab notebook. The entry begins with the date of the transfer. For the culture you are expanding, it should list the strain number or mushroom type, the origin code, the generation code and substrate, and the date the culture was made. For the expansion you are making, it should list the span of codes you use to label the batches you make. For example:

Today's Date: 7-15-13
Pink oyster (PDJAMOR1) 2-B / Grain / 7-1-13
Transferred to 3-A1 to 3-A10 sawdust (formula)

This has all of the information I need to know. I used pink oyster grain spawn coded 2-B (originating from the isolate coded PDJAMOR1), which itself was created on July 1, 2013, to make ten bags of sawdust spawn. I labeled these ten bags 3-A1 through 3-A10. If all of the bags turn out to be extremely good-quality spawn, I can confidently go back and pull another jar of grain spawn from the 2-B pink oyster culture to create more spawn, knowing that it has turned out to be productive. Tracking stock in this manner helps spawn producers and cultivators decide which stock cultures to use for spawn production. Tracking is the best means of understanding the history of your laboratory expansions, what methods are working, and what levels, or generation numbers, you should be stopping at to ensure that you are consistently producing quality spawn.

CHAPTER 19

Storing Your Cultures

Many of the storage methods discussed in this chapter can be performed by anyone just about anywhere with basic refrigeration or freezing capabilities. For those of you who do not have access to electricity, storing cultures indefinitely can be tricky, and the methods you will most likely employ are those offered in chapter 12.

I recommend pursuing the goal of purifying your culture—make sure it is free of contaminants—before storing it. While it's impossible to purify spores—contamination and genetic variability just come with the territory—you can be selective about the media you use to limit unwelcome guests. Mycelium and liquid cultures, on the other hand, can be cleaned up a bit before storing. We'll focus here on storage; for a good discussion of how to rescue cultures from contamination, see the discussion in chapter 18.

AGAR PLATES

You can seal agar plates with tape or Parafilm and keep them in a refrigerator for months, sometimes up to a year. I like to store my sealed cultures in plastic ziplock bags not only for extra protection against drying out, but also for maintaining the cleanliness of the plates when moving them from the fridge to the laminar. The bags, like the plates, can be cleaned with alcohol. Just be aware that alcohol typically dissolves any labeling you may have done with permanent marker, which can be critical in identifying the plate, the contents, the type of media, the date, and other important information.

MAKING BACKUPS

Making backups (that is, duplicates) of your cultures is essential. If something goes wrong with your primary culture—contamination or spillage—having a backup means you won't lose the entire culture. Think of it as backing up your computer hard drive. These are your most precious copies of mycelium on the planet.

Your backup cultures need to be the purest form of any particular fungus you maintain. If a particular culture is not suitable for expanding into grain spawn for production, don't store it. Keep working with it on agar until you have determined it to be as pure as it can be before committing it to storage.

It is a good idea to store your collection of cultures in multiple forms at multiple locations. A prolonged power outage, fire, or other natural disaster can ruin years of work spent collecting and culturing fungi with little or no warning. For my own operation, I store duplicates of some of my most important cultures in liquid form in refrigerators belonging to friends and family. These small tubes do not take up much space, hidden in the back of crisper drawers, and they give me peace of mind that in an emergency I won't lose everything.

SUBMERGED AGAR CUBES

The most common problem with longer storage on agar plates is that the culture can lose viability, aging from beautiful fluffy mycelium to a flat, discolored mat that appears dead. One way around this is to take a fully colonized plate, slice the agar into small cubes, and transfer the cubes into a test tube half filled with sterile water. Refrigerate the test tube, making sure the cubes stay fully submerged. Whenever you need the culture, you can simply use a sterilized tool to fish out one of the cubes. Cap the test tube and return it to cold storage. The cubes last much longer than agar plates because the water helps protect and hydrate the mycelium. It can typically last for over a year like this. I have used the same process with sterilized mineral oil in place of the water, with an even higher rate of success, although for best results you must rinse the oil away in sterile water before you use the culture.

AGAR SLANTS

Agar slants are a common short-term storage method. To make slants, you will need to gel the agar at an angle inside a test tube. To do this, mix your agar formula together (best done with warm water) and fill test tubes one-third full. Place them in a test tube rack or small jar to keep them upright and sterilize them for twenty to thirty minutes. After autoclaving, transfer the tubes to the laminar station to cool down under sterile airflow, but place them at an angle (slanted) by tipping the test tube rack or jar to create a long slope of agar inside the tube. Allow the tubes to cool. Gelling the agar at a slant in the tube allows the greatest surface area. Transfer cultures to the gelled agar by cutting a small wedge of mycelium from a purified plate; once they have adequately colonized it, store the tubes in the refrigerator, where they'll keep for up to a year, sometimes longer. I store blank or extra poured slants, ready for instant backup, so that I do not have to crank up the sterilizer or autoclave every time I need just a few to work with.

COLONIZED GRAIN

Many species can be stored for years on colonized grain. I have come to love this method. Submerge the colonized grain in sterile water in a large tube or jar for easy retrieval, and store in the refrigerator. I've found that one 50 ml centrifuge tube filled with colonized grain is enough to last me for a year, allowing me to pull out a few grains when I need them. This saves many steps and critical time. Many cold-sensitive species prefer this method of storage, since the coating of water helps insulate them from cold damage. Water can also be replaced with autoclaved mineral oil to protect difficult-to-store species. This can be extremely useful if you are trying to maintain cultures of cold-sensitive strains such as paddy straw (*Volvariella volvacea*) and other tropical mushrooms that can die if they are stored under general refrigeration or temperatures below 50°F (10°C) for longer than a week.

You can store colonized grain submerged in water to slow growth and prevent dehydration. Refrigerating this grain can extend its storage time even more.

The best use of colonized grain may be to prepare liquid cultures for use in spawning a great quantity of substrates (see the discussion on page 184).

FREEZING

Freezing is an excellent option for storing backups of your cultures almost indefinitely. A word of caution, however: Freezing cultures without a protectant can damage and kill the mycelium, and the protectant itself can be toxic if this procedure is not exacting. This is chemistry, not home cooking, so follow the guidelines precisely and exercise caution.

Testing Frozen Cultures for Viability

Before transferring your entire collection to liquid backups and freezing them, begin by testing different species to determine their viability under freezing. Prepare a set of identical liquid cultures of a single fungal strain, and freeze them. Remove one tube after one month, transfer the frozen specimen to a fresh agar plate, and allow the petri to incubate. Follow a protocol for thawing frozen species, such as placing the vials in a warm water bath of 77 to 86°F (25–30°C) for two to five minutes to allow them to acclimate before transferring them. Observe and photograph the culture, evaluating both its recovery and the speed of that recovery. If the growth of the first transfer doesn't look healthy, try tranferring it to a fresh agar plate. You're looking for uniform mycelial growth that best resembles that of the original culture before it was frozen. You can also view the mycelium under a microscope to verify the presence of clamp connections or other characteristics that can validate the identity and purity of your culture. (A clamp connection is a small "elbow" that wraps around dividing hyphae. It appears like a bump located near the septa, or wall between dividing cells. It can verify that you have a basidiomycete, or gilled mushroom, and not a mold or other contaminant.)

Repeat the process after two months and every consecutive month thereafter to gauge the longest interval your cultures can withstand being frozen and thawed without incurring damage or compromising your collection.

Preparing Cultures for Freezing

One way to freeze a culture is as colonized agar cubes or grain (both are discussed above) submerged in a 5 percent glycerol solution. Glycerol, also known as glycerin, keeps the mycelial cells from rupturing due to crystallization when they are frozen. As when you're preparing agar cubes for refrigeration, you'll need to first sterilize water, and glycerol now as well, refrigerating the liquids before mixing them (if you don't chill the liquid culture before you add the glycerol, it can damage the mycelium of some species, so to be sure to chill it first.) Fill a series of test tubes half full with the sterile water, and add enough glycerol to make a 5 percent solution. Agitate the test tubes to distribute the glycerol, and drop in the colonized agar cubes or grain. To allow for expansion of the water as it converts to ice crystals, keep the caps loose for now. Place the test tubes (in a rack to keep them upright) in a sterile autoclavable bag with a filter patch, seal it, and put it in the freezer. Once the tubes are frozen, you can remove the bag. Cut it open in front of your laminar flow hood, screw the caps on tight, and seal them with Parafilm. Place the tubes in plastic ziplock bags and return them to the freezer.

Another method, and probably the one preferred by professional laboratories, is to make a slurry using colonized agar and sterilized water. Transfer the slurry to a liquid broth containing a weak solution— roughly 10–20 percent—of the original antibiotic and nutrient formula you plan to use when preparing the transfer plates. Agitate it for one to three days, stirring it continuously, using a laboratory plate stirrer with a sterilized magnetic stirring bar inside, to allow the mycelium to recover and repair the damage incurred by the blender when you made the slurry. After a few days, pre-chill the mixture either in the refrigerator or on ice, and then add enough glycerol to the broth to make a 5 percent glycerol solution, such as 95 ml of your culture slurry mixed with 5 ml of glycerol (again, if you don't chill the liquid culture before you add the glycerol, it can damage the mycelium). Transfer the mixture to sterile storage tubes and freeze as described above.

CONSIDERATIONS FOR COLD-SENSITIVE SPECIES

Cold-sensitive species are particularly hard to maintain under cold storage since they can die if temperatures drop below 50°F (10°C) for as little as one week. Storing spores can be a good idea in this case. Simply take a spore print, and keep the spores dry and cool; generally they'll remain viable for at least a year. You can also store spores in water under refrigeration for up to two years. To keep things as clean as possible, I take a spore print on a sterilized piece of aluminum foil, setting the cap of the mushroom on a disinfected metal grill to keep it from touching the foil. I flood the print with sterile water from a syringe and then draw the water back into the same syringe for storage. This technique is commonly used for psilocybin mushrooms, but it also works well for many cold-sensitive strains. When you want to prepare a culture from the spore solution, you can inoculate an agar plate with a small drop of the solution, preferably on one side of the plate and not swirled over the entire surface, to allow the spores to germinate and outrun potential competitors.

Another method is to use cultures of the cold-sensitive species to prepare grain spawn using larger grains, such as milo, and to refrigerate that grain spawn at a moderate, not cold, temperature—about 50 to 60°F (10–16°C). This spawn may remain be viable for up to a year. In the entrance to our walk-in cooler, there is a separate space that we keep at 50°F (10°C) year-round, much like a root cellar, and it is a perfect place to store grain spawn of tropical species (and also a great place to brew beer, culture sauerkraut, and store vegetables).

DEHYDRATION

I have experimented with dehydration as a long-term storage method with some success, finding that some dehydrated cultures will remain viable for up to a year. For dehydration, I like to use grain spawn. I transfer a small amount into a large, sterilized jar and seal it with a lid that has a micron filter disk. The mycelium dries naturally and quickly, typically within a week or two, at which point I transfer it to a test tube for storage in a dark, cool location. Do not refrigerate dried spawn, regardless of its cold tolerance; it may not survive.

If you are using dried grains, you can easily retrieve the grains one at a time whenever you want to start a new petri plate culture. Dried myceliated grain is slow to leap off, but it is typically viable and recovers a week or so after being wedged into fresh agar. I sometimes place a small wedge of the dehydrated culture on the middle of a fresh agar plate and then cut and flip some fresh agar over and on top of the wedge to make a "sandwich." This allows more contact points with the fresh agar and provides some stimulation by gas deprivation to kick-start the mycelium into reviving.

CHAPTER 20

Advanced Cultivation and Research Strategies

Don't let the "advanced" in the title scare you away from applying the techniques covered in this chapter. You do not have to have a PhD to employ these ideas. They're here for your experimentation and for you to build on and get creative with once you develop your skills and fully understand your operation. Without compromising the seven basic stages of mushroom cultivation outlined in chapter 2, try experimenting with some of the information and ideas in this chapter to see how they positively or negatively affect your cultivation. Report, publish, or otherwise share your results so that others may benefit from your experiences and so that we can all build on one another's knowledge—which will help us move forward not only with mushroom cultivation per se, but in related fields such as agriculture, medicine, and alternative energy as well.

CASING SOILS

Casing soils are designed to cover colonized bulk substrates, such as straw or compost, to stimulate mushrooms that benefit from an interface between them and the surface. Casing soils give primordia more time to form in a high-humidity environment, helping them to avoid drying out. They can also benefit mushrooms that require interaction with certain microbial communities before they can fruit. In fact, you can supplement casing soil with the microbes found in the soil of their native growing environment to facilitate their fruiting. Casings can be used indoors or out but are almost universally applied after colonization of the substrate is complete, to reduce the chance that competitor organisms will take over the substrate.

Casing soils are most commonly made of materials such as peat moss, coconut coir, vermiculite, perlite, sand, lime, and gypsum. A high-quality casing soil has high water retention and good aeration; it's also pH neutral to slightly acidic. A mixture of these materials can be formulated to meet an individual mushroom's needs.

Experiment with the moisture content, depth, and density of the casing—these are all factors that impact yields. To make a casing soil, moisten and mix the ingredients together and allow the water to drain, until the casing is moist but not wet. If you can squeeze just a few drops of water from a handful of wet casing, it is perfect. If you can squeeze out a steady stream of water, it's too wet; if no water drips, you need to wet it more. Bear in mind that it may take some time for the media you use in the casing soil to absorb the water into their fibers, which can be hydrophobic at first. The casing cannot be too wet or the aerobic mycelium will suffer, colonizing the casing weakly, and possibly creating a haven for competing molds and bacteria. A dry casing can be just as bad, allowing the mycelium to dry out and preventing the development of primordia at or below the surface. If the primoridia do not have enough humidity, the mycelium will not commit to spending energy on a fruiting cycle. After it's spread, you can maintain moisture in the casing soil by misting it periodically. Casing soils are spread directly on a colonized substrate at a depth that is proportional to the depth of the biomass below. As a basic rule, a casing soil can be applied 1 inch deep for every 6 to 8 inches of colonized media.

The casing structure should be somewhat loose, with minute air pockets, not packed down tight. It should be level overall, though you should rake or dibble the surface to give it a series of ridges. Primordia can form in the hollows between the ridges, developing their caps and swelling with the surrounding water to rapidly mature. As the mycelium rises from its relatively dense substrate to encounter the less dense casing soil, it becomes rhizomorphic, or "strand-like," bundling itself into ropes that channel its resources upward. When it reaches the surface, the mycelium shifts its growth horizontally to form a dense mat and cross connections.

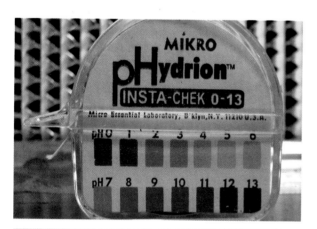

Soil microbes operate within narrow pH windows, so test the pH of all soil or media you plan to use for casing in order to best replicate the pH levels that the mushroom you are growing prefers in nature. You can use pH paper, which is inexpensive and portable, or a digital pH meter, which is more expensive but also more accurate.

Casing soils should not contain mycoparasites such as *Trichoderma* spp. and other pathogens that could impede or pose a threat to the mycelium, the mushrooms, or your own health. For this reason, many growers pasteurize their casing soil. You can use low-heat pasteurization (130–145°F/54–63°C) for this, which preserves the heat-tolerant microbes that will guard the casing soil and prevent competitive pathogens from multiplying. Load the wet casing soil into a steam chamber (in pots or barrels for small batches) and steam for at least six to eight hours, monitoring the temperature periodically to keep it within the temperature window. I don't recommend sterilizing casing soil unless you plan to reintroduce a microbial community, namely one that has been shown to trigger fruitbody formation for the particular species you are cultivating (see below).

Most wood-decomposing mushrooms, such as shiitakes and oysters, do not require microbial interactions to stimulate fruiting, so casing is generally reserved for the other species, such as those that grow on composted manure, grasses, and different types of agricultural wastes. These are primarily terrestrial mushrooms when found in the wild. The pH of the casing should best represent the chemistry and biology of the native environment from which the mushroom was isolated. Casings are traditionally nonnutritive, but modern casings at commercial mushroom operations now often include time-release fertilizers and supplements high in fat and protein that act as yield boosters.

Casing with Microbial Slurries

Many species of mushrooms require or benefit from the presence of certain microbial activity, and you can introduce those microbes via casing soil to induce or enhance fruiting. Although these complex interactions are not yet fully understood, the most common species of mushrooms that require or benefit from microbial casing include but are not limited to *Agaricus* spp. (white button, portabella, and almond portabella), *Lepiota* spp. (parasol), *Stropharia* spp. (king stropharia), *Clitocybe* spp. (blewits), and *Morchella* spp. (morels). What *is* understood is that often a fungus will remain in the colonization stage until it encounters a selection of compatible microbial partners. It's possible that once we learn more about the compatible microbes for these species, our understanding will change the methods people have used to cultivate mushrooms for centuries. So at this point there's a lot of knowledge to be gained from experimenting with microbial casings to try to induce or improve the fruiting of difficult species. After all, it may not be the presence of a certain species of microbe, but rather the input or removal of a particular chemical or molecular product that triggers fruiting,

Many *Agaricus* growers over the years have learned that casing is critical not just for the moisture and microclimate created for the developing primordia, but for the microbial partnerships as well, although the compatible species still remain something of a mystery. What *Agaricus* growers have discovered is that *sterilized* casing soils—in which all microbes have been destroyed—results in no yields, but that *pasteurized* casing soil almost always induces fruiting. Again, the exact mechanism is still in question, but we do know that it must involve either chemical or physical aspects of the casing that signal the mycelium to devote its energy to fruiting. The bigger picture is that we are not just cultivating mushrooms. In order to cultivate mushrooms successfully, we must also focus on the other kingdoms of life.

So if you want to make a microbial casing, how do you know which microbes to add and how to add them, especially if you are working on a small scale or without access to microscopes and a laboratory? The best method I can recommend is to collect soil from the stem base of either a wild specimen or a recently harvested commercial mushroom. Chances are, the soil at the stem base will contain the right combination of microbes for that particular mushroom, embedded in or adhering to the mycelial hyphae. You can then prepare progressive dilutions of that soil to dose sterilized casing soil to see at which dilution the microbes have the best effect on the mushroom you are trying to cultivate. In other words, dilutions help you "chase" the right microbes; once you catch them, you have to figure out how best to put them to work.

King stropharia (*Stropharia rugoso-annulata*) fruiting from sterilized sawdust topped with a casing treated with a slurry of microbes isolated from wild stropharia stem bases. Casing with native microbes improves your chances of successfully "taming" many wild mushrooms.

If you don't have a lab where you can prepare and test dilutions, it's still worth taking soil samples from the stem bases of wild or freshly harvested cultivated mushrooms and perpetuating them as a microbial slurry that you add to your casing soil. Start with using this untested slurry on small quantities of colonized media, such as a few bags, buckets, or even cups; if your slurry happens to contain microbes that will compete with your mycelium, rather than aiding it, you'll want to find that out at a small scale, rather than after you've used it in the casings for your entire operation.

Dilution to Extinction Step-by-Step

Step 1. Harvest a fresh mushroom. It can be either wild or cultivated. The base of the mushroom, which includes the anchoring mycelium, will be covered with microbes matched to that species' microbial

The stem bases from mushroom species that benefit from or require microbial triggers can be harvested and soaked in water to make a dilution, preferably to extinction, to isolate the causal organisms.

preference in a convenient package for your dilution assay. You want to use the soil and substrate attached to the base of the mushroom, clinging to the mycelial thread, not the mushroom itself. If no mushrooms are available, use a small sample of the native soil from or near where the species of mushroom fruits naturally. Each gram of soil can harbor more microbial cells than there are species of life on the planet, so chances are reasonably high that a native soil sample will have the microbial life your mushroom needs.

Step 2. You will need about 250 ml of sterilized water. You can boil or autoclave it, but distilled water is best. If you boil or autoclave the water, keep the container covered while it is cooling. You will also need to sterilize a dispensing tool (pipette, baster, dropper,

or graduated cylinder), twelve jars or tubes, and one large quart jar or similar vessel.

Step 3. When the sterilized water has cooled, you can add a small sample of the soil from the stem base of the mushroom, at least a few grams or teaspoons. The microbial population is there and you will be diluting it, but what is more important is to include soil that has mycelium in it. Include any binding rhizomorphs (hyphal strands fusing substrate together) to ensure that the microbial species adhering to the mycelium are included. Label this sample 0. Shake this original soil and water sample vigorously to break apart the bits of mycelium and adhering soil particles. Allow the particles to settle for 10 minutes before transferring and starting the dilution.

Step 4. Fill each of twelve smaller jars or test tubes (which have been sterilized) with 9 ml of sterilized, distilled water. Cap each with a sterile screw-on lid to minimize airborne contaminants. Label each consecutive jar from 1 to 12.

Step 5. Shake the jar containing the original 0 slurry and, using a sterile pipette, transfer 1 ml into the tube labeled 1, which had 9 ml of sterilized water, which now has a final volume of 10 ml. Shake tube 1 vigorously, or vortex it in the lab, and then allow the slurry to settle for a few minutes as the particulates fall out of suspension. Resterilize your dispensing tool or use a fresh pipette tip. From jar 1 take a 1 ml sample, or 10 percent, and transfer it to tube 2. Repeat the process: Sterilize your tool or replace the pipette tip, transfer 10 percent of tube 2 to jar 3, and shake it, until you have finished with tube 12.

Step 6. Cover and store the samples in a clean place, or refrigerate them briefly, while you prepare the casing soil in which you'll test the dilutions. You are now ready to perform your casing dilution assay.

Casing Dilution Assay Step-by-Step

You now have twelve individual samples, a collection of possible microbial populations, theoretically

Dosing king stropharia (*Stropharia rugoso-annulata*) with serial dilutions to isolate the fruiting trigger point.

decreasing in number down to a single individual cell in jar 12 (which is why it is called a dilution to extinction). Now you are ready to determine which dilution you will want to use for inoculating the microbial slurry for your casing soil, performing an assay to identify the lowest common denominator that will trigger a successful fruiting. The dilutions of interest will be the one that results in fruiting and the adjacent one that does not result in fruiting. (You don't need to identify the particular microbes in this process; what matters is the point at which the microbes trigger fruiting.) This will leave you with a sample of microbes that you can further perpetuate by additional stem base harvesting and dilutions or purify further using sterile culture to pinpoint keystone species. If you have a laboratory, examination of the casing soils using polymerase chain reaction (PCR) or denaturing gradient gel electrophoresis (DGGE) to extract and isolate the DNA from the samples can reveal the identity of the microbial communities present in your slurries.

Once you have your twelve dilution samples, it's time to prepare samples of growing media to test each of the twelve dilutions in replicates of at least three containers per dilution—so you will need thirty-six fruiting containers that have been colonized with pure mycelium, preferably in the lab, and treated with sterilized casing. And you will need six controls (three with an untreated casing and three with no casing at all) to eliminate the possibility of an abiotic factor such as a chemical or physical trigger present in the casing soil itself.

It is important that you try to use identical containers for all the samples, keeping the shape, weight, and quantity of the mycelium as constant as possible for comparative study and consistent results. The following is a basic casing dilution assay using king stropharia (*Stropharia rugoso-annulata*) as a model.

Step 1. Prepare your casing soil as described earlier in this chapter, mixing up approximately 1 gallon of this moist soil. Autoclave or steam the soil for one hour and allow it to cool, keeping the soil in a clean or sterile environment while it cools.

Step 2. While the casing soil cools, prepare forty-two individual samples of colonized growing medium packed into identical containers at identical weights. The easiest way to do this is to use quart jars that have colonized with mycelium of a pure culture. This way you do not have to sterilize empty containers and try to fill them, risking contamination. The jars should also already be fitted with a breathing filter. The containers need to be large enough so there is space to add casing soil on top, along with an air space to allow for mushroom formation. For example, a regular quart mason jar can be filled halfway with growing medium and a pure culture of mycelium, then cased with 1 inch of sterilized soil, and still have 2 to 3 inches of head space below the lid. You can use any size container, but smaller containers are more economical, take up less space, and are easier to handle. You don't need a lot of biomass to test for a fruiting trigger.

Step 3. Cover the colonized substrate in all but three of the containers with an identical weight or volume of your sterilized casing soil, gently tamping it down level or lightly compressing it with a sterilized object. Uncover and cover the cultures as quickly as possible while you work to minimize the risk of contamination.

Step 4. Number thirty-six of the jars as dilutions 1 through 12, in groups of three. (For example, the first three jars would be numbered D1-1, D1-2, and D1-3.) The remaining six bottles are two sets of controls—one group with casing and one without—and can be labeled as such (for example, C1-1 through C1-3 and C2-1 through C2-3).

Step 5. All of the treatments will receive an identical amount of solution; I use 1 ml. Perform the control exposures first, adding sterilized or distilled water, not your dilutions, to the surface with a sterile pipette.

Resterilize your pipette or use a fresh one. Then add 1 ml of dilution 1 to each of the three jars labeled D1-1, D1-2, and D1-3. Proceed, transferring 1 ml of each dilution to each of the three jars in its group, until you have finished with D12-3. As before, uncover and cover the cultures as quickly as possible while you work to minimize the risk of contamination.

Step 6. If no micron filter lids are available, fit each jar with a secure but not airtight seal, such as a foam plug or a wad of cotton or polyester fill (like the kind you'd use to stuff a pillow), covered loosely with aluminum foil. This protects the jars from contamination but allows the mycelium to breathe. Transfer all of the covered samples to an incubation area that is temperature-appropriate for the species (check part 4), keeping them in darkness as much as possible. A laboratory incubator or Styrofoam cooler works well to minimize temperature fluctuations.

Step 7. Observe the behavior of the mycelium on a daily basis, checking each group for signs of hyphae streaming upward into the casing. Note any morphological features (strand-like, rhizomorphic, threading, knotting) that are visible as the cultures colonize the casing and team up with microbes for primordia formation. This is the moment when the mycelium is recruiting its symbionts or not, depending on what's present or absent—which depends on the given dilution. If you're using clear jars, which will allow you to see the mycelium beneath the casing soil, photograph and circle with a permanent marker any small, growing masses so that you can follow the progression. Record your observations in a lab notebook.

Step 8. After a few weeks of casing soil colonization, transfer the jars to a fruiting area that is a few degrees cooler and offers some diffuse natural or fluorescent light to trigger primordia formation and fruitbody development. Now you will be able to see the dilution point at which fruiting is triggered. That point—the dilution of the samples in which fruiting begins, next in line from the dilution whose samples do not begin fruiting—is the one you'll want to trial as a microbial slurry for your casing soil.

Preparing and Using a Microbial Slurry

Once you've determined the targeted level of dilution that triggers optimal fruiting, you can harvest the casing to produce a mother culture capable of expansion for use on a much larger scale. Do this as soon as you notice mycelial knotting or primordia formation, rather than waiting for the mushrooms to mature, which can invite contaminants. Remove the casing soil from the three jars in the group you identified as having the targeted level of dilution, and blend that soil into a larger volume of sterilized water and casing soil to expand it. Mix it thoroughly and allow for drainage and gas exchange.

Now you'll need to experiment to determine the proportion of microbial slurry to use for a given volume of biomass to maximize primordia formation and fruiting. Trials must be performed to substantiate the microbial population needed to trigger a fruiting response that meets the criteria or goals of the experiment. It may be possible to use less, applying a thin layer or spraying of the solution onto the casing surface, or at localized points using a few drops spaced evenly over the surface to position mushrooms magically across the casing soil in a perfect pattern. Original dilutions can be used to create additional casing treatments, and when this volume is spent, fruitbodies from the next flush or cycle can be used to create a casing soil that becomes increasingly more calibrated to the substrates and techniques you are using. One word of caution: You may also be perpetuating potential pathogens, some mycoparasite *Trichoderma*, which can quickly destroy casing soils run through with mycelium. Commercial cultivators should screen their dilutions using a streak test (see chapter 18) prior to using large volumes or committing to using this technique on a large scale for quality control.

On a commercial scale using microbial slurries in casing soils could prove extremely valuable, since the volume and close proximity of cultures in fruiting rooms increases the risk of cross-contamination from undesirable microbes in casing soils. Disease-carrying vectors such as flies, mites, and harvesting hands can move contaminants about at an alarming rate. Adding these slurries may help create a resistance or defense by "exclusion of species," setting up a protective force of beneficial microbes.

To perpetuate the microbial community that enhances your casing soils, you have only to harvest healthy mushrooms from these "clean" growing beds. You can use the substrate that collects around their stem bases to prepare microbial slurries, as described above, for the next batch of casing soil.

Commercial cultivators can modify aerobic bioreactors (similar to the ones used in the beer brewing industry, with the addition of a mechanism to inject filtered air or oxygen) to accommodate batch production of the casing microbial culture. (Batch reactors, which have a fixed volume and are used for creating precision mixtures in an enclosed vessel, may be more accurate for producing the casing cultures than continuous feed reactors, which can accumulate undesirable secondary and tertiary metabolites.) In this case, you should add an enrichment solution that promotes the division and proliferation of the targeted species. Figuring out the optimal mix of ingredients for this enrichment solution will take some microbiological exploration. You can culture the microbial isolates on agar plates that contain specific nutrients to determine their metabolic requirements; this can also aid in identification of the microbes. You'll also want to investigate temperature thresholds so that you can figure out how best to store the isolates without loss of viability. Performing storage trials with dilutions will be critical to maintain your microbial populations from year to year to sustain their fruiting efficiencies and viability, using techniques outlined in the chapter 19, such as refrigeration or freezing in water.

Eliciting Novel Metabolites from Fungal Biomass

Fungi are enzymatic factories, supplied with the biological machinery to produce an array of primary, secondary, and possibly tertiary metabolites. And fungi are capable of shifting their gene expression

Mushroom metabolites harvested from galleries can be added to agar formulas or used for inhibition studies.

to manufacture a different spectrum of enzymes in response to different chemical and biological triggers. Understanding this, researchers can observe, construct, and harvest from fungal enzyme production units that can provide significant volumes of unique metabolites. As a result, fungi have many applications beyond culinary ones, and there are many ways to experiment with fungi that can lead to discoveries of new antibiotics, antifungal compounds, enzymes with industrial uses, and complex metabolic mixtures that can be used for mycoremediation.

I use simple laboratory plating techniques to test how different species react to various stimuli, such as positioning a wedge of mushroom mycelium an inch away from a specific bacterium and observing their behavior over the next few hours to several days. This tells me a great deal about their relationship, and particularly whether the two organisms are inhibitory or facilitative. This test will quickly show which organism is dominant or submissive, which is stronger or weaker, and can give researchers clues about potential applications. Organisms that prevent the linear growth of another may be diffusing an inhibitive substance into the agar gel, creating what is sometimes referred to as a zone of inhibition. Others may sweat metabolites specific to the threat of an oncoming biological attack, manufacturing defensive molecules capable of deflecting or devouring the opponent.

Fungi that exhibit a strong metabolic response when interacting with other organisms can be cultured in larger volumes to allow extraction of those metabolites. This is easiest to do when you cultivate the fungus on a "block" of substrate enclosed in a plastic bag. Once the fungus has colonized its growing substrate, shape to form a small depression or well on the upper surface, using your hands or a rounded object such as a spoon, without opening the bag. If the bag is too inflated to allow you to reach the block, wipe the bag's surface with 70 percent isopropyl alcohol; at a laminar flow hood, poke a small hole in the bag, press the bag to partially deflate it, and then tape or seal the hole shut. Now you should be able to make a depression in the substrate.

Allow the bag to incubate further and the well should completely seal over with a nice, thick mat of mycelium. Because the chitinous mycelium is hydrophobic to any substance other than pure water, like a pond liner it will prevent any solution from leaching into the body of the substrate. At this point, dose the block's well with the contaminant of interest. This will trigger the fungus to manufacture large volumes of metabolites along the sides of the bag, away from the contaminant. You can then harvest these metabolites with a sterile syringe.

In my experiments, the medicinal mushroom *Fomes fomentarius* (iceman polypore) has demonstrated activity

Forming a depression on the top of a pure culture cultivated on sawdust without opening the sealed bags. These depressions, or wells, seal over with a mat of mycelium in just a few days.

Dosing the well with a known biological or chemical contaminant. The mat of hydrophobic mycelium keeps this fluid pooled, rather than absorbed throughout the media.

Two blocks of *Fomes fomentarius*. The left one was dosed with *E. coli* and the right one (a control) with water. Mycelia can create different metabolites in response to their interaction with biological or chemical compounds.

The *Fomes fomentarius* blocks dosed with *E. coli* each produced over 100 ml of metabolite-rich fluid over the next few days. I drew out the metabolites with a syringe, poking it in just above the liquid surface and angling it down to avoid leakages. I then taped the bag back shut.

against *E. coli* and other bacterial pathogens. This led me to develop a procedure for manufacturing metabolites that are unique to not just human pathogens in general, but even the strains or ecotypes of the pathogen that have customized themselves to our bodies and immune systems. In other words, this process can create novel chemical cocktails specific to an individual's needs, be it an anticancer compound or an antibiotic for a rare infection. I have also

used this procedure to elicit metabolites from several mushroom species that typically do not produce metabolites at all in culture.

Comparing the metabolites a fungus produces when triggered by a specific contaminant against those that it produces without that contaminant allows the researcher to eliminate those that are in compound and thus to ascertain the shift in metabolic function. The researcher can also screen strains of fungi to determine which are the most proficient at producing the compounds of interest at a rate that would satisfy industrial demand, should large-scale production ever be called for.

Using Primordial Hormone Extracts to Induce Fruiting

Mushrooms, like plants, produce hormones that orchestrate their fruiting cycles. These hormones are concentrated in the primordia, and you can harvest these hormones to make an extract and apply it as a mist or liquid fertilizer to a colonized biomass to help induce fruiting. A primordial hormone treatment could be useful in controlling the location and timing of fruiting, focusing primordia development at the

sites where you want the fungi to fruit, or inducing fruiting to complement your production cycles. Of course, the myceliated biomass still needs the same time to charge up enough nutrients before you induce it with a fruiting hormone or the premature induction will result in reduced yields (Mau and Beelman, 1996).

To create a primordial hormone extract, harvest a few primordia from your cultivated mushrooms, cleaning away any substrate debris. Surface-sterilize the harvested primordia, swirling them gently in ethanol for thirty seconds, then rinse with distilled water; repeat three times. Blend these primordia with water in a sterilized blender to create a slurry, and vacuum-filter it to remove the particulates. Transfer the slurry to a centrifuge, and spin the sample at 10,000 xg for five minutes. Withdraw the supernatant into a sterile syringe, which is then fitted with a syringe filter

Tiger sawgill (*Lentinus tigrinus*) primordia initiated on agar at a locus of hormonal stimulus. Hormone extracts that stimulate fruiting can be used to initiate site-specific mushroom formation, which could help cultivators control the placement and timing of mushroom production.

to sterilize the fluid. Collect the sterilized fluid in a sterile vial. You can dilute this solution, which will be a very small amount compared to what you started with, once you've experimented to determine the rate of application for your species, based on the volume of primordia you used to make the extract.

Boosting Production with Electrical Pulses

Many biology students, sooner or later, are exposed to a basic principle in muscular contraction while dissecting frogs: They can induce a "twitch," or contracting effect, in the frog's leg muscle by touching it with an electrical probe. For years I wondered whether electrical stimulation could similarly stimulate mushrooms, only to discover recent research reporting on

mushrooms responding favorably to a mild electrical field (Ohga et al., 2004).

At Iwate University in northern Japan, a team of scientists performed a four-year study to determine the effectiveness of low-level electrical stimulation on mushroom formation during the cultivation of a variety of species. Out of ten kinds of mushrooms that were tested, eight responded favorably by growing at an increased rate when jolted with only a single pulse of electricity between 50,000 and 100,000 volts for one ten-millionth of a second. The electrical stimulation was shown to double the yield of many species, including shiitake, nameko, and reishi, in comparison to the controls. The researchers hypothesized that the electrical charge, which may be an imminent threat to the survival of the fungi, could trigger a defense mechanism that causes the fungi to fruit more prolifically. This may (or may not) have something to do with the

Electrical pulses can turn cottony mycelium into rhizomorphic strands (a preliminary precursor to fruiting) in mushrooms thought to require microbial interactions to trigger fruiting.

known phenomenon of mushroom emergence immediately following an electrical storm, which may be
due solely to a drop in barometric pressure but could
also be related to static charge, which shocks the fungi
into producing fruit.

More research is needed to understand these
mechanisms, although re-creating these low-level
charges and introducing them to the mushroom cultivation systems in a way that is easy and safe for the
cultivators may well be the next challenge.

CHAPTER 21

Morel Cultivation: Research Update

Morels are some of the most coveted mushrooms on the planet, but they are extremely particular about their growing environment, which makes them not only difficult to forage for, but even more difficult to cultivate. When I'm teaching, this is the mushroom that I get asked about the most, usually by beginners who think that since we can grow shiitakes and oysters on logs, growing morels must be equally simple. Unfortunately that is far from the truth. Unlike other cultivated mushrooms, these fungi are extremely variable, so each strain isolated from a particular region can have cultivation parameters much different from those of the next.

Most cultivation manuals that propose strategies for successful morel cultivation offer the information as suggestion rather than fact, and this book is no different. Balance any broad claims about morel cultivation alongside your own observation of local strains and any tests you conduct in the lab. Caveats aside, recently there has been some great research and even some success cultivating morels on a site-specific basis. Although cultivation is not yet consistent, we are learning more about how to support and trigger morel fruitings both in outdoor beds and, more infrequently, indoors. So keep those dreams alive and experiment with the following parameters. Using the information here, you may be able to uncover clues specific to your location and indigenous strains—and help crack the code of morels.

ECOLOGY AND LIFE CYCLE

In order to consider cultivating (or finding) morels, it's important to try to understand their ecology and life cycle—an understanding that is, for everyone, a work in progress. Morels have been classified as both saprophytic and mycorrhizal fungi, and in some cases as both at different points in their life cycle, making them debatably obligate, or dependent, on living trees at times. Morel mycelium has evolved to interact with soil microbes that could be specific to the root zones, or rhizosphere, of these particular host trees instead of interacting with the tree directly. There is also evidence of bacterial endosymbionts living inside morel hyphae as part of the Hartig net, the site of nutrient exchange between plant and fungus, where the mycelium enters the root cells of compatible plants and trees, typically with ectomycorrhizal fungi, further suggesting a great deal of complexity and dynamism in the soil ecology associated with this genus.

Morels are not gilled or pored but instead have a series of cups arranged in every direction on the

A beautiful blond morel (*Morchella americana*) fruiting wild in the southeastern United States. Note the signature hollow stem glowing from the sunlight passing through.

upper "head" or cap surface above the stem. These cups are lined with tightly bundled tubes along the inner surface, called asci. Once the mushrooms mature, the asci "hiss" or vent out concentrated bursts of spores into the environment, looking for suitable habitat, which is usually in the general vicinity of the morel itself. That said, spores are weightless and can travel on wind, float on water, and stick to clothing. Wildlife that feed on morels can also transport the spores considerable distances. Morel spores germinate quickly, much more rapidly than those of white rot fungi such as shiitake and oyster mushrooms by comparison, with a growth rate that actually rivals that of

some molds. The mycelium itself can grow at a rate of an inch or more per day, rivaling the paddy straw (*Volvariella volvacea*) as one of the fastest-growing fungi on the planet. (Fast-growing mycelium does not always equate to fast-growing fruitbodies or fast life cycles; it is simply a characteristic of the vegetative stage and not representative of the overall production cycle.) To put that into scale, imagine lying down on the ground and dividing yourself end-to-end nonstop at a rate of twenty miles per hour!

Morel mycelium typically does not form fruitbodies without some sort of environmental stressor. This prompts them to produce sclerotia, or resting

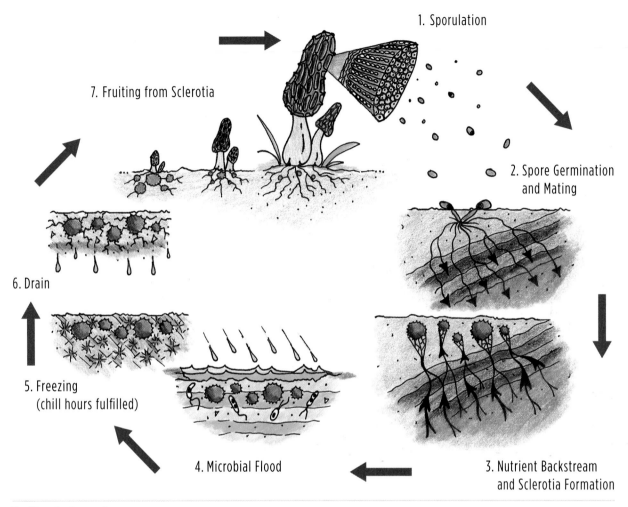

1. Sporulation

7. Fruiting from Sclerotia

2. Spore Germination and Mating

6. Drain

5. Freezing (chill hours fulfilled)

4. Microbial Flood

3. Nutrient Backstream and Sclerotia Formation

The life cycle of a morel.

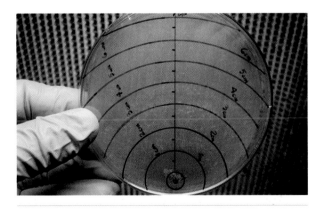

Morel mycelium can grow at a rate of over 1 inch per day.

Morel sclerotia, the structures the mushroom uses to stockpile nutrients for fruiting, are seen here covering a petri plate.

"microtubers," which serve as energy reserves while the mycelium waits for favorable water and temperature conditions to trigger the small, hardened masses to develop into mushrooms. If you look carefully at the harvested bases of unearthed morels, you can sometimes see these sclerotia where they formed just under the surface of leaves and debris. The sclerotia typically form in the summer and fall, before dropping temperatures slow the mycelium, and develop into fruitbodies the following spring.

The mycologist Thomas Volk found that when he inoculated plates that were half nutritive agar and half nonnutritive agar with morel culture, mycelium would form sclerotia only on the nonnutritive side. The hyphae were able to "find" the nutritive agar and translocate the nutrients back to the nonnutritive side, where the mycelium produced and stored all of the sclerotia. In re-creating these trials, I duplicated Volk's process but added an additional gallery to observe morel mycelium on the middle of a plate, at the interface between the nutritive and nonnutritive agar. The culture in the middle produced sclerotia almost immediately on the nonnutritive side—within two to three days—whereas the culture that traveled a great distance from nonnutritive media to nutritive media backstreamed nutrients and produced sclerotia into the nonnutritive side in seven to ten days. Adversely, all replicates of galleries where morel mycelium was placed on the nutritive side streamed to the nonnutritive and continued to colonize the entire plate, without ever producing sclerotia, even three months after incubation. What I am suggesting is that this simultaneous habitat capture and nutrient transport—backstreaming, as suggested by Volk—is essential; it saves time, increasing the efficiency of morel sclerotia formation. Placing the mycelium between the nutritive and nonnutritive layers seemed to hasten the process, a factor that may eventually change the way morel mushrooms are cultivated, potentially cutting as much as a week off production time by simply changing the way a cultivator inoculates the substrates.

After sclerotia form in nature, they can be subject to environmental fluctuations in soil such as drought, flood, freezing, and sometimes heat in the form of forest fires scorching the forest floor above. In some cases these occurrences—particularly fire—seem to encourage a flush of morel development. One of the most common questions I hear from both hopeful foragers and cultivators is whether a burn or ash is helpful or essential to the morel life cycle. Although many people argue that it is essential, I think the phenomenon is site- and species-specific, at least in North America. East Coast morel sites do not experience a massive bounty following a burn the previous year. Different ecotypes evolve to fulfill their particular environmental niche, which means you have to allow for some variability.

One phenomenon, however, that is fairly consistent with most morel strains is the need for a certain number of "chill hours" to induce the mycelium to develop. A chill hour is an hour in which the soil at a depth of 4 to 10 inches is below 32°F (0°C). You can often get this information from local weather stations, which monitor soil temperatures with probes that extend as far as 8 feet into the ground. Morels that develop during years with fewer chill hours could be considered "low-chill-hour morels," much like fruit trees that have been selectively bred to produce in warmer or more tropical climates. Having a good sense of these chill hours is probably necessary to successfully establish a morel patch or to cultivate most strains indoors.

Whether you plan to cultivate indoors or outdoors, your first step will be to clone or isolate local strains. Unlike exclusively saprophytic mushrooms such as oyster and shiitake, which can fruit on a variety of wood types just about anywhere within the right temperature windows, morels form complex biological and chemical relationships with their surrounding soil. Your success in growing morels relies heavily on the strain that you find, which you'll want to screen for a few basic properties.

For ten years I have cloned and taken spore isolates of many wild morel collections, thanks to friends such as Tim Geho and Judy Roberts, personal collectors and supporters of my morel research, who mail me samples from across the East Coast every spring. In recent years I have tried to collect spores from morels

Morel mycelium on a gallery plate. The mycelium was placed on the nutritive side (upper half) and encouraged to travel across to the nonnutritive side (lower half). No sclerotia ever formed on any of these replicated plates, even after three months.

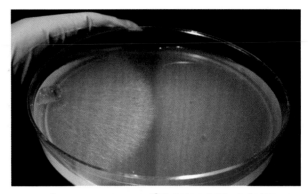

Inoculated on the nonnutritive side, the morel mycelium is seen here running hyphae toward the nutritive side. All of these plates began producing sclerotia in the nonnutritive side in about seven to ten days once the entire nutrient capture was complete.

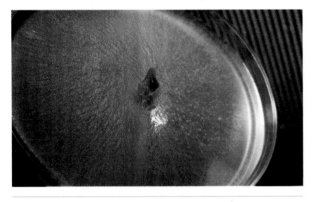

The plates inoculated at the nutritive and nonnutritive interface colonized and formed sclerotia almost immediately, even before completing habitat capture. This observation could change the next generation of morel cultivation techniques.

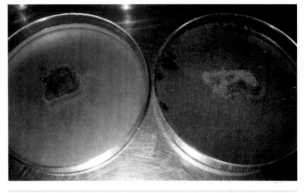

Morel mycelium apparently places sclerotia in the most energy-depleted region, even if that is its point of origin. The plate on the left contains a more nutritive solution of potato dextrose than the inoculating cube in the middle, forcing the sclerotia to stay put. The plate on the right contains a less nutritive media, and in turn the sclerotia form as far away as possible from the inoculating cube.

that don't seem to fit the "mycorrhizal model," meaning that they aren't found near a logical tree partner, which in my area would be a tulip poplar, ash, or elm. My aim is to isolate an anomaly: a morel strain that does not form a direct association with a specific tree host or the bacteria that thrive in a specific tree's rhizosphere. Such a strain could potentially eliminate a major challenge factor for cultivation: the tree.

Another important factor is the availability of the raw materials that your local morel strains will recognize as a viable substrate. There is almost no use in isolating a morel far away from your location and attempting to grow it on a food source it can't connect with or in a climate it doesn't recognize. The best strategy for cultivating all difficult species is to understand that they are extremely sensitive about their new homes—so make them feel as comfortable as possible.

Once you locate the morels you want to use, it's important to collect them in a precise way. Don't just cut them and throw them in your basket for taking spore prints or tissue samples later. You will be losing valuable insight about the environmental conditions the morels need and how to try to re-create those conditions. First, understand that unless the morels are fused at the base or in close proximity to each other, it's probable that the strains you are collecting are genetically different from each other. Second, since each strain forms associations with the specific microbial community in the soil where it grows, you'll want to take a soil sample for morel you collect.

I collect morels that are close to maturity and firm. I make a conical incision around the morel, about 2 inches into the soil, to keep the morel and the attached soil plug intact. I then separate the fruitbody from the soil plug and place each sample into a waxed-paper sandwich bag, which I label with the date, the location, and a sample number.

The sample numbers are important for tracking the progress of both the morel mycelium and the soil sample, especially as I attempt to recombine them later in the process. In a notebook, I record the sample numbers as well as basic data such as the temperature, the soil pH (which I ascertain later with a pH meter or test strips), the morel's distance from and orientation from neighboring trees, and the size and species of those trees. I record the last few observations as a combination, such as "3 meters west of 2-meter-diameter ash" since I have collected morels many times in the same spots. I also note any understory vegetation, the proximity to water or evidence of surface water flow and hydrology, any simultaneous wildflower blooms—anything that makes the location of this morel unique in relation to that of other samples.

I place the samples in a cooler with a few ice packs to keep them cool but not frozen. Freezing can damage the fruitbody tissue and make sterile sampling much more difficult. When I return home, I store the fruitbodies under refrigeration until I am ready to culture them. I have also allowed the fruitbodies to dry completely and still had success with both sterile and nonsterile spore germinations. As for the soil, I either refrigerate or freeze it to preserve the microbial community.

Once morels are collected, the processes for indoor and outdoor cultivation differ. Outdoor cultivators will need to make spore slurries and prepare beds for woodland gardens. Indoor cultivators will need a few more steps, isolating the strains using sterile culture in a small laboratory and preparing the soil for reintroduction to the mycelium at a later time. In either case, however, not only are we cultivating mushroom mycelium; we're making a multidimensional interkingdom effort to re-create a complex environment, using soil microbes as an additional feedstock and biostimulant for the morel mycelium. Whether you are trying to grow indoors or outdoors, each of the methods offers insight to the other, so I encourage you to read both sections to help increase your chances of a successful yield.

CULTIVATING MORELS OUTDOORS

Most mushroom growers don't have the equipment to isolate morels in pure culture in a laboratory. The upside of not having the space or equipment is that morels are notoriously unpredictable anyway when you separate them from their microbial partners and native environments, so natural cultivation methods have proven to be more successful and are worth the time and expense to try. On the other hand, morels fruit within a very narrow soil temperature window, from approximately 48 to 56°F (9–13°C). They also require the chill hours referenced above, so if you live in warmer or tropical climates, outdoor cultivation may not work for you. I tell potential morel growers that if you have never seen or heard of morels being collected in your immediate area, then a morel bed probably will be a waste of time.

To extract the spores from a morel you have collected, bring 2 cups of water to a boil, pour the water into a clean jar, and let it cool to room temperature. Place the head of one morel into the water and swirl it around for thirty seconds. Remove the morel (and cook and eat it, no?). Cap the jar, allow it to sit at room temperature for twenty hours, and then place it in a cool or cold location (refrigeration is ideal) for at least three weeks. (Culture jars can remain refrigerated for months, so prepare them whenever the morels are available.) Inspect the jar every week. You should notice the water darkening, an accumulation of floating biomass that turns into a jelly-like biofilm, and strands of morel mycelium floating around in the solution. Once you notice the water darken and a biofilm on the top, drop in a small sample of the original soil you collected from the stem base of the same mushroom to complete the microbial partnership. This concoction is a spore slurry of both morel spores and the complex of soil yeasts and bacteria that were colonizing the fruitbody, resembling a kombucha-like layer gelling above the solution. Be sure to keep the jar refrigerated, or else bacteria and yeasts may outcompete the morel. The morel itself is better able than the other microorganisms to grow at the cooler temperature of refrigeration, and that cooler temperature and saturation in water mimics a winter flood and chill scenario, decreasing the morel's need for oxygen and forcing it into an anaerobic state. After a few weeks, transfer this slurry to the freezer for at least a day to crystallize the matrix. Store it frozen until you are ready to use it.

When you are ready to construct and inoculate your outdoor bed, remove the slurry from the freezer and thaw it. (Making the bed is a quick affair and it needs to be inoculated immediately, so don't start to prepare it until your slurry is prepared and in the freezer.) Place the liquid in a blender and process it to homogenize the solution and biofilm layer. Dilute this slurry in a clean 5-gallon bucket of water. (Keep your soil sample—the one you collected along with the morel that you've made this slurry from—refrigerated for now, since you won't be using it yet.)

Constructing Your Morel Bed

Preparing the outdoor morel patch is easy, but be sure you have all your materials ready before you begin. As far as timing goes, I prefer to duplicate the natural cycles of the morel sporulation and germination process, constructing beds in spring at the time that they are fruiting. The most important component is the spawn or slurry, so if you have not made it yet, you will need to do so. (You can also purchase morel

spawn and prepare it as a slurry, although it may be from a different region and not contain the microbial population to partner with to induce fruiting.) One of the most interesting differences between cultivating morels and traditional mushroom culture is how the spawn is used in relation to the prepared medium. Recall that morels produce sclerotia in nonnutritive soils, sending out hyphae to find and transfer nutrients to the nonnutritive site. With this in mind, you don't mix morel spawn into the entire growing substrate, as you would with king stropharia, for example. Instead you apply it as a shallow, topical application, allowing the morel mycelium to find its way down into the nutritive substrate below. Morel mycelium understands that sitting on a soil surface that lacks nutrients puts it in danger of not being able to produce a fruiting body to release its spores. It responds by searching outward and downward for nutritive resources. The mycelium colonizes the rich nutritive layer of the substrate in an elaborate web, much thicker than in the top layer of soil, building volume and absorbing as many nutrients as possible. It then ships the nutrients upward to store them in dense masses that develop into sclerotia, which will eventually give rise to fruitbodies. Inspect your beds periodically. Finding a few fruitbodies is a good reason to feel like you are on the right path.

Step 1: Collect Your Materials

To build a morel bed, you'll need the following equipment and materials. Be sure to have everything on hand before you begin, because the bed comes together quickly and must be inoculated promptly.

- **Newspaper.** Enough for two 2-foot by 8-foot layers (one at the bottom of the bed and one on top of the nutritive layer).
- **Fresh or dried hardwood sawdust.** Soaked overnight and drained. Preferably from a tree that matches the host tree for your morel strain. You'll need about 8 cubic feet for the 2-foot by 8-foot bed described here.
- **pH testing strips (or a digital pH meter).** Testing strips are inexpensive and available from pool supply company; a pH meter is more accurate.

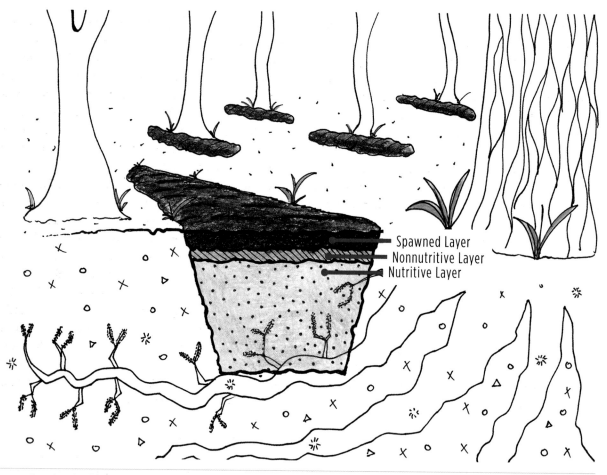

Cross section of outdoor morel cultivation. You can place the spawn between the layers or on the top surface. Just be sure not to mix your spawn into the nutritive layer or sclerotia may not form.

- **Garden lime and aluminum sulfate.** To adjust pH, if necessary.
- **Wheat bran or soy meal.** From a farm supply or feed store. You'll need 5 pounds.
- **Hydrated limestone.** You'll need 1 pound.
- **Dried peat moss or coconut coir.** You'll need 3.8 cubic feet (one compressed bale).
- **Sand.** Extra fine, not coarse. You'll need about 50 pounds.
- **Logs or untreated lumber for the frame.** Optional.
- **Chicken or poultry wire.** For covering and protecting the bed from digging wildlife.
- **Spawn.** Use your own spore slurry or purchased black or blond morel sawdust spawn (one 5-pound bag can cover up to 200 square feet). Using both

would be ideal, giving you a mass of mycelium for colonization and a slurry with the matched-up microbes to trigger fruiting.

Step 2: Build the Bed

Locate your bed in a shaded area—away from anyone who might find the morels! A spot along a creek bank or a wet, boggy area that typically dries out a little in the summer, with trees that morels like to associate with, such as poplar, ash, cottonwood, or spruce, is best.

Dig your bed about 2 feet wide, 8 feet long, and 8 to 10 inches deep. It's actually best to build a series of beds, small and spaced a few feet apart, to increase your chances. Separate beds can develop differently from one another since they will host distinct colonies of organisms. Rather than gambling the entire biomass on one location, dividing your resources allows the different beds to explore different environmental triggers in their slightly different surroundings.

Line the bare soil at the bottom of the bed with a single layer of wet newspaper.

Step 3: Fill the Bed with Nutritive Medium

Soak the sawdust overnight to hydrate it. Squeeze out a small sample of water and test its pH. Your target is a pH between 7 and 8, so adjust accordingly by adding garden lime (to raise pH) or aluminum sulfate (to lower pH). Make this nutritive layer a day in advance of adding it to your bed so that you'll have time to adjust the pH if needed, draining just before it is ready to use.

When you're ready to fill the beds, mix the nutritive supplement, approximately 10 percent wheat bran or soy meal by volume, into the soaked sawdust. Then fill the prepared bed with 6 to 8 inches of the nutritive medium, leaving 1 to 2 inches of depth for the casing soil. Level the nutritive layer with a rake and tamp it down tightly. Morels need a very dense medium, not a fluffy one. Once it's level and packed, broadcast pelletized lime at a rate of 1 cup per 16 square feet evenly across the surface of the nutritive layer. This will act as a slow-release pH buffer (the bed will acidify slightly as the mycelium eats it up) and will also supply the mycelium with the extra calcium it loves in order to produce abundant sclerotia. Water the bed thoroughly to wet the nutritive layer and soften the wheat bran. After watering, level the bed and pack it firmly again to remove air pockets. Line it with a single layer of wet newspaper to separate it from the nonnutritive casing you will add next.

Step 4: Add the Nonnutritive Casing Soil

Mix the dried peat moss or coconut coir and sand in equal parts in a wheelbarrow or mixing tub. You'll need enough to cover the nutritive layer to a depth of 1 to 2 inches. Wet the mixture thoroughly, allowing it to hydrate but not become supersaturated. I use the "squeeze test": When a fistful of wet soil is squeezed

it produces a few drops of water, not a stream. Mix in additional dry media if it is too wet or add more water, then repeat the squeeze test, and test the pH of that water, just as you did for the nutritive layer. You want the pH between 7 and 8. This is critical, since most morel strains are very specific about pH for optimum metabolism. If your soil is acidic, you will need to add lime, a cup at a time, mixing it into the wet slurry. Repeat the squeeze test until the pH rises to between 7 and 8 and stays there for at least an hour. Then drain the casing soil, pouring away the excess water, and use it immediately to cover your nutritive layer.

Add the wet casing soil to your bed at a depth of 1 to 2 inches. Use a ruler to check your depth as you apply it. Compress the casing, then check the depth and add more casing soil if needed to keep it at 1 to 2 inches deep. Using a a long broom or rake handle, or a wooden dowel, press small indentions into the casing soil to create a textured surface for the morel spawn. (If the upper layer is smooth, the mycelia will have a difficult time acclimating to the immediate and possibly drier environment.)

Step 5: Inoculate the Bed

Sprinkle your spawn sparingly but evenly, at a rate of ¼ cup for every square foot of the bed, over the casing soil. You only need to dust it.

Cover your bed with some poultry or chicken wire to keep wildlife (squirrels, pigs, turkey, armadillos) from digging into and destroying it. Consider also laying a wooden lattice over the bed to limit the exposed surface area and promote microclimates, and also help protect the bed against digging animals. Cover the bed with leaves or wheat straw to a depth of 1 to 2 inches to shade the bed and provide moisture protection for developing primordia.

Step 6: Water

Water the bed well. For the next two months water the bed well enough to keep it moist, but not wet and waterlogged. Reduce watering to once a month from late summer through mid-autumn (that would be August through October in North America, or December through February in the Southern Hemisphere).

Olga likes to dry morels on aluminum foil, covered with paper towels, which can all be washed in a bucket of water to reintroduce the spores into the woods around host trees such as elm, poplar, and ash in our area.

During this time, healthy morel mycelium can grow about 1 inch per day!

Step 7: Flood and Freeze

When a winter freeze is predicted to hit your area, the night before the freeze, slowly flood the bed. This will stimulate the morels to feed on beneficial bacteria essential for fruiting.

Step 8: Spring Thaw and Fruiting

As the spring thaw progresses, you should see sclerotia that have formed in the nonnutritive layer swell with water and begin to form primordia. Keep a careful eye on your bed and on local soil temperatures. Morel fruiting generally occurs when soil temperatures are in the 48 to 56°F (9–13°C) range. Good luck—you will need it!

Incentive and Hope for Improved Methods

Successful methods for cultivating morels are more highly guarded secrets than morel hunting grounds. I have often dabbled with different methods, taking into account reports of both accidental and intentional outdoor fruiting efforts. When time permits, I run trials of methods that other growers report both online and in scientific journals. In my own attempts to replicate these reported methods, few if any have succeeded. In some cases, I suspect that the original cultivators replicated their experiments many, many times so that even if only a few morel beds produced, they could claim success. But to me, successful cultivation means reliable repeatability (and financial results, for a commercial enterprise).

One person in particular comes to mind when I think of morel cultivation—a gentleman from Mississippi who contacted me about growing shiitake mushrooms in his climate. He told me he was in his mid-seventies and rambled on a bit about how he cultivated his own morels. The man told me he didn't care much for making money at it since he was "getting old anyway," and that numerous visitors and professors who had witnessed the fruiting were relentlessly bothering him to divulge his methods. But on this day he must have felt at ease with my gentle questioning and was willing to share his secrets—and he said that I could publish his methods in order to benefit others.

I was not convinced of his expertise until he began to describe, using layman's terms, what for me was a very familiar procedure. He told me the exact method for outdoor cultivation that I have described here, with a minor substitution in the base ingredients in the nutritive layer of the growing substrate. The gentleman said that he would trench out a bed 1 foot deep, 2 feet wide, and 40 feet long, fill it with sawdust from ash trees, from a local sawmill, and a "protein source" that he would not identify, and then cap the bed with a layer of his native soil. After capping the soil he added a slurry of morel spores and soil that he had saved from the morel cultivation bed of the year before, mixed into solution with water and stored in the freezer until he planted the row in the fall. He thawed the slurry and drizzled the mixture down the center of the row into a depression in the casing soil. He covered the entire bed with nearby leaves and chicken wire to prevent damage from turkeys and other wildlife and then left it alone. The prepared trench, conveniently located in a low-lying wooded area, similar to a floodplain, provided the shade and moisture necessary to drive the morel mycelium down into the nutritive layer below.

In midwinter, he would walk the row and check under the leaf litter in multiple spots for sclerotia, which he said were always plentiful, like small cinnamon-colored stones lining the entire trench. The trench would also flood in the fall and winter, either naturally with adequate rainfall or with help from the farmer, who would completely saturate the nutritive layer. This flooding would encourage the sclerotia into a fruiting cycle as the ground temperature approached 40°F (4°C). A visiting University of Illinois professor, who had heard of the gentleman's success, stopped by during one fruiting cycle and described morels "fruiting shoulder-to-shoulder in a row forty feet long like a 'row of corn.'" Amazed, the professor "begged and pleaded" for the farmer to divulge his methods, but the farmer felt nervous about the professor's (and other people's) intentions with the discovery. "I felt like they were just after my process and that it could be either stolen or used by larger companies," he said, "so I kept it to myself all these years. Now I don't see much use in keeping it a private process."

Although we came to our knowledge in different ways, both my own and the gentleman's methods represent an improved process. By adding spawn only to the top nonnutritive layer, cultivators can force the mycelium downward and promote healthy sclerotia formation—just as the mycelium's behavior on the agar plates suggested. Using this technique, every year we harvest about four hundred morels from our outdoor beds using a wild isolate of the tulip morel (*Morchella deliciosa*), a small species that is less reliant on host trees than the larger yellow or ash morels in our area.

CULTIVATING MORELS INDOORS

Although few people have succeeded in year-round indoor commercial morel production, the good news is that over the past few decades there has been a lot of progress in the scientific understanding and experimental trials. I have never felt comfortable making claims about how close we are to consistently fruiting morels indoors, except to tell excited growers that I feel like we are "on the edge of completing a process." What I mean is that we—meaning everyone who is currently researching or experimenting with this—can learn from the many different successes and failures and try to find the hints, clues, and least common denominators that are embedded in all of the research and experimentation. The key is sharing information, which many growers and researchers are hesitant to do unless they have protected their particular unique component of the process by patenting—there is just too much money at stake here, so many advances remain as guarded as, well, wild morel patches. But without a full body of research and trials it is difficult to isolate a particular series of triggers that stimulate morel primordia to initiate and develop into mature fruitbodies.

Outdoor cultivation is simpler than indoor cultivation (though still not easy) in part because the "key players" are most likely to be in the soil, where the morel mycelium is free to wander and locate the nutrients and microbial partners it needs to eventually fruit. But for those of you who dream about growing morels indoors, I have outlined a method that has worked for other commercial morel growers, with a few minor changes. At our farm we are currently investigating these growing conditions and techniques in the hope of fruiting our first indoor morels cultured from native strains. I believe that the techniques developed in this book, such as microbial slurries added to casing soils, will be a critical component in solving the morel cultivation equation, both indoors and out.

There are basically two methods researchers use for indoor morel cultivation. The first method is derived from a patent filed in 1986 by Ron Ower (U.S. patent no. 4,594,809), which involves harvesting, chilling, and then replanting the sclerotia in a nutrient-deprived medium, such as sand and peat moss, then flooding it with cold water and allowing it to percolate slowly, as air temperature in the growing room is slowly raised to mimic natural growing conditions. The second involves flooding the sclerotia with a microbial slurry while they are in bags or trays, and then draining them to a level just below the casing soil, or at the

interface of nutritive and nonnutritive, without any physical separation whatsoever, which is the method I am developing for our farm. Although I have not yet fruited morels indoors with this method, every trial has resulted in abundant sclerotia, which shows me that I am on the right track—just a few factors such as chill hours and microbial populations remain for me (and you) to solve. My technique is experimental, of course. Every strain has unique nutrient, metabolic, and environmental requirements, so there is really no one-size-fits-all method for perfect morel cultivation. Since most claims about fruiting morels indoors are just that—claims—I offer this protocol only as a collection of evidence based on existing patents, observational evidence from my laboratory experience, and principles taken from years of fruiting mushrooms. I simply want to offer as much insight and evidence as possible to advance our understanding.

My advice is to experiment and work toward developing small, incremental changes in this system, in an effort to carry the body of knowledge forward. Remember that this is much, much more difficult than fruiting oysters. All the stars have to align for a cultivator to succeed. If you find that you are producing morels consistently indoors on a year-round basis, congratulations, you can retire. Needless to say, I advise anyone reading this to wade into commercial production with extreme caution. Millions of dollars in research have been spent on unlocking this perfected system and many have failed. Enthusiasts have filed patents with the hope of making millions of dollars a year, only to have their system fail early in large-scale production trials. Documenting the precise environmental parameters and formulas and, most importantly, maintaining the health and vigor of the original morel culture and soil microbes will be your main focus; otherwise the biological aspect of the process shuts down.

A main difference between my procedure and the other published or patented procedures is the lack of physical separation of the nutritive and nonnutritive layers. I leave the sclerotia intact in the top layer of the casing soil, instead of removing them from the casing soil as in Ower's patented method, where the entire casing including the sclerotia is removed and replanted. To me this seems quite unnatural. Leaving the sclerotia intact removes a few steps in the process, making it less laborious. The nutritive and the nonnutritive layers aerobically separate when you flood and freeze them, submerging the entire biomass for the recommended number of chill hours. The floodwater, which eventually drops to just below the nonnutritive casing, renders the nutritive layer anaerobic. This method results in aerobic mycelium above and waterlogged anaerobic biomass below, with communication and interplay between the two, resulting in the same biochemical triggers and associations found in the morel's natural habitats.

Phase 1: Screening and Pairing Isolates with Microbial Partners

First you need to isolate a pure culture of the morel strain you collected, along with a slurry of the soil you collected with it. Once you have isolated the morel culture, you can transfer it to grain or sawdust for storage and use as inoculum. Spore germinations and tissue cultures (using tissues taken directly from the interior of the morel fruitbody) are the best methods for establishing pure inoculum. It is critical to keep the culture and growing media as pure as possible in this first phase. If you have success, it's wise to maintain your morel culture in pure form for the inoculation of your substrates, but also harvest the stem bases of fruitbodies and capture the associated microbial populations. Some cultivators hybridize the outdoor slurry method with the indoor preparation parameters rather than using pure cultures for the morel inoculum, but without using a pure culture and a pasteurized substrate you always run the risk of introducing contaminants such as the mycoparasite *Trichoderma,* which can quickly spread through an indoor operation.

Since this method is largely experimental, you should build mini trials into your process in order to gather as much information as possible. Use replicates of three to five identical fruiting substrate formulas and spore dilutions to decipher the combinations that work and the components that are functional versus

nonfunctional. (Refer to chapter 20 for discussion of dilution and testing techniques for wild soil samples or stem base isolations.)

Phase 2: Preparing and Inoculating the Containers

This stage of indoor growing resembles the outdoor method, in that it requires the growing media to be layered and it's best to use horizontal fruiting containers. Morels will not fruit en masse in vertical containers like you would use for oyster mushrooms. You're going to need to sterilize the growing media. If your containers are heat-tolerant, you can fill them with the growing medium and then sterilize in an autoclave. If they're not heat-tolerant, you can sanitize the containers, sterilize the two types of growing medium (nutritive and nonnutritive), and then add the media in sequence to the containers.

Use opaque containers, since sclerotia are light-sensitive and form more readily in the dark. If you like, you can also use one clear bag, or "spy bag," that you can see into to help you gauge the morels' progress, allowing you to orchestrate your interventions in an efficient manner. You'll fill each container with a nutritive layer at a depth of 3 to 4 inches and then top it with a nonnutritive layer at about half that depth.

Adding the Nutritive Layer

Almost all nutritive layer formulas that have been published or patented describe a seed or grain mixture. The first published patent on morel cultivation described an annual ryegrass seed mixture for this layer, although most cereal grains have produced abundant sclerotia in my lab. You could use any smaller grains or particles such as ryegrass seed, millet, or wheat bran, mixing them into a volume of sawdust, preferably harvested from the same species of host tree from which the morel strain was originally isolated. Since morel mycelium is extremely fragile and fine compared to other edible mushroom mycelium, and it does not built a dense mat on its own, it needs a high-density growing medium. The sawdust particles, if fine enough, fill in the gaps between the larger particles of grains, building density for the mycelium.

I further supplement the sawdust with small percentages of sugars and protein such as soy meal, which has a particle size and density that morels seem to like. Adding a small amount of sugar can help the mycelium feed rapidly once it reaches the nutritive area, giving it the boost of energy it will need to begin decomposing and feeding on the sawdust and supplements.

Check the nutritive medium's pH; it should be neutral (7) to slightly alkaline (up to 8.1), and corresponding as closely as possible to the pH of the native soil from which you collected the morel. Add small amounts of gypsum or agricultural lime to adjust the pH as needed.

Hydrate the nutritive medium to between 45 and 50 percent moisture content. For small quantities you can submerge the entire mixture in water and then let it drain for a few hours, until a handful of the mixture produces only a few drops, not a stream, when squeezed tightly. For larger quantities I recommend soaking and mixing the nutritive medium in drainable tubs, opening the drain valve to release the excess water, then using a press to squeeze out the moisture to achieve the desired moisture level.

Mark a depth of 3 to 4 inches inside each container. Fill the containers to that mark with the moist nutritive medium. Compact it lightly and level the surface. Your containers are now ready for the nonnutritive layer.

Adding the Nonnutritive Casing Layer

The term *nonnutritive* implies exactly what it means: depleted of organic food sources and essentially lacking all the nutrients that the morel mycelium will be looking for. You can use many materials for this casing layer; it just has to hold moisture, have a fine particulate composition, and be pH-neutral (7 to 8.1). As with the nutritive layer, the density of the nonnutritive layer is critical. Avoid materials with large particulates, which can create air pockets and cause the mycelium to struggle in forming intricate connections with its surrounding biomass. Fine sand is an excellent density-building addition, which helps sclerotia formation, and also aids with drainage, with its nonnutritive properties and lack of moisture-holding ability, which becomes an important factor during phase 6 (thawing and draining).

A lot of people use a nonnutritive casing formula similar to what you would use with *Agaricus* species, without any added supplements. It is a mixture of peat moss, sand, and lime or fire ash. If peat moss is not available or you would prefer not to use it, you can substitute a more sustainable material such as coconut coir, which is sold in compressed bricks. My basic formula for the nonnutritive layer is 75 percent peat or coconut coir and 25 percent sand. I soak the mixture in water (as described above for the nutritive medium) and let it drain, until squeezing a handful of the mixture produces only a few drops, not a stream. I use agricultural lime to adjust the pH as needed to keep it between 7 and 8.1 and some gypsum to aid in particle separation.

Spread the nonnutritive casing soil over the nutritive layer evenly, at about half the depth of the nutritive layer. Compress the casing soil lightly and evenly to eliminate air pockets and areas of varying density.

Inoculating the Containers

Once you have filled your containers with sterilized media, it is time to add the morel spawn. Only a small amount is needed. If you are using grain spawn, which is more nutritive than sawdust spawn, about one grain for every square inch is more than enough. (As a side note, for morel grain spawn I use larger grains such as wheat or rye berries.) If you are using sawdust spawn, you can broadcast it a little more generously across the surface as a very light dusting.

Once you have spawned the containers, you should cover them and place them in a warm, high-humidity environment for optimal colonization and nutrient transfer.

Phase 3: Colonization

Colonization lasts for two to three weeks. During that time, keep the containers at around 70 to 80°F (21–27°C), and the humidity at around 90 to 95 percent. Do not let the containers dry out; moisture fuels the internal hydrology of the mycelium, allowing it to divide and explore. Without this precious resource, the mycelium is at risk of drying out and failing to reach the nutritive layer, so mist or lightly water as necessary to keep the surface of the casing soil and spawn fuzzy and active.

The containers should be kept in complete darkness, although to be sure the spawn is leaping off, you can shine a light on them to inspect the surface for a slightly fuzzy appearance. Remember, however, that all of the colonizing activity occurs underground, and exposing your containers to too much or even periodic light can disrupt their metabolic triggers, making them shift gears midcycle. To prevent this, light in wavelengths that may reduce photosensitivity (black or red, for example). After about one week the entire container should be lightly colonized to the bottom. At that point supercolonization of the substrate becomes very evident, with much more visible fuzzy, brown growth, similar to animal fur. The hyphae are now feeding voraciously, awaiting environmental cues to trigger the production of sclerotia in the casing layer. Keep an eye on the containers to monitor their progress and prepare for the next phase. Timing is critical.

Phase 4: Sclerotia Formation

After two to three weeks, when supercolonization is complete, drop the temperature of the room in which you're incubating the containers to 55 to 65°F (13–18°C) to signal to the morel mycelium that cold weather is imminent. Maintain high humidity conditions. During this phase, the hyphae channel nutrients into the upper casing soil, streaming as much energy as possible up to the oxygen interface to strategically position its reserves for the fruiting cycle. Sclerotia that form deep in the casing soil, rather than on its surface, may indicate that the casing layer is too dry or has desiccated since application, so be sure to maintain extremely high humidity during this phase.

Sclerotia are hard and vary in color from species to species, from whitish to cinnamon to dark brownish black. Their size is also variable depending on the strain and techniques you use. These nodules are fat reserves, storing the lipids that will fuel the development of the bulk of the fruitbody. Once the sclerotia have formed and then begin to slow in development, usually after a period of two to three weeks, the cultures are ready for the next phase, a soil drench and freeze to simulate winter floods and a massive drop in temperature. Consider easing the temperature down as the morels approach the next phase, mimicking dropping fall temperatures.

Phase 5: Microbial Flood and Freeze

Now comes the time for the soil sample that you collected along with the morel strain you are now

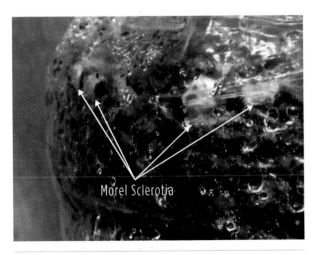

A close-up of morel sclerotia, which are much larger on sawdust than on agar and typically form where the nonnutritive layer is inoculated, activated by light and oxygen.

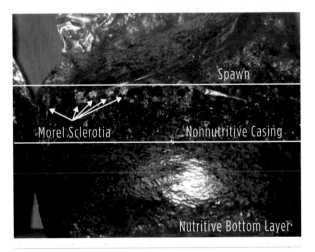

Our indoor trials of casing formulas and media treatments. These bags were flooded days later and frozen for several days to achieve the specific number of chill hours needed for the strain based on climate data.

cultivating. Thaw it completely and dilute it in water to a volume sufficient to flood the containers. You'll need approximately ¼ teaspoon of soil per gallon of water for this slurry. Add enough water to each container to completely submerge the soil in the slurry, and let it sit for twenty-four hours at room temperature to allow the morel mycelium time to interact with the soil community. Then freeze the containers for the number of chill hours that is needed for the strain you have collected to crystallize the matrix. This data can be collected from weather stations that have probes at depth intervals, and you are looking for a total number of hours below freezing at a depth between 2 to 4 inches for the winter season in your region. The freezing process not only signals to the mycelium that winter has arrived but also plays an as-yet unknown role in microbial soil ecology, possibly killing or injuring a portion of the microflora, making them available for the morel mycelium to feed on or providing some sort of environmental trigger. The length of time below freezing temperatures of 32°F (0°C) may also be proportional to temperatures below this mark, meaning less hours are needed to accomplish the chill hour requirement. Targeting this requirement will prove to be critical in a production process and also lower costs associated with chilling equipment to avoid unnecessary overexposures if they are not needed. This stage will require no light since the media are frozen, nor humidity if the blocks or trays are kept in an enclosed unit. Lids on the trays can be used to decrease the dehydration of the frozen blocks since no gas exchange is occurring while the blocks are frozen.

Sclerotial Chill Hours

17 to 21 days (400 to 500 chill hours) for cold strains
13 to 17 days (300 to 400 chill hours) for intermediate strains
9 to 13 days (200 to 300 chill hours) for warm strains

Phase 6: Thaw and Drain

Now that the sclerotia and microbial soup have crystallized together, you'll be "separating" the layers

(making one aerobic and the other anaerobic) by thawing and draining the containers down to the level where the nonnutritive casing layer and nutritive layer meet. This stage will last approximately one week, and during this time you should keep the temperature at 45 to 50°F (7–10°C) and expose the containers to indirect light. Switch out solid lids for similar covers fitted with small airholes that will allow for a gradual acclimation to the gas exchange, while also maintaining a high humidity inside the container environment.

If you are using disposable or pliable containers, poke a number of small holes at the interface between the nutritive and nonnutritive layers to allow for the slow and gradual drainage of water from the casing layer. If you are using solid trays, I recommend popping out the frozen blocks and transferring them into identical trays that have had holes predrilled in their sides at the appropriate height (making this technique ideal for mass production).

Allowing the casing layer to drain while maintaining the nutritive layer in a waterlogged anaerobic state re-creates the morel mycelium's natural habitat in springtime, without subjecting it to the abrupt physical removal of the entire casing layer, as other methods and patents have suggested doing. Morel mycelium is quite active in aqueous solutions at cooler temperatures, which may suggest the bottom layer's significance in facilitating communication and transport throughout the entire mycelial system. The mycelium in the aerobic nonnutritive layer can transport oxygen down to the submerged hyphae in the aerobic layer, keeping the bulk of that biomass still in the metabolic loop, while that biomass is still capable of sending nutrients up to the aerobic layer. With the slightly warming temperatures and the separation of the soil layers into anaerobic and aerobic states, the morel mycelium and sclerotia feel safe (or threatened enough) enough to devote their energy into fruitbody formation.

Phase 7: Fruitbody Formation

Raising the temperatures to 50 to 55°F (10–13°C) in the final phase is critical for rapid fruitbody formation. Morels are generally triggered to fruit when soil temperatures reach 48°F (9°C) or higher, so air temperatures slightly higher than that allow for the gradual warming of the substrate in the containers. Keep humidity at a maximum; a fog-like environment is ideal. But be careful not to overwater or supersaturate the developing mushrooms, which can cause bacteria blotch and rot. Artificial light should be subtle, natural fluorescent or blue-green LED, to mimic the gentle bathing of reflected light on the forest floor, as these primordia are forming under and in between layers of leaves and clumps of grasses. Light will need to be increased over the period of ascocarp maturation, simulating the availability as the mushrooms mature and enlarge into and toward the light source.

Morels mature over a period of seven to ten days during this phase. Observe their formation to determine whether they have the biomass they need to optimize fruiting (or suffer a complete fruit failure due to lack of nutrients). Based on your observations and the yields from this phase, you can recalculate your formulas, container dimensions, ratios of casing to nutritive layers, and any environmental parameters to try to improve the outcome in your next batch.

Few have tried and succeeded at indoor morel cultivation, but you could use this chapter as a foundation for getting started on trials that could eventually provide some critical information or variables that will improve the process to make it commercially viable. Given the complexity of the morel life cycle, and particularly the need to freeze morels to fulfill a calculated chill hour requirement, commercial cultivation may simply not be as profitable as we hope for. Unless we are able to isolate low-chill-hour morels to reduce the time and energy needed to complete the cycle, we could be stuck hunting them in the wild or simply planting outdoor patches that can be harvested seasonally. I don't mind that at all as a consolation prize. I hope this chapter serves you well and keeps your gears turning and your morel cultivation dreams alive and hopeful.

CHAPTER 22

Introduction to Mycoremediation

The term *mycoremediation* can be broken down as *myco* (fungus) and *remediation* (to clean, resolve, or correct), and indeed, mycoremediation is the use of fungi, specifically mushrooms, for creating simple yet effective biomass capable of breaking down environmental and industrial pollutants. The mycelium is a sort of self-healing filter that targets specific organic compounds and pollutants. Research has proven the efficacy of using fungi to degrade contamination such as PCBs, aromatic hydrocarbons, and oil spills. Biological pollutants, especially *E. coli*, have been of special interest in recent years, and a wealth of data now supports the benefits of mycoremediation in reducing or eliminating such pathogenic organisms.

Mycoremediation can be used clean polluted soils and water, and it can also be used prepare sites for potential contamination, such runoff from proposed farming or industrial sites. However, it should not be used as a license to pollute, and you should never assume that a mycoremediation project will always be successful; as with all cultivation efforts, it is your own responsibility to practice common sense as you experiment with these ideas. I am confident that mushrooms are going to make a difference in our sustainable future. I hope this chapter will spark your interest so that we can all build on the collective knowledge to solve our environmental problems in an inexpensive, efficient, sustainable way.

How It Works

In nature, mushrooms are decomposers and constructors, the agents of habitat renewal. With their enzyme systems sweating into the environment, fungi swim through their own fluids, or waste, breaking down materials into smaller molecular units that they can absorb through their cell walls and use as a food source. Fungi need to interact with their environment for a period of time (called the factor time) in order for the chemical reactions necessary to mycoremediation to take place. (This is also how disinfectants work.)

Mycoremediation can address two main forms of pollution: microbial contaminants and chemical contaminants. In microbial mycoremediation, the main mode of "removal" of a pathogenic microorganism from a system is by inactivation, by disrupting replication or reproduction, preventing cell synthesis and division, and eroding cell membranes. Fungi can also enforce species exclusion (gobbling up territory in advance to prevent entry by pathogenic organisms), or they may alter chemical gradients such as pH or the availability of nutrient resources needed by other

Olga found this strain of oyster mushroom (*Pleurotus ostreatus*) on our way to dinner one night in town. Here it is being "trained" to remediate the herbicide atrazine at concentrations three times the labeled application rate. The fungi stopped its linear growth after two days of getting a "taste" of its new food source.

The same plate eight days later, showing the mycelium's ability to adapt its enzyme production, rapidly shifting genetic expression to initiate the disassembly of a previously unencountered chemical compound.

organisms. Microbial mycoremediation is especially useful in cases of fecal coliforms in water: soil, manures and other livestock waste, and failing septic systems. With chemical remediation, the fungi use many types of enzymes, secreted into the environment, to cleave the molecules of a chemical compound into smaller units, making them more easily degraded by other organisms in a species sequence approach. (But just because the molecules are smaller doesn't make them any less dangerous; testing is always critical to confirm contamination removal.) Chemical remediation can be used against herbicides, pesticides, fertilizers, dyes, many other kinds of chemical residues.

Many fungi also absorb and retain specific elements, namely heavy metals, in their biomass, a process called hyperaccumulation. This process makes the mushrooms themselves toxic, but the biomass can be used to filter or treat soil, water, or air and is then removed and disposed of or concentrated for reuse.

All fungi are tooled to produce a different narrow or wide spectrum of "chemical keys" that they use for specific functions when they encounter a substance in nature. If you think of mushrooms like factories, consider that some factories are small and others large, some primitive and others high-tech, some produce in small volume and others in large volume, and at the end of the day they all produce a different product.

Fungi vary in their ability to produce cellulases, ligninases, peroxidases, and laccases, which are all used in combination to attack and decompose woody plant material, primarily lignin, cellulose, and hemicellulose. Mushrooms can be generally divided into two categories based on their main mode of decomposing these materials: brown rot and white rot. Brown rot fungi tend to be more aggressive in breaking down cellulose. White rot species, on the other hand, are highly effective at breaking down lignin and any complex polymers architecturally similar to lignin, such as many of the chlorinated compounds found in pesticides. The beauty of fungal enzymes is that they are cell-free, or extracellular chemical catalysts. Fungi secrete them out into the environment, and they begin their metabolic work well ahead of the mycelium, and even ahead of the advancing hyphae, so that the fungi

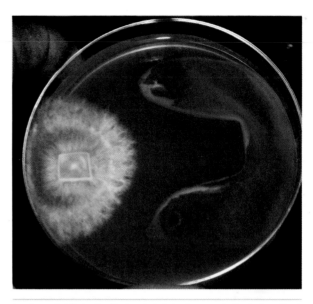

A culture of shiitake (*Lentinula edodes*) chasing away two colonies of bacteria. Creating galleries to screen fungi for their ability to inhibit pathogenic organisms is a good starting point for any full-scale mycoremediation effort.

are beginning to digest their food before they are even in contact with it. Bacteria, by contrast, must come into intimate contact with a compound in order to begin to degrade it.

While a fungus breaks down complex compounds into more simple ones that it can digest, those metabolites that it does not digest can be utilized by associated microorganisms, such as bacteria, which degrade the compounds even further, with the end goal of recycling the elements back into a self-perpetuating food chain. Fungi may play some role in regulating the balance of microorganisms in soil or water, enacting the rise and fall of certain populations to achieve a tolerable level of harmony. All organisms are opportunistic in some way, whether passively or aggressively, and mushrooms are masters of taking charge in uncharted territories.

SIMPLE SMALL-SCALE MYCOREMEDIATION PROJECTS

There are many simple mycoremediation projects that you can undertake on a small scale, even at home,

Using mycoremediation barriers to manage site runoff from open areas, parks, and playgrounds can be a simple and sustainable way to reduce erosion and improve local water quality.

and that can have a profound effect with just a few easy-to-follow steps. For example, with myceliated wood chips or bales of wheat straw, you can improve the water-holding capacity of the soil and reduce erosion due to rainfall or water drainage on a sloped area. If you have pets who relieve themselves in your yard, place a barrier of myceliated wood chips at the farthest downside portion of your property; the rain will wash fecal coliforms into the fungal filter, where they will be trapped and remediated by the mycelium before passing into the local watershed. Have fun experimenting with these simple, small-scale projects, but practice common sense as you do it; don't assume that any contaminant is remediated just because you've run it through a fungal filter.

Dog and Cat Waste

According to the Humane Society of the United States, there are an estimated 86.4 million domestic cats in the United States, or one for every third household—which translates to a lot of cat litter. Litter that is made of compressed newspaper, wheat gluten, or finely powdered corncob debris will work, wheat and corn being the best because they clump, rather than just absorbing urine, which makes it easier to scoop out the soiled litter. Although pine-based litters are fairly incompatible with most cultivated fungi, newer fungal isolates (of *Neolentinus lepideus*) that will be suitable for use with pine are being developed and will likely be available in a few years. In general, the best species for remediating feline feces are king stropharia and oyster mushrooms.

Scoop the soiled litter into a bucket with a lid to prevent odors from escaping, and periodically cover it with clean, wet biodegradable litter that has been inoculated with fresh spawn. The mycelium will thread its way into the substrate, decomposing and deodorizing the soiled litter in as little as twelve weeks. Run this

Spent myceliated sawdust—here with *Lentinus strigosus*—can be used for any pet litter. The fungus helps absorb urine, and when the litter is removed and placed into a bin, any bacteria and parasites are remediated by the fungus.

myceliated litter through a vermicomposting cycle next; don't mix it into your regular garden compost until it has been thoroughly decomposed by the worms.

You can also use spent mushroom substrate from primary decomposers such as oyster mushrooms as your litter; allow it to dry, and then shred it to make a fluffy bedding for the litter box, on its own or mixed in with the existing litter. When you are ready to change the litter, dump it into a bucket and place it outdoors in a well-ventilated spot for a couple of weeks to reduce the ammonia. Add just enough warm water to the litter to rehydrate the substrate, and the mycelium will soon come back to life and consume the substrate and waste. After it has fully colonized, you can vermicompost the mixture. Do not mix it into your garden compost until it has been thoroughly decomposed by worms.

There are about 74.3 million dogs in the United States according to a 2013 SPCA assessment, and—not to blame everything on cats—they produce a lot

of waste, which often ends up in small plastic bags bound for a landfill or even polluting local waterways, as runoff from sidewalks and yards. One way to handle dog waste is similar to the cat litter remedy described above—designate a 5-gallon bucket with a lid for collection, and periodically add a spawned substrate to it, using moistened hardwood sawdust, wood chips, or spent coffee grounds. You can even use wet layers of newspaper, lightly sprinkled with your spawn. Occasionally pick up and drop the bucket onto the ground from a few inches high to tamp the mixture and increase the density and contact points for the mushroom mycelium. Once you've stopped adding materials, it will take a few weeks for the mycelium to fully colonize its contents, but then the bucket will become a firm block of mycelium, smelling sweet (not that you would want to put your face right in there, at least until the very end) rather than putrid. You can start a new waste bucket while the other one finishes colonizing. The finished mycelium cake can be removed—invert the bucket and gently tap its sides and bottom to release the biomass into a worm composting bin—in a shady location to decompose safely into rich topsoil and worm castings for your garden.

City Chicken Coops and Runs

Although chickens have long been a fixture on rural homesteads, urban chickens are gaining popularity and acceptance, allowing city and suburb dwellers to take advantage of the rewards of fresh and naturally raised eggs. There are a few obstacles to this, however, including odor control and the coliform bacteria that can accompany poultry droppings. So how can you plan to manage this when you are petitioning for the right to homestead a few chicks? Look no further than king stropharia.

Stropharia mycelium has a sweet smell; it's one of my favorites. I make a point to route visitors to our farm past our chicken fortress so they can see not only how we manage it but the complete lack of smell. In fact, a reporter recently published an article calling ours "the best-smelling chicken run" he'd ever been

Chickens can make a mess in a small space. Luckily mushrooms are here to keep the coop clean and smelling sweet.

around, a great compliment to our mushroom mycelium's hard work. And of course mushrooms are great at using the excess nitrates and phosphates that accumulate in poultry waste, attracting earthworms and other organisms that collaborate to build an incredibly rich and sweet-smelling soil amendment. Our living fungal filters are easy to install and typically last about a year, at which point they are a wonderful garden amendment filled with rich worm castings that can be used to build soil and fertilize garden plants.

If you're keeping chickens and you'd like to set up a similar fungal filter to remediate your poultry waste and build a great soil amendment, follow the steps below.

Getting Ready

- Measure the area of your coop and/or run, and calculate the volume of wood chips you will need to cover that area, assuming a depth of 8 inches. When you measure the length and width of the coop, give yourself an extra foot on each side. For example, a coop that is 7 feet by 9 feet should be calculated at 9 feet by 11 feet, or 99. Multiply that by 8 inches (0.75 foot) and you get 74.25 cubic feet. Divide cubic feet by 27 and you have cubic yards, which is the unit of measure most mulch and wood chip companies use to sell their product. For this example, 74.25 divided by 27 gives you 2.75 cubic yards. Go ahead and round up to 3 cubic yards so you can have a little extra. You can always spread the chips a little deeper.

To build the fungal filter, you'll need the following:

- **Hardwood chips.** Make sure your chips are super fresh (harvested within the past four weeks). Be sure to get only hardwood chips, without any conifer mixed in. Power-line companies, college campuses, and cities sometimes stockpile chips to make mulch from their own tree trimming crews. Some even deliver! Get a tarp ready and tip or reward the delivery guys.
- **Poultry or chicken wire.** This wire mesh is used to separate the base layer from the upper layer of

chips, allowing the chickens to scratch about 2 inches deep, but not all the way down to the point where they'd destroy the mycelial biomass. You'll need enough poultry wire to cover the entire area of the coop or run.

- **King stropharia (*Stropharia rugoso-annulata*) sawdust spawn.** Available from a mushroom spawn supply company, or make your own (see chapter 18). A 5-pound bag will cover about 1 cubic yard of chips.
- **Garden tools.** You'll need a wheelbarrow, a shovel, a rake, a hoe, and other garden tools to move, dig, mix, and level your wood chip installation.

Installing the Fungal Filter

You can either dig down or build up, depending on your situation. In my opinion raised beds are easier to install, maintain, and clean out. Transfer the birds to another area so you can work without scaring them or allowing them to get involved, as they tend to get in the way. Rake out the coop or run and lay down a layer of wood chips about 1 inch deep. Massage your

bag of sawdust spawn so it is broken up well. Then sprinkle the spawn over the surface of the chips and rake it in gently. Repeat, until the inoculated wood chips have reached a depth about 2 inches less than what you want in total. Now install a poultry wire over the entire surface of the chicken enclosure. This will keep the chickens from digging down too deep and destroying the mycelium, but allows them to scratch and harvest worms in the top 2 inches. Cover the wire with another 2 inches of wood chips and spawn. You'll also need to install and inoculate a 1-foot-wide wood chip bed around the perimeter of the enclosure. This bed will act as a barrier to catch any excess runoff from the chicken enclosure. You can either mound up the wood chips against the enclosure wire or fencing, or build an enclosed bed with untreated landscape timbers or even mushroom fruiting logs (with shiitakes, reishi, et cetera) to make a medicinal mushroom landscape.

The only thing you really need to remember is that the mycelium needs a little moisture to spread properly and colonize. For the first month or so after inoculating the wood chips, it would be a good idea to

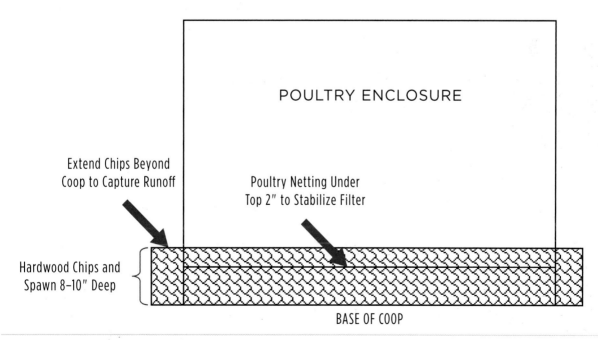

Cross section of the fungal filter base. The hardwood chips should be deep enough to handle the waste above. The netting a couple of inches below the surface keeps the chickens from scratching down into the mycelium.

gently water the bed every few days if it has not rained or if the area is under cover. This will also wash some of the poultry droppings deeper into the chips to feed to the waiting mycelium below.

Every year it is a good idea to start over and make the filter anew. The wood chips will have decomposed mostly into a rich humus and will not be able to filter bacteria and odors as well as they did when they were fresh. Locate an edge of the buried chicken wire and lift, pulling and raking the chips/debris into a pile. Examine the chips underneath and harvest any patches that are fully colonized with white mycelium by placing them into buckets (you can use this as spawn for your new wood chip bed if you have enough). Then remove all the wood chips/compost to prepare the enclosure for a new wood chip bed. In the fall, use a flat shovel and wheelbarrow to carry the compost to a finishing area, where it rests for the winter to allow for the mycelium and worms to complete the remediation process. Finally we bring it to our vegetable garden to top it off every spring. The compost is rich and absolutely beautiful, full of worms and priceless castings, just what the garden is begging for. You'll notice the rewards by the abundant yields of veggies and fruit that you receive each year. I hope you and your chickens enjoy the coop upgrade and mushroom fruitings as much as ours "doo."

Cardboard Manure Capture

Inside a chicken coop you may notice a generous buildup of odiferous chicken litter directly under the roosting platform. To put mushrooms to work in this situation, simply place strips of cardboard under the roosting area to catch the litter. Allow the litter to collect for anywhere from a few days to a few weeks, depending on the number of chickens and the accumulation. Removing this sheet of cardboard is much easier than raking or scraping it out with a flat shovel

We use cardboard just under the roosts to capture the falling manure. It makes cleaning the coop a five-minute task. We then inoculate the cardboard with spawn, layer it, and leave it in the woods, covered with wood chips, to remediate the bacteria and produce mushrooms.

and carting the loose material away for composting. I used to sprinkle shredded wheat straw directly under the roost, which helps clump the litter together for easier cleaning, but the cardboard is much easier and dust-free! Once removed from the coop, the litter-covered cardboard makes an excellent substrate for mushrooms. You can use it in a wood chip mushroom cultivation bed (see chapter 5) or on its own (see chapter 10). A very easy method of remediating this litter-covered cardboard is to inoculate the layers with king stropharia spawn, watering lightly and stacking them into a tight tower, then covering it loosely with a tarp or plastic to maintain humidity and offer a little gas exchange for the mycelium to colonize it and consume the entire biomass. Once the tower has fully colonized, you can use it to line your coop, garden, or other areas in the yard where you plan to add wood chips, using the remediated cardboard as spawn.

Spent golden oyster mushroom substrate piled in our composting area makes great inoculum and biomass for creating biological filters and for use as a topdressing for contaminated-soil sites.

These techniques can be used for many kinds of domesticated animals that are kept in enclosures. Rabbits, birds, reptiles, gerbils, et cetera, all share one thing in common: What goes in must come out. The solution here is the same as for chickens—position cardboard at the animals' preferred litter site, or install a wood chip bed below their enclosure, and use fungi to remediate the litter, filtering out potential pathogens and preventing runoff into local waterways.

EVALUATING SPECIES FOR MYCOREMEDIATION POTENTIAL

Not all mushrooms are created equal, and even different strains of the same species can exhibit profound differences that enable them to undertake particular activities in their unique environment. Some mushrooms are extremely "picky eaters," while others, like oysters, will devour just about anything you give them. Many can be trained to adapt to different environmental conditions, or to remediate high levels of certain contaminants, even beyond the thresholds they normally tolerate in their native environment.

To find the more talented fungal strain that will do best for a proposed mycoremediation project, you first must determine the field conditions of your project. What are the native soil or water conditions, in terms of pH, nutrients, soil density, force of the water flow, and so on? Will any contaminants be present? Are you proposing microbial remediation or chemical remediation? Or do you simply want to reduce soil erosion?

Once you've identified the scope of your project, you can select a panel of likely fungal candidates to see which seems most suitable.

Testing Species Against the Contaminants

If you're proposing remediation of a known biological or chemical contaminant, you'll want to culture a fungal strain known to be active against that contaminant. Test a series of tissue cultures from different strains in agar mixed with a trace amount of the

Debris and trash can be used for transforming devastated habitats into environmental edens using mushrooms native to the area. Here a local wood and leaf litter drop-off site is composting wood chips back into soil, but it could easily be retrofitted to include mycelium for expanding the fungal biomass needed for mycoremediation projects.

contaminant. Of those cultures that appear to do well in the contaminated agar, attempt to refine their ability by scaling up to more concentrated formulas. Keep records, and keep your protocols consistent for all the strains that you test.

Another option would be to perform spore germinations on agar that has been mixed with trace amounts of the contaminant. Since spores are genetically variable, some may be better suited to tolerate the contaminant, and you should be able to identify them on the plate as they develop and outcompete their "cousin" strains.

Consult part 4 for help choosing mushroom species for your given climate and target pollutant. Be sure your mushroom also matches the growing substrate that you have available for expanding the biomass efficiently. You may have to match up a mushroom with the media in order to achieve your remediation goals.

Matching Species with the Site Conditions

The site of every mycoremediation project is unique, with varying climate, hydrology, and naturally occurring microbial ecology, which makes setting up a small-scale trial critical for determining how much biomass of what particular fungal species will be needed.

To begin, you'll need to be sure the fungus is compatible with the substrate. If you intend for the fungus to proliferate in the ground, it will need to be compatible with native environmental and soil conditions. If you intend to construct a freestanding filter unit, you'll need to be sure that the fungus will thrive in the substrate you intend to use in the filter—or, conversely, that the substrate your chosen fungus prefers will contribute to, rather than be detrimental to, the filtering process.

For example, if you're building a water filtration unit comprising several sequential modules of fungal biomass, with the aim of cleaning the water of a particular contaminant, you'll need to determine how much contact time the fungus needs with the targeted contaminant to inactivate or degrade it. The particle size of the growing medium becomes extremely important, since large particles will allow the passage of water quickly, reducing contact time, while finer particles can bind together tightly, preventing water flow altogether. You'll need to experiment with the particle size and perform simple drainage tests as part of design process for the filtration unit.

Likewise, all mushroom species bond their biomass in different strengths and densities. Mycelial tenacity,

A homemade tabletop mycofiltration unit.

We have designed a simple tabletop mycofiltration unit that can be built for about $30, using a three-tiered, plastic module with a filter to constantly recirculate the water. This unit can give you a wealth of information about your proposed mycoremediation before you commit to any on-site work. It will help you collect data about such factors as the effectiveness of the mushroom species you propose to use, the volume of biomass needed, the optimal rate of water flow, and probable installation costs. (Such data can, coincidentally, help you write a successful grant proposal.)

Constructing a Test Unit

Most mycoremediation projects employ some sort of water mycofiltration or enzyme harvesting system, since contaminants can be treated directly in water or percolated into soils. The simplest filtration unit is one that uses the force of gravity to feed water through it. If you're constructing a prototype with more than one unit, set them up so that they are stacked, or so each is downhill of the previous unit, to maximize efficiency.

Building a test unit will give you valuable information that will guide and shape the design of your in-field installation. By constructing a test unit, you can determine the following:

- **Rate of flow.** The rate of flow determines how much contact time the mycelium will have with the contaminant. With the contact time required being a known quantity (something you will have established in your earlier testing), you can adjust the rate of flow in various ways to optimize contact time, such as the volume of fungal biomass, the substrate, or the pitch of the filtration unit. Or you can adjust the number of modules in the unit to increase or decrease the total contact time.
- **Biomass needed.** The amount of fungal biomass you need will be based on the mushroom's efficiency in degrading or inactivating the contaminant, the contact time, the rate of flow, and the concentration of the contaminant.
- **Efficiency of the fungal species.** Though you'll have tested various fungi for their activity against the contaminant, a prototype allows you to test

as you might call it, becomes an important variable to match up with substrate density. Mushrooms that intensely bind organic matter, such as polypores like reishi (*Ganoderma lucidum*), turkey tail (*Trametes versicolor*), and multicolor gill polypore (*Lenzites betulina*), can be of value when you're dealing with strong water flows or you need to provide structural support to a biofiltration medium. Conversely, water processing units that are overly slowed by dense mycelium run the risk of overflowing and distributing unfiltered water down the line.

CONSTRUCTING SMALL-SCALE PROTOTYPES

Developing prototypes or small pilot studies is essential for any proposed mycoremediation project. Since every remediation site is unique, we cannot discuss the specifics here, but in general, your strategy should be to develop the best method on a small scale to test whether the treatment you propose will work. In a laboratory setting, you can do this with test tubes and flasks; for in-the-field testing, you can begin by building a tabletop unit that simulates the site conditions and remediation processes.

Building a tabletop mycofiltration unit.

them under in-the-field conditions. You may test multiple species or strains with protoypes, or you may use a species sequence approach, with multiple filtration units in one prototype, each with a different species or strain.

You can easily measure the rate of flow with a large graduated cylinder or flow meter. Start by running the system without the contaminant present. The rate of flow out of the system must equal the rate of flow in. If the biomass and/or substrate offer a great degree of resistance to the flow, decrease the density of the filter by adding a substrate with larger particles, such as volcanic rock or sand. If the water flows through the unit so quickly that the mycelium doesn't have

adequate contact time with the contaminant, you can add filtration units to the system, so that the water is cycled through multiple filters.

The amount of fungal biomass needed to adequately remove the contaminant from the water is determined by the rate of flow and the fungus's efficiency. Run water with measured amounts of the targeted contaminant (in concentrations equivalent to in-the-field conditions) through the filtration system, and collect and test the effluent. Calculating the effluent's post-filtration contaminant levels, the time of exposure to the fungal filter, and the total volume of fungal biomass in the filter can give you some idea of the quantity, in volume, of the biomass needed for the full-scale installation. (You'll need to test the water continuously to determine the life of the filter to estimate when it must be replaced, since the biomass will eventually become exhausted.) To calculate the volume of the biomass, use water displacement measurements: Submerge the biomass in a netted bag in a measured quantity of water in a graduated container of some sort. The volume of water displaced by the biomass is equal to the volume of the biomass. You can also weigh a a small sample of the biomass, measure its water displacement in a small container, and then weigh the entire bulk of biomass and calculate its total volume using the ratio of weight to volume you established for the smaller sample.

Once the prototype filtration system is up and running, testing the effluent will determine the rate of filtration, or efficiency, in either chemical or biological units, depending on the contaminant, for an established rate of flow and contaminant concentration. The system should be tested continuously, and effluent samples should be collected and tested at regular intervals, from the first flush of contaminated water.

Your end goal will be unique to your remediation project, of course. But for guidance on the total maximum daily load limits for various biological and chemical contaminants, consult the EPA, which sets forth standards as directed under the Clean Water Act.

Using Your Prototype to Harvest Enzymes

A tabletop filtration unit such as a prototype for calculating the efficiency of certain strains of mushrooms in removing or biodegrading pollutants can also be used to capture enzymes by "washing mycelium." This setup is most suitable for commercial or laboratory applications, where the entire process can be carefully controlled in a sterile environment. It can provide a method for concentrating specific enzymes by exposing mycelium to a particular chemical or biological contaminant.

As we've discussed earlier, fungi are able to produce a wide spectrum of enzymes and other metabolites, and they can retool their production of metabolites in the presence of new environmental factors, such

as chemical compounds or biological entities. These metabolites can take the form of novel antibiotics that defend the fungi from other microbes, or chemoattractants to attract microbes that can assist in a complex degradation process. Dosing fungal filtration units with a known compound, whether chemical or biological, allows researchers to coax fungi into responding, metabolically speaking, to these compounds. To dose the biomass and harvest metabolic products you will need the mycelium and the biological or chemical trigger to come into contact, yet avoid the washing or introduction of this contaminant into the harvester unit. To do this simply insert small sterile, open-ended tubing into the biomass, accessible from the exterior, typically at an angle that can be capped, where solutions can be added to infuse into the block of mycelium. Allow the tube to sit for one week in the mycelial block before adding a solution to allow the mycelium to heal and seal off any holes in the filter and make it watertight. If done correctly the solution, once added, should not flow, but rather sit and maintain a level throughout the experiment. Biological or chemical contaminants can be used to elicit a metabolic response that can be harvested in the form of novel antibiotics or enzymes specific to the mechanisms needed in response to the stimulus.

Basic Tests for Biological and Chemical Contaminants

How do you know whether a mycoremediation project is working? You can rely on expensive laboratory equipment to test the results of any mycoremediation effort, but I prefer the holistic approach, using organisms such as beans, worms, and minnows, which have extremely low thresholds for toxin exposure. These methods are low-tech but highly accurate in signaling toxicity. They are also low-cost and accessible, requiring no special equipment, and appropriate for everyone from young students and home cultivators to commercial operators and college professors demonstrating the basic concepts of mycoremediation.

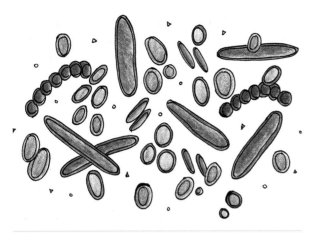

Gram stains can help identify unknown bacteria by sorting them into classes based on cell membrane composition, shape, and metabolic functions. It's important to identify and quantify contamination before, during, and after a mycoremediation experiment to monitor the ongoing efficiency of your filtration system.

Testing Mycoremediation Results: Biological Contaminants

Cell Counting

Cell counting is an important tool for tracking the growth rates of organisms such as bacteria. Determining the initial cell counts in a mycofiltration experiment is critical to calculating the efficiency of the filter. Most labs would prepare a stock solution the day before doing a cell count, allowing the bacteria to run through a rapid log phase (exponential growth), followed by a stationary phase (leveling off).

Step 1. Perform a gram stain test on a small sample of your bacteria to help you validate the identity and purity of the culture. Gram stain kits can be purchased from many science supply stores and are relatively inexpensive (under $30).

Step 2. The bacteria are best counted using dilutions of the bacterial culture that can be spread across an agar plate and incubated overnight, resulting in individual colonies that you can actually count. Use the dilution-to-extinction method described in chapter 20 to prepare a series of dilutions of your bacterial culture. Prepare three plates from each dilution, and incubate them overnight.

Step 3. Bacterial colonies should form on the plates, appearing mostly as clear, yellow, or rose-colored dots on the surface that if left to develop will radiate outward over the next few days. In lesser dilutions you may see a proliferation of overlapping colonies; those won't be helpful in counting. Instead, take note of those plates on which there are thirty or fewer colonies. Record the dilution rate for those plates, so that in the future, when you want to count cells, you can dilute directly at that ratio, without having to go through the whole dilution-to-extinction series.

Step 4. Count the colonies on the selected plates, and calculate the average on the three plates for each dilution. Multiply that number by the inverse of the dilution to calculate the number of bacteria cells in a given volume of the original culture.

I have also been using the EasyGel coliform detection kits, since they do not have to be autoclaved and you can perform the test anywhere. The coliforms are conveniently stained in different colors depending on type and are easily counted. The kits are inexpensive and make the identification of coliforms without the use of microscopes very simple.

You will want to calculate bacterial loads by counting cells before and during the mycoremediation process at specific time intervals according to your project design. Since bacteria are prolific—many, such as *E. coli*, can double in population every 20 minutes at their optimum incubation temperatures—you may choose to pull samples at initiation, thirty minutes, one hour, three hours, six hours, twelve hours, and every six hours thereafter until the experiment has run its course. Once you've pulled a sample, dilute it to the appropriate level (which you will have determined in your original cell counting), plate it, and allow it to incubate overnight. Perform another gram stain test on the bacteria and examine the morphology of the specimen under a microscope to make sure you are counting and following the pathogen of interest.

Testing Mycoremediation Results: Chemical Contaminants

For a low-tech but effective test for chemical contaminants, there are two good options: the "bean test" and the "worm assay." Anyone can perform these tests, at home, in school, or in a cultivation operation.

The Bean Test

For this test, you'll need at least 70 bean seeds of known viability and germination potential, preferably from the same stock and not from different sources or suppliers. You'll also need fourteen small jars or large test tubes. Be sure to wear gloves and eye protection for this entire process, since you will be using herbicides or other chemicals that can be harmful. The containers should be large enough to accommodate the beans after they have been soaked and begun to swell, so small-diameter test tubes may be a problem.

Step 1. You will be making twelve serial dilutions, with one control and one full-strength solution. Label one container "control," the next 0, and the rest 1 through 12. If you're using test tubes, set them in a rack or jar to keep them upright.

Step 2. Add five beans to each container.

Step 3. Add 10 ml of distilled water to the control, making sure the beans are at least partially submerged.

Step 4. Add 9 ml of distilled water to each of the remaining containers.

Step 5. Prepare a 1 percent solution of the chemical compound you are testing. You only need to make enough to cover the beans in container 0. Add enough of the solution to container 0 to cover the beans. Shake well or vortex to distribute the solution.

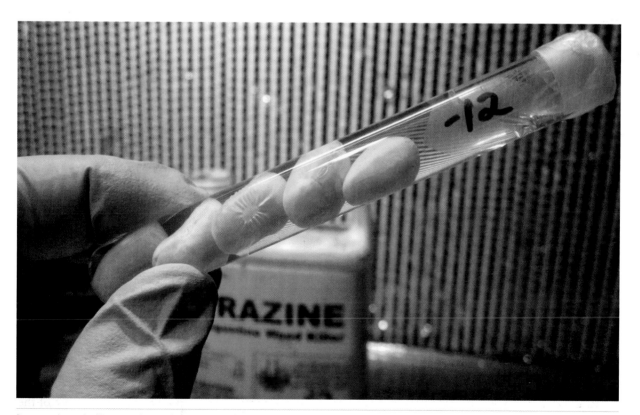

Beans can be soaked in contaminant solutions of varying concentration and then planted to determine the safe thresholds. Similar tests can be conducted using water from a mycofilter.

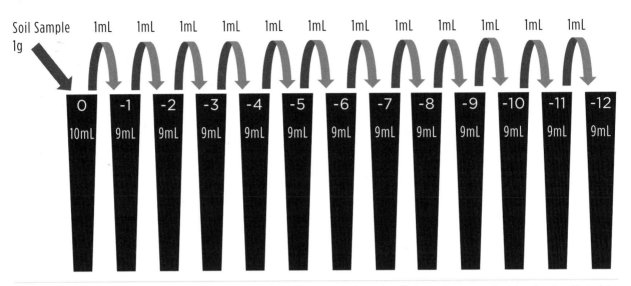

Dilution to extinction prepares samples of a contaminant in solution in progressively greater dilutions, allowing you to determine the effect of the contaminant at different concentrations.

Step 6. Using a sterile pipette, transfer 1 ml from container 0 to container 1. Shake or vortex well to distribute the chemical.

Step 7. Repeat the process, transferring 1 ml from container 1 to container 2, and so on, until you have reached container 12.

Step 8. The original stock solution in container 0 has a strength of 1 percent, meaning it contains one part per hundred. Since you tranferred one-tenth of the solution to every container down the line, each dilution is magnified by a factor of ten, giving you one part per thousand (container 1), one part per ten thousand (container 2), one part per hundred thousand (container 3), and so on. Calculate and record this information for each container.

Step 9. Soak the beans for 24 hours. This will completely hydrate the seeds with the dilutions you have created, along with the varying levels of herbicides. Strain them and save the water in its appropriate vial to be used for watering the seedlings as they sprout.

Step 10. You'll want to plant the five beans from each dilution group together, but separate from the beans of the other dilution groups. So you'll need fourteen good-size containers, sterile soil to fill them, and a growing area where light and temperature conditions are conducive to bean sprouting and growth. Be sure to label the containers with their dilution number. Plant the beans in the sterilized soil, and water them with a dropperful or two (being consistent for all the containers) of the dilution solution you saved from their soaking container.

Step 11. In a few days seedlings will begin to emerge (or not). Inspect all those that do emerge for any damage: abnormal leaf tissue, color variations, wilting, or any factor that differs from the appearance and vigor of the control seedlings. Over the next few days and even weeks, record the sprouts' rates of growth and general appearance and health. (At the conclusion of the experiment, unearth the seeds that did not emerge to see if any of them managed to open.) You may wish to photograph the sprouts as they grow to provide you with visual reference for your records.

Your goal here is to draw a correlation between the concentration of the contaminant in the soaking and watering solution and the appearance and growth of the bean seeds. The level that can generally be labeled "safe" is the highest concentration at which bean

Red wigglers enjoying their new chemical-free home courtesy of oyster mushroom mycelium that had been eating motor oil for the past three months.

seeds grow normally, behaving in all manner the same as the control group.

Water harvested from a mycofiltration system can then be run through the bean test without dilution and the results compared against the spectrum of response you have established with this original bean test. This should allow you to determine with general accuracy the concentration of contaminants in the water, and thus the efficiency of the mycofiltration system.

If you are attempting to strip herbicides out of a water source, you may want to consider administering the dilution as a foliar spray rather than a seed soak and drench. You can also do a similar bean test for contaminated soils or composts, where the persistent herbicides can take months and sometimes years to degrade to safe levels suitable for gardening. This test is as simple as planting herbicide-sensitive seeds, such as beans, corn, and tomatoes, into your soil or compost sample (try this before you buy bulk composted manure) and watering to see if the plants are stunted or have foliage damage, such as twisted or burnt leaves. Many organic gardeners relying on manures for their fertilizing have been devastated by total crop loss due to persistent herbicides in composted manures, which even after the forage or hay has been digested and decomposes the herbicides can persist for four to six

years (see Chlopyralid). Many countries are attempting to ban or inform the public on the use of these herbicides and the dangers of introducing them into other streams of gardening or industry before they have a chance to break down.

The Worm Assay

Worm assays can be used to determine levels of soil toxicity, where worms are added to soil samples with incrementally increasing degrees of contamination. Worm assays can provide information that bean tests do not, since worms are highly sensitive to a broad range of hydrocarbons and other toxins that beans may be able to withstand.

Your goal here is to determine the maximum levels that the worms can survive without sickness or injury. You're creating visual and written records here for your own use, but worm toxicity information is actually available for chemicals on the manufacturer's product information materials or under the EPA registration documents (which are available online). Locate the worm toxicity information before beginning your worm assay, as a complement to your experimentation.

Step 1. Using sanitized plastic cups that will hold approximately 1 cup (250 ml), fitted with lids perforated for gas exchange, fill containers with collected soil samples moistened with distilled or deionized water. If you are creating an experiment using a contaminant, hydrate the soil with a solution based on parts per million dilution that you are testing based on the active ingredient or chemical compound of interest. Samples should be moist and not wet to the point of dripping or leaching water. Be sure to prepare cups for controls with no contamination present. A good benchtop replicate number is typically five replicates of each test (five controls, five herbicide 10ppm, five herbicide 100ppm, et cetera).

Step 2. You'll need twenty worms per sample assay (either *Eisenia fetida* or *Lumbricus terrestris*). Use cultivated and not wild-harvested worms, all of the same species and approximate size. Inspect the worms and record their original color, reaction to light (will they dig when place on wet soil?), and sensitivity to touch (healthy worms will twitch when touched in the middle). Add the worms to the soil samples, cover with a perforated lid (be sure the holes are small enough that the worms cannot escape), and incubate the cups at room temperature, between 65 and 75°F (18–24°C). A small light above helps keep the worms from trying to leave the container. Leave these containers alone, opening once daily to observe and record any noticeable changes in worm behavior, color, or mortality.

Step 3. The test will span fourteen days, of which you will have daily observations and notes. On the last day you will need to dissect the soil sample and sift through it to find and count the worms, again noting and recording their final color, reaction to light, and sensitivity to touch. Compare these results to the original observations and to the control worms placed in soil without chemical contamination.

As with the bean test, once you have determined the effect of the contaminant at different concentrations on worms, you can use that information as a reference point for evaluating contaminated soil that has been through a mycoremediation effort. Worms can also be analyzed for their properties, such as coelomic fluids and blood, to determine how and where the bioremediation or chemical tolerance took place, or not.

Follow Up with Professional Testing

Once you have begun the initial testing of a mycoremediation project, you might consider having the results—the remediated water or soil—analyzed by a professional lab to determine what has become of the pollutants. Did the mycoremediation enact a complete inactivation or degradation of the contaminant, or did the contaminant simply undergo biotransformation, converting different compounds? There's even a chance that the results could be worse than the original contamination—the compound could have biotransformed into a form more toxic than the original.

Any data you can collect about the action and effectiveness of your mycoremediation project, whether labwork, cell counting, bean tests, or worm assays, will be important, guiding you in recalibrating the next series of trials. Chase the results with good data, and you will be able to fine-tune the different aspects of the project up front, before performing a full-scale installation in the field, where more is at stake. Your success depends on good data.

PART IV

Meet the Cultivated Mushrooms

Introduction

There is no all-encompassing list of cultivated or edible mushrooms, since growing conditions, the availability of spawn, spores, or tissue for culturing, and even what people consider to edible can vary throughout the world. That said, I have tried to compile the most commonly cultivated mushrooms for several climates and regions, as well as species that I think people can, or soon will, have success with. (I have not included the very-difficult-to-cultivate morels (*Morchella* spp.) here since I dedicated a chapter to them in part 3.) If there is a mushroom you want to try to grow but it isn't included here, you can try applying the protocol for a mushroom within the same genus, or with similar climatic and habitat preferences. New mushroom species are being isolated and cultivated all the time from different regions around the world, with growers and researchers analyzing them for commercial potential in regard to improving yields and increasing their nutritional content. After all, all cultivated mushrooms started out in the wild, which means that it took growers years of trial and error to establish successful cultivation techniques for every mushroom you have ever heard of.

For medicinal properties and referencing, there are too many properties to list for each species, and I would strongly encourage pairing this growing guide with *The Fungal Pharmacy* by Robert Rogers. Rogers has included the medical referencing in extensive detail.

DIFFICULTY RANKING

Each species has its own unique challenges, potential yield, and cultivation preferences. For this reason it's difficult to offer a straightforward ranking of how easy or difficult it is to grow certain mushrooms. The ease of cultivation depends on many variables such as where you live, what potential substrate materials you have available, whether you'll grow indoors or out, and your experience. That said, some mushrooms are easier to fruit than others (compare oysters and morels, for example), so I've developed a simple ranking to help guide you toward species that match your skill level. But don't give up if you have failures. It may just take time to develop your skill and adjust your methods to get it just right. The ranking is broken down as follows:

1 — beginner (easy to fruit)
2 — intermediate (likely to fruit)
3 — advanced (difficult to fruit)
4 — expert (requires precise methodology to fruit)
5 — experimental (unlikely to fruit)

Since there are challenges to assigning a difficulty ranking—at a certain point it becomes arbitrary—I've opted to organize the profiles in this chapter alphabetically by genus, rather than by their difficulty ranking. Below, you'll find the ranking for each of the mushrooms profiled in this section, which should give you a sense of where to start. Flip directly to the full profile and you'll find much more information that will help you both decide what to grow and succeed at growing it.

Difficulty Ranking by Genus
Agaricus (portabella mushrooms and relatives):
    indoor 3 / outdoor 2
Agrocybe (black poplar): indoor 2 / outdoor 1
Auricularia (wood ear): indoor 2 / outdoor 1
Clitocybe (blewit): indoor 3 / outdoor 2
Coprinus (shaggy mane): indoor 3 / outdoor 1
Fistulina (beefsteak): indoor 4 / outdoor 2
Flammulina (enoki, velvet foot): indoor 2 / outdoor 1

Fomes, Fomitopsis, and *Laricifomes*
 (iceman polypore): indoor 4 / outdoor 3
Ganoderma (reishi and other varnished polypores):
 indoor 2 / outdoor 1
Grifola (maitake, hen of the woods):
 indoor 4 / outdoor 2
Hericium (lion's mane, pom-poms):
 indoor 2 / outdoor 2
Hypholoma (brick top): indoor 2 / outdoor 1
Hypsizygus (elm oyster, shimeji): indoor 1 / outdoor 1
Laetiporus (chicken of the woods): indoor 5 / outdoor 3
Lentinula (shiitake): indoor 3 / outdoor 1

Macrocybe and *Calocybe* (giant macrocybe,
 giant milky): indoor 3 / outdoor 2*
Macrolepiota and *Lepiota* (parasol):
 indoor 4 / outdoor 2
Pholiota (nameko): indoor 2 / outdoor 1
Piptoporus (birch polypore): indoor 5 / outdoor 3
Pleurotus (oyster mushrooms): indoor 1 / outdoor 1
Sparassis (cauliflower): indoor 4 / outdoor 3
Stropharia (king stropharia, garden giant, wine cap):
 indoor 2 / outdoor 1
Trametes (turkey tail): indoor 2 / outdoor 1
Volvariella (paddy straw): indoor 3 / outdoor 2

---

* Calocybe is probably more in the range of indoor 2 / outdoor 1.

The Genus *Agaricus*

Common Species

A. augustus (the prince)
A. bisporus (white button, portabella, and crimini)
A. bitorquis, A. campestris (forest mushroom, field mushroom, pink bottom, rose bottom)
A. blazei, A. brasiliensis, A. subrufescens (almond-scented agaric, almond portabella, royal sun agaric)

Difficulty Level

Outdoor cultivation—2
Indoor cultivation—3

General Description and Ecology

Agaricus species are common worldwide, growing naturally on composted livestock manure, grassy fields, wood chip piles, and forest leaf litter. The species are usually specific to their food sources, with very little adaptability to different substrates. The most recognizable of all the mushrooms in this genus are the white button and portabellas (*Agaricus bisporus*), which are the same species, as is the cleverly marketed "crimini"—white button is harvested young, portabellas later, and crimini during the second flush when mushrooms are smaller. The almond portabella, or royal sun agaric, is a tasty and meaty button mushroom with the smell and sweet flavor of almond extract or anise. It is highly sought after but rarely cultivated on a large scale, due to the lack of availability of spawn and the higher fruiting temperature it requires—making the almond portabella an ideal candidate for smaller-scale outdoor cultivation in more temperate or tropical climates. In temperate climates, the royal sun agaric has a unique affinity for compost piles and can naturalize easily if compost is added to its bed periodically throughout the year, and the heat generated by the composting process creates a safe haven for mycelium during the colder months.

All mushrooms in this genus function primarily as secondary decomposers; they are not very efficient at lignin degradation, so it can be beneficial to grow a primary decomposer such as oyster mushrooms first, and then almond portabellas on the spent media, after the oysters have broken down the lignin. (This also works well with paddy straw mushroom, another secondary decomposer.) *Agaricus* mushrooms require interaction with the microbial community in their natural environment to initiate primordia formation and fruitbody development, and so they require a microbe-enriched casing soil to be applied after you have colonized a fruiting substrate.

Mycelium and Spawn

The mycelium is white and linear, with irregular margins and cottony in pure culture but rhizomorphic when it collects microbial partners, and exudes a light yellow metabolite on overcolonized spawn or fruiting substrates. The mycelium's rate of growth is much slower than that of other genera, such as *Pleurotus* (oyster) and *Volvariella* (paddy straw). The fragrance of the mycelium is specific to the species, ranging from musty to sweet, like almond extract. Commercial spawn is normally available in the form of cereal grains or supplemented sawdusts; grain is the carrier of choice for most commercial cultivators. Sawdust spawn is slow to leap off unless it is supplemented with wheat bran or cereal grains but tends to be more useful for home cultivators who are unable to completely pasteurize their compost, in which case insects can eat up the grain spawn before the mycelium has a chance to spread through the growing substrate.

Fruitbody Development

The cap is centered on the stem and has free gills (not attached to the stem) that are pinkish white at first,

A close-up of an almond portabella from an outdoor patch spawned at Sharondale Farm in Virginia, showing its beautiful veil and pink gills. This mushroom is a member of a large genus whose members share similar growing conditions and are cultivated primarily on composted livestock manure or aged sawdust. *Photograph by Mark Jones, Sharondale Farm.*

darkening at maturity as spores develop and stain the gill blades. The cap is white, covered with a fine matting of fibrils, and the stems of a few species stain bright yellow when cut or bruised. The spore print is chocolate black.

Common Strains and Ideal Fruiting Conditions

Agaricus mushrooms are terrestrial, needing to grow from a horizontal growing medium, such as prepared beds, and are also dependent on microbial interactions for fruiting. Fruiting temperature varies among species, as does the preferred growing substrate, so choose a species that fits your climate and production needs. Strains grown indoors generally follow a well-orchestrated timing of higher spawn run temperatures, a lower "middle" temperature phase to initiate primordia following colonization, and a slightly higher temperature to promote mushroom formation and maturation. Outdoor bed cultivation is best accomplished by choosing the strain that matches your climate and substrate availability, since you are basically relinquishing environmental controls. Outdoor beds generally produce two flushes a few weeks apart, depending on the temperature fluctuations in your area, which could shorten or lengthen the fruiting cycles, making the harvests much more unpredictable from a timing standpoint than in indoor, climate-controlled cultivation, where fruiting occurs much sooner and in more regular and predictable cycles.

- **The prince (*A. augustus*).** The optimal spawn run temperature is 70 to 80°F (21–27°C). The optimal fruiting temperature is 75 to 85°F (24–29°C).
- **White button, portabella, and crimini (*A. bisporus*).** The optimal spawn run temperature is 70 to 80°F (21–27°C). The optimal fruiting temperature is 60 to 65°F (16–18°C).
- **Forest mushroom, field mushroom, pink bottom, and rose bottom (*A. bitorquis, A. campestris,* and related ecotypes).** The optimal spawn run temperature is 70 to 80°F (21–27°C). The optimal fruiting temperature is 60 to 65°F (16–18°C). *A. bitorquis* is the most heat-tolerant commercial strain;

its optimal spawn run temperature is 77 to 91°F (25–33°C), and its optimal fruiting temperature is 75 to 91°F (24–33°C). A good candidate over *A. bisporus* for tropical climates
- **Almond-scented agaric, almond portabella, and royal sun agaric (*A. blazei, A. brasiliensis, A. subrufescens*).** The optimal spawn run temperature is 70 to 80°F (21–27°C). The optimal fruiting temperature is 75 to 85°F (24–29°C). These species form deeper under the compost surface than most other agarics, and many growers are now using a perforated plastic barrier placed strategically in the upper third of the casing to reduce the sites where primordia can form, thus increasing yields by forcing the fungi to direct their energy to these specific fruiting spots.

Wild Spawn Expansion Techniques

A compost-loving mushroom, agaricus relies on soil microbes to initiate and stimulate the fruiting cycle, making it a difficult genus to perpetuate using natural expansion methods. You can transplant the rhizomorphic stem bases or harvest plugs of colonized compost to start new beds (though multiple transfers will result in strain degeneration). Likewise, you can use the stem bases to make a slurry to mix into fresh compost, bringing all the key microbial players needed to initiate mushroom formation. I recommend storing the stem bases under refrigeration until you need them to make microbial slurries to seed the casing soil.

Laboratory Isolation and Spawn Culture

This entire genus is slow to grow in culture. I prefer to cultivate it on potato dextrose yeast agar, which is created simply by adding 2 to 3 grams of yeast per liter of boiled potato water, then mixing with agar. The pH of the agar mixture should be closer to neutral (7.0) for agaricus species (other mushroom species generally prefer a pH of 5.8 to 6.2). Because it grows so slowly on agar, I prefer to blend the petri cultures with sterilized water and add the liquid culture directly to sterilized grain. Grain spawn is enhanced by the addition of calcium, in the form of limestone or gypsum powder.

Store tissue cultures and spore germinations of tropical strains at room temperature or frozen in a glycerol solution for prolonged storage, since refrigeration lasting several weeks can kill the mycelium. *A. blazei, A. brasiliensis,* and *A. subrufescens* are extremely sensitive to prolonged refrigeration, so I would recommend keeping both petris and grain of these species out at room temperature. We keep them in our walk-in cooler entrance, which maintains an even 50 to 55°F (10–13°C) year-round, and subculture from these, even when they dry out. Dehydrating grain spawn also works well for storage of these tropical strains.

Preferred Fruiting Substrates

Agaricus species in general prefer composted mixtures of manure and cereal straws, with cotton hulls or any other agricultural by-products common in your area, such as bean hulls, cornstalks, cotton waste, beet pulp, and sugarcane bagasse. Almond portabellas (*A. blazei, A. brasiliensis, A. subrufescens*) prefer a complex media matrix over a basic one, so be sure to give them a nice variety of supplements to enhance the fruiting and development.

Casing Soil

A microbial casing soil is required. When the entire surface of the substrate, or just below if you scratch around in a few spots to check, is a continuous white mat, dress the bed with a microbial casing soil, which stimulates primordia formation (see chapter 20).

Outdoor Cultivation Notes

Media preparation. Follow the composting procedures discussed in chapter 6 to prepare the growing medium. Fill rows or raised beds approximately 8 to 10 inches deep in a shady location, and be sure to cover the beds with chicken wire or plywood, which helps deter digging animals. I use supplemented sawdust spawn outdoors to minimize the risk of bugs eating all the spawn before it has a chance to colonize. If you have low pest pressure, upgrade to grain spawn for increased yields. Dress with microbial casing soil just before or as soon as the substrate appears to be colonized.

Preferred spawn type. Sawdust, plug, cardboard, or grain spawn.

Inoculation to fruiting time. From 80 to 100 days, depending on temperature and inoculation rates.

Expected yields. Fair to very good. The genus of this mushroom is broad, and yields vary depending on species and growing substrate. Commercial spawn grown on supplemented composted manure-based substrates indoors is high-yielding, while most growers appreciate moderate to occasional harvests seasonally in their gardens and landscapes, typically fruiting 2 to 3 pounds of mushrooms per square foot for outdoor planted beds.

Indoor Cultivation Notes

Media preparation. Fill containers or trays with the growing medium 8 to 10 inches deep, poke narrow holes in a diamond pattern every 6 to 8 inches across the surface, and fill the holes with spawn. Cover the spawned holes with wet compost to keep the spawn from drying out and to keep insects from attacking it immediately. Cover the substrate with perforated plastic or with plywood that has had holes drilled through it to preserve moisture while still allowing for modest gas exchange. Dress with microbial casing soil just before or as soon as the substrate appears to be colonized.

Preferred spawn type. Grain.

Inoculation to fruiting time. From forty-five to sixty days, depending on temperature and inoculation rates.

Expected yields. Good. You could expect 5 to 7 pounds per square foot.

Harvesting

These mushrooms form singly or in small groups scattered about the casing surface. Many times a few clusters fuse just below their bases, rising from an underground mass of mycelium. Cut the mushroom as close as you can to the casing soil with a sharp knife, and trim the very base, where growing medium is attached, into a scrap bucket.

Storage

When fresh, these mushrooms will keep for up to two weeks at 38 to 42°F (3–6°C). They do not rehydrate well but can be dried and powdered. Agaricus mushrooms pickle well and can be canned for prolonged storage.

Marketing

This is a meaty, flavorful mushroom that is excellent when sautéed. This one needs no special help in marketing; have a few cooking at your booth for sampling, and the aromas are sure to lure a crowd. Common *Agaricus* species can sell for $3 to $5 per pound, while the less common (and in my opinion far superior) almond-flavored species can fetch $15 to $20 per pound—which can be a hard sell, so lay out the nutritional benefits (see below).

Nutritional Value and Medicinal Uses

Agaricus mushrooms have an average moisture content of around 93 percent. Common species, such as the white button and portabella, contain 25 to 33 percent protein (dry weight). The almond-flavored agarics are some of the most nutritional of all gilled mushrooms, with protein concentrations ranging from 35 to 49 percent (dry weight), depending on the strain and growing conditions, making it one of the most protein-rich of all cultivated mushrooms, comparable to the paddy straw (*Volvariella volvacea*). As is true for many medicinal mushrooms, studies suggest that many species of *Agaricus*, specifically the almond-scented ones, can help boost and modulate the immune system, regulate blood pressure and sugar levels, lower cholesterol, and provide cardiovascular and digestive support. They contain antiviral and antibacterial properties as well as anti-inflammatory compounds, and they also have been shown to suppress many cancer cell lines.

Uses in Mycoremediation

Mushrooms in the genus *Agaricus* have been shown to hyperaccumulate cadmium, copper, and zinc. The almond-flavored agarics in particular also contain high levels of antibiotic properties.

The Genus *Agrocybe*

Common Species

A. aegerita, A. cylindracea, and *A. praecox* (black poplar, pioppino, and swordbelt)

Difficulty Level

Outdoor cultivation—1
Indoor cultivation—2

General Description and Ecology

The black poplar mushroom, commonly sold in Europe and now the United States under the name pioppino, is relatively easy to grow on hardwood substrates. Though it often grows in the wild on poplar trees, it has a broad spectrum of hosts, making its range much wider than its common name would suggest. My only warning for anyone growing this mushroom is that it is a little more difficult to identify than some of the other cultivated mushrooms, like shiitake and oyster. When you are cultivating *Agrocybe* species outdoors on logs or buried rounds, where uninvited species may fruit alongside your planted *Agrocybes*, you will have to take care to distinguish which species you are harvesting. I recommend thoroughly studying and becoming confident in the identification of this genus before you eat it.

Agrocybe aegerita, black poplar mushrooms, fruiting from sterilized hardwood sawdust supplemented with wheat bran, showing their thin, fragile veils and the dark spore deposits that have landed on the shoulders of the stems.

Mycelium and Spawn

The mycelium is white and linear, exuding a light yellow metabolite on overcolonized spawn and fruiting substrates. Its smell can be sweet to slightly anise-like. Commercial spawn is usually available in the form of supplemented sawdusts and wooden dowels for outdoor cultivation. Grain spawn is more difficult to find, making lab culture helpful for cultivating this one commercially indoors on supplemented sawdust.

Fruitbody Development

The cap is centered on the stem and is attached to slightly descending gills that are smoky gray at first, darkening at maturity. The gills are never yellow or greenish, as they are in poisonous species of wood-inhabiting mushrooms like *Hypholoma fasciculare* (sulfur tufts), which are more brightly colored, with yellow to orange caps, compared to *Agrocybe*'s typically cream-colored to light brown caps. The spore print of *Agrocybe* is dark gray to black, as is that of *Hypholoma*, so be sure of your identification.

Common Strains and Ideal Fruiting Conditions

Strains are available from many spawn laboratories, or specimens can be cloned from identification tables at reputable foray events, such as those sponsored by the North American Mycological Association. Fruiting temperatures are on the cool side, between 50 and 65°F (10–19°C). Indoor fruitings are generally not as abundant as those of oyster or shiitake mushrooms, yet the flavor is distinctly nutty, making *Agrocybe* a good addition to your spawn collection and cultivation efforts. Outdoor fruitings generally produce two flushes a few weeks apart. *Agrocybe* typically fruits in the spring in temperate climates, so if you're cultivating indoors, it may benefit from a mild cold shock to initiate fruiting. If your containers are not fruiting, try placing them in a cooler environment just after colonization to allow them the chill hours they need to begin the fruiting cycle. Allowing the containers to rest at warmer temperatures for two weeks after harvest, then rehydrating them with cold water and returning them to the cooler environment, will stimulate their fruiting cycles more efficiently. Outdoors, soaking logs in ice-cold water can also trigger fruiting in temperate climates.

Wild Spawn Expansion Techniques

The black poplar is a hardwood-loving species. Since this mushroom fruits from stumps and logs but does not reproduce easily from stem base samples, you will need to pull back the bark and find some of the elongated stem base or colonized wood near the attachment point to have any luck harvesting viable mycelium for expansion. Try experimenting with natural cardboard expansion methods (see chapter 12) for preliminary isolations.

Laboratory Isolation and Spawn Culture

Spores of this genus germinate rapidly and outcompete most mold spores on agar, which makes taking a spore print a valuable way not only to identify this group, but to store spores from year to year. Adding screened sawdust, or fine particulates of the fruiting substrate or wood type, helps these spore germinations diverge into high-quality strains that will fruit well in your area. This group of species prefers malt extract agar over potato extract agar.

Preferred Fruiting Substrates

These mushrooms love hardwood but are omnivores and can be tried on any type of non-coniferous wood. For indoor cultivation I would recommend sterilized, supplemented sawdusts. This mushroom is not very fond of agricultural waste, such as wheat straw, cotton hulls, and other dried vegetable matter; to grow on these substrates, you'd have to inoculate heavily with grain spawn, which is not economically viable.

Outdoor Cultivation Notes

Media preparation. Fresh logs or stumps will do. Make sure logs or rounds are fresh (cut no more than one month before inoculation) to ensure the mushroom has a chance to outcompete any wild fungal species. *Agrocybe* mushrooms benefit from ground contact, so use a submerged log technique, covering them with wood chips. You can also grow them on wood chips alone, using supplemented sawdust spawn, perhaps in between rows of vegetables in the garden or along woodland pathways. You can even cultivate *Agrocybe* species outdoors on cardboard.

Preferred spawn type. Sawdust, plug, or cardboard spawn.

Inoculation to fruiting time. From eight to ten months, depending on temperature and inoculation rates. These mushrooms typically fruit in cooler weather, which is usually spring in temperate climates.

Expected yields. Good. *Agrocybe* mushrooms cluster nicely and can be harvested in tight bundles from outdoor fruitings when they are young. Expect two flushes a few weeks apart. The patch will keep fruiting for two to three years.

Indoor Cultivation Notes

Media preparation. Sterilized, supplemented sawdust is the ideal growing medium for *Agrocybe*, since it is very representative of the mushrooms' native substrate in terms of nutrition and density. Containers should be ready to fruit three to four weeks after inoculation. Tight clusters of mushrooms, rather than rangy groupings or single specimens, will have a longer shelf life, so if you're growing for market, encourage the primordia in this direction with narrow openings in the bag, rather than large exposed surfaces. Once colonization is complete, poking a few supplemental holes in the bag's extra overhead space can help trigger fruiting. Once primordia have formed, cut the bags and roll them down to just above the primordia, where the mushroom caps can flatten and mature immediately.

Preferred spawn type. Grain or supplemented sawdust spawn.

Inoculation to fruiting time. From four to six weeks, depending on temperature and inoculation rates. Incubating the bags in warmer temperatures during colonization and then moving them to a cooler environment for fruiting will trigger mushroom formation much sooner. After fruiting and a two-week rest period, return the bags to cooler temperatures for a period of time equal to the time from completed colonization to fruiting (typically a few weeks), to stimulate another flush.

Expected yields. Good. *Agrocybe* mushrooms cluster nicely and can be harvested in tight bundles from indoor fruitings when they are young. Expect two flushes a few weeks apart.

Harvesting

These fleshy mushrooms form singly on logs and in small clusters when grown in bags. Typically they are mature and ready to harvest five to eight days after emerging, depending on temperature and humidity. Cut any singletons at the base, or pull entire clusters out together, trimming off any wood debris or attached substrate before storing.

Storage

Fresh mushrooms will keep for up to two weeks at 38 to 42°F (3–6°C). When dried, the caps of these mushrooms rehydrate well, and the stems can be used to make powders.

Marketing

This is a crunchy, nutty mushroom that is excellent in stir-fries. The common name can vary depending on the venue; black poplar is more common, but pioppino is often used at upscale restaurants. Loose mushrooms are quickly damaged, so sell these in bundles wrapped in paper with just the caps exposed for viewing.

Nutritional Value and Medicinal Uses

Mushrooms in this genus are on average 27 percent (dry weight) protein. Studies have shown that *Agrocybe* species contain antiviral properties, for both human and plant viruses (tobacco mosaic), as well as antibacterial properties (against *E. coli*, *Streptococcus aureus*, *Bacillus cereus* and *subtilis*, and *Salmonella typhimurium*). Extracts of these fungi have high levels of anti-inflammatory compounds and have been shown to suppress many cancer cell lines, including some forms of leukemia. In Chinese medicine this mushroom is used to treat disorders associated with the spleen.

Uses in Mycoremediation

Mushrooms in this genus are known for their wide spectrum of activity based on laccase, an oxide reductase enzyme. They are known for their ability to break down endocrine disruptors, like BPA, and could be useful for mycofiltration projects.

The Genus *Auricularia*

Common Species

A. auricula, A. fuscosuccinea, A. polytricha (wood ear, tree ear)

Difficulty Level

Outdoor cultivation—1
Indoor cultivation—2

General Description and Ecology

Wood ears don't get the respect they deserve. They are thin and rubbery little mushrooms, crunchy and flavorless. So why do we grow them? Because this texture is desirable in many dishes, especially in the cuisine of Asia and the tropics, where these mushrooms thrive on many types of hardwood logs and stumps. (On a recent trip to Haiti, I found a few dried pieces of wood ear on a landscape timber made from a mango tree trunk.) The wood ear is a crunchy treat, raw or cooked.

Mycelium and Spawn

The mycelium is white and linear, growing brown with age, and it exudes a light yellow metabolite on overcolonized spawn and fruiting substrates. Its smell can be slightly unpleasant and musty. Commercial spawn is usually available in the form of cereal grains, supplemented sawdusts, and wooden dowels for outdoor cultivation. Grain spawn should be used exclusively for indoor cultivation, since many insects are attracted to and will eat it.

Fruitbody Development

This is a jelly fungus that is cup-shaped, or sometimes lobed. The mushrooms are gelatinous and rubbery, sometimes forming as firm, roundish nodules that appear hazy when they are young, then growing quickly and developing the characteristic cup shape, and becoming more clear and transparent in the process. The spore prints are white in this genus, and the spore are released from the incurved cup side of the fruitbody.

Common Strains and Ideal Fruiting Conditions

Fruiting temperatures for most strains are in the more temperate to tropical range. For outdoor growers, cultivating in the warmer months is fine, even with no insect protection, since these mushrooms are not highly sought after by insects, slugs, and other fungus-loving organisms. You can cultivate indoors year-round, at temperatures that depend on the strain but usually are in the range of 60 to 80°F (16–27°C). It's wise to rotate among strains with different temperature fruiting windows during the year in temperate climates to maximize your harvest periods and yields without having to invest a lot of energy to heat or cool a fruiting room.

Wild Spawn Expansion Techniques

When you find a wood ear that you want to collect, peel the bark away from the log in which the mushroom was growing, insert a small piece of wet cardboard, and replace the bark firmly. In a few days the mycelium can be seen leaping off onto the cardboard, and you can then transfer that mycelium to an agar plate or make layers of inoculated cardboard. You can also pull back the bark to find some of the whitish attachment points, or even chip away small chunks of colonized wood near the attachment point, for isolation. I use a sterilized dental pick to gently pull small pieces of mycelium from the attachment point, soak it in 3 percent hydrogen peroxide for several minutes, and then transfer the small bits to cardboard or antibiotic agar for culturing and sectoring.

Fresh wood ear mushrooms emerging from cracks and wounds in the bark of a log inoculated with sawdust plugs. The mushrooms form under the bark as whitish, rubbery nodules and turn brown once they steer their way out into the light. Wood ears can dry completely, rehydrate, and continue to grow and produce spores over several months.

Laboratory Isolation and Spawn Culture

Like all jelly fungi, wood ears are much more difficult to isolate directly from fruitbodes in sterile culture than fleshy gilled mushrooms are, since the rubbery bodies are thin and seem unwilling to provide mycelium without associated contaminants. I prefer to take spore prints from mature wood ears for use in germination or to use the primitive method listed above to at least get the fungus into a myceliated state. Once I notice mycelium spreading on wet cardboard, I then transfer those hyphal tips to antibiotic agar to clean up the culture, getting rid of hitchhiking molds or bacteria. Once you have the strain purified on plates, things proceed quickly; getting this culture onto grain spawn takes only a few weeks.

Preferred Fruiting Substrates

The wood ear is a hardwood omnivore and can be tried on any type of non-coniferous wood. For indoor cultivation I recommend sterilized, supplemented sawdusts. This mushroom likes to form on the open face of stumps, squeezing out of the space just under the bark surface, making half-buried rounds an excellent choice for outdoor cultivation. Since insect pressure on this mushroom is extremely low, transferring colonized bags of sawdust from indoor incubation to outdoor fruiting structures actually works just fine. This mushroom is not very fond of agricultural waste, such as wheat straw, cotton hulls, and other dried vegetable matter; to grow on these substrates, you'd have to inoculate heavily with grain spawn, which is not economically viable.

Outdoor Cultivation Notes

Media preparation. Fresh logs, stumps, or wood chips will do. Make sure logs or rounds are fresh (cut no more than one month before inoculation) to ensure the mushroom has a chance to outcompete any wild fungal species.

Preferred spawn type. Sawdust, plug, or cardboard spawn.

Inoculation to fruiting time. From ten to fourteen months, depending on temperature and inoculation rates.

Expected yields. Very low. Wood ears are not high yielding outdoors and are slow to mature compared to fleshy, gilled mushrooms.

Indoor Cultivation Notes

Media preparation. Use sterilized or super-pasteurized supplemented sawdust. The supplements of choice are wheat or rice bran, at 5 percent (dry weight) of the sawdust mixture. After three weeks of colonization, containers can be fruited indoors or out. Bags are best with this species. To initiate fruiting, make just a couple of long (3- to 4-inch) slices in the bag, either on opposite sides of the bag, or as an X on the side of the bag where you wish the mushrooms to form (such as if you are stacking the bags and creating aisles of fruiting walls, which makes harvesting easy). Prolonged pasteurization of wheat straw also has been shown to produce fruit from grain spawn, but the yields are not as high as those cultivations using supplemented sawdust formulas.

Preferred spawn type. Grain or supplemented sawdust spawn.

Inoculation to fruiting time. From fifty to sixty days. You can expect three or more flushes two to three weeks apart.

Expected yields. Up to 1 pound for every 5 pounds of supplemented sawdust substrate.

Harvesting

Wood ears attach firmly to logs, so you'll need to cut them free. Sawdust-grown mushrooms pull away readily, along with some of the growing substrate, which you'll need to trim away before storing the mushrooms.

Storage

Fresh wood ears will keep for up to three to four weeks at 38 to 42°F (3–6°C). They also dry well.

Marketing

These mushrooms ship well since they are virtually indestructible as long as they are not stored too wet under refrigeration. They also reconstitute from a dried, cracker-like state back into a fresh state in a matter of minutes if you soak them in warm water, and so drying them is a good option for extending their shelf life. Be sure to include recipes and demonstrate the rehydration method if you're selling wood ears at markets, since many consumers will not know how easy they are to use. A clever way to sell these, as I learned from Greg Carter at Deep Woods Mushrooms in North Carolina, is to soak dried wood ears in a flavorful broth, like teriyaki or any sauce you can fathom. The wood ears absorb the fluid; drying them then produces a crunchy flavorful cracker, a perfect trail food.

Nutritional Value and Medicinal Uses

The moisture content of *Auricularia* species is around 90 percent. The protein levels in dried wood ears range from 8 to 10 percent. These mushrooms are well known for their anticoagulant properties (they are commonly used as blood thinners) and anti-inflammatory actions. In homeopathy they are used to treat pancreatic imbalances.

Uses in Mycoremediation

Auricularia mushrooms may be best used for habitat renewal, particularly for devastated ecosystems that are rich in lignicolous debris, such as logged forests. Given their ability to thrive with more sun exposure than other wood-decomposing mushrooms, these fungi can be useful to help build soil in arid, sunny environments.

The Genus *Clitocybe*

Common Species

C. nuda (blewit, wood blewit)
C. saeva (blue legs)

Difficulty Level

Outdoor cultivation—2
Indoor cultivation—3

General Description and Ecology

The arrival of blewits in the fall and early winter means that we have had a cold snap, since these mushrooms are triggered by a heavy frost or freeze. It is unclear exactly what else triggers the fruiting response, although we do know that the mushrooms form complex alliances (or battle) with microbial

The beautiful blewit (*Clitocybe nuda*) loves to fruit in colder weather on composted manure and leaf litter and can grow quite large, sometimes hiding under the leaves, only visible as a bump or mound on the forest floor. The color quickly fades from purplish blue to tan or beige when exposed to light and as the mushroom expands.

communities in the soil. There's nothing like spotting a lone blewit as it lifts up the recently fallen leaves, revealing a sliver of purple or blue, then upon closer examination finding dozens more scattered about, visible only by as slight bumps and mounds on the leaf-littered forest floor. Once I've found one, I tend to kneel down and try to get closer to the ground to spot more. The large-diameter cap seems shy, preferring the carbon dioxide–rich microclimate it has created under its small bubble of leaves. Very few of these mushrooms make it out into the open air to sporulate, which may limit their spore dispersal; then again, perhaps they rely on other mechanisms of distribution such as insects. Blewits can resemble the blue species of *Cortinarius* mushrooms, some of which can be potentially toxic, but their spore print colors are remarkably different, so there should be absolutely no mistake here. Blewits produce creamy pink spores, while those of *Cortinarius* are rusty orange. I recommend thoroughly studying and becoming confident in the identification of this species before you eat it.

Mycelium and Spawn

Blewit mycelium is bluish and linear, and it does not exude many metabolites except in the presence of contamination or during colonization on pasteurized substrates. One native strain I isolated actually luminesces bright orange under black light—it was absolutely beautiful!

Fruitbody Development

This is a stemmed mushroom, with the cap positioned in the center and attached gills that generally lift and become wavy at maturity. The outer margin of the cap sometimes has a distinct color zone, a different shade of watermarked purple or cream when compared to the center of the top of the cap. This feature allows me to identify a blewit from yards away. The spore print is creamy pink.

Common Strains and Ideal Fruiting Conditions

Wild strains are common and should be tested for their fruiting capabilities; you'll need their stem bases

to make a microbial slurry to trigger the fruiting cycle. Since most strains require a significant cold shock in order to fruit, most cultivation is done outdoors in temperate climates and is reliant on the natural cycles. I typically do not see any blewits in our beds until temperatures have dropped below freezing for at least a few days. After the cold shock, if the temperature rises for just a few days above freezing this can trigger mushroom formation, bringing the blewits out in full force. With blewits, local strains are always better, so track some down and start cloning them with cardboard to make your own spawn.

Wild Spawn Expansion Techniques

Cardboard culture works well with these mushrooms; wrap up the purple stem bases into "blewit burritos" and place them in a plastic bag in the refrigerator for a few weeks. Blewits are pretty much indestructible. They can freeze solid, thaw, and then continue to mature, a quality that led me to invent a "blewit bomb": Place the stem bases, soil and all, in a blender with a little water, pulverize the mixture, pour into a ziplock bag, and freeze for a few days. Place the frozen chunk on top of a compost pile, where it will slowly melt, coating the compost beneath with not only blewit mycelium, but also possibly the microbial triggers that it will need later. When I did this myself, the following year a nice patch of blewits appeared just feet away from my compost pile. Blewits prefer primarily hardwood tree bark and leaf compost. I have seen massive fruitings on composted horse manure in a neighbor's yard—some of the largest blewits I have encountered. They clearly liked what they were getting from that pile. An intuitive cultivator will try to figure out and mimic the conditions responsible for a prolific fruiting like that, which appeared naturally on its own.

Laboratory Isolation and Spawn Culture

Blewits are easy to clone from tissue located inside the stem base, which is usually fluffy and sometimes hollow. Using a sterile scalpel or dental pick, pull out a small chunk of this mycelium, and flip it upside down onto an agar plate for best results. Spore germinations are possible but are prone to contamination. This mushroom's mycelium is slower than that of most cultivated mushrooms, meaning that you will need to make sure you use very good sterile technique and antibiotic agar media to give it enough time to generate enough mycelium on petris to transfer to grain. Blewit strains store well on colonized grains submerged in water.

Preferred Fruiting Substrates

Blewits prefer composted hardwood leaves or composted mixtures of manure and cereal straws or any other agricultural by-products common in your area, such as bean hulls, cornstalks, cotton waste, beet pulp, and sugarcane bagasse. These cold-weather fruiters enjoy a complex growing medium, such as a composted mix of hardwood leaves, twigs, and bark debris.

Casing Soil

A microbial casing soil is required. When the entire surface of the substrate, or just below if you scratch around in a few spots to check, is a continuous white mat, dress the bed with a microbial casing soil, which stimulates primordia formation (see chapter 20).

Outdoor Cultivation Notes

Media preparation. Use composted leaves, spent oyster mushroom substrate, composted hardwood sawdusts, or manure-rich composts of cereal straws with cotton hulls or other agricultural by-products. This mushroom prefers a complex matrix over a basic one, so be sure to give it a nice variety of media to support vigorous fruiting. Spread the compost in shallow rows or beds approximately 8 to 10 inches deep, allow the mycelium to colonize the substrate, and then dress it with microbial casing soil to stimulate primordia formation.

Preferred spawn type. Supplemented sawdust spawn.

Inoculation to fruiting time. The time to colonization will be from six to twelve months; the time to fruiting will depend on when the cold weather sets in.

Expected yields. Good, but sporadic. Expect 4 to 5 pounds for every 16 square feet of beds, prepared to a depth of 8 inches. You may need to replace the beds every two years. For more intensive production, replace the beds at the end of each winter to allow enough time for the new beds to establish themselves and fruit the following fall.

Indoor Cultivation Notes

Media preparation. Although blewits are typically cultivated outdoors, some people have reported success fruiting them indoors on composted mixtures of sawdust and rice bran or composted manure and straw formulas. My strategy is to maintain a sequenced flow of trays that have been colonized, cased with a microbe-rich solution from previous fruitings, and then frozen briefly before being brought to the fruiting room. Fill the trays with the growing medium 8 to 10 inches deep, poke narrow holes in a diamond pattern every 6 to 8 inches across the surface, and fill the holes with spawn. Cover the spawned holes with wet compost to keep the spawn from drying out and to keep insects from attacking it immediately. Cover the substrate with perforated plastic or with plywood that has had holes drilled through it to preserve moisture while still allowing for modest gas exchange. The cold shock at the end of the colonization cycle is important; blewits will not fruit without it. You might also consider adding beneficial nematodes as a food source for the mycelium; their presence may prompt the mycelium to react as it would in the wild, possibly triggering a feeding or fruiting cycle.

Preferred spawn type. Grain or supplemented sawdust spawn.

Inoculation to fruiting time. Once colonization is complete and a cold period is rendered, usually by refrigeration or flash freezing, the mycelium is triggered to fruit. A typical indoor fruiting cycle for blewits is to allow three to four weeks for colonization and casing, and an additional three weeks for cold shock treatments and mushroom maturation. Estimate about six weeks, minimum, for completing this cycle.

Expected yields. Low and unpredictable. More research on indoor cultivation is needed to improve the methods and yields suggested here. Expect 2 to 3 pounds of mushrooms from every square foot of fruiting area.

Harvesting

Blewits, picked indoors or out, form tight buttons that quickly enlarge—at which point they store poorly. Mushrooms that are large and firm, with their cap margins still rolled just slightly inward, toward the stem, are most desirable.

Storage

Firm, fresh mushrooms will keep for up to two weeks at 38 to 42°F (3–6°C). Dried blewits do not rehydrate well but can be used to make powders. Olga found that sautéing and then freezing these mushrooms works best for long-term storage.

Marketing

When I sell blewits, I describe them as "sweet and silky shiitakes." When sliced, the flesh is smooth, but sautéing it can make the mushroom unappealing to some people, since it tends to soften and becomes a bit slimy where the smooth layer cooks off. The solution is to add sautéed blewits to soups and gumbos, where they blend in perfectly with soft cubes of potato, leeks, and a splash of sherry.

Nutritional Value and Medicinal Uses

Blewit mushrooms are about 90 percent water and have a high protein content, in the range of 24 to 26 percent by dry weight. Blewits contain compounds that inhibit both gram-positive and gram-negative bacteria and have shown high activity against tuberculosis. *Clitocybe* species also contain antifungal and antibacterial compounds that have been shown to suppress the growth of *Candida albicans* and *Serratia aureus*.

Uses in Mycoremediation

Some studies suggest that extracts from the fruitbody of blewits can be used as a foliar bactericide and insecticide. Blewits are also well known for their ability to hyperaccumulate metals in their mycelium, making them well suited for mycofiltration of contaminated soil and water.

The Genus *Coprinus*

Common Species

C. comatus (shaggy mane, inky cap)

Difficulty Level

Outdoor cultivation—1
Indoor cultivation—3

General Description and Ecology

The shaggy mane is a cosmopolitan mushroom, gracing yards and lots all over the world. It even pops up in construction sites and other places you would think mushrooms simply couldn't grow. Shaggy manes are not highly prized due to the extremely brief existence of their fruiting bodies (one to two days at the most), but that's what makes them so intriguing to grow at home—if you are able to closely monitor their progress, your chances of actually seeing, picking, and eating them improve.

Mycelium and Spawn

The mycelium on agar is white and cottony, with irregular and asymmetrical growth, and may develop brownish tones as it ages. The smell of the mycelium and spawn is sweet and pleasant. Commercial spawn

Harvested shaggy mane mushrooms, which were cultivated on composted rabbit manure and wheat straw, spawned strategically beneath a raised outdoor rabbit cage. The mushrooms are ideal for consumption at this button stage; if not harvested, they can enlarge and liquefy in a matter of hours, so check your patches often.

is usually available in the form of cereal grains, supplemented sawdusts, and wooden dowels for outdoor cultivation.

Fruitbody Development

Shaggy mane is a stemmed mushroom, with the cap centered on the stem and free gills (not touching the stem). The mushroom resembles a drumstick popping out of the grass—it emerges from the soil with the entire cap covering the length of the stem. Fruitbodies develop relatively fast if the substrate is producing heat (as compost would). This mushroom can tolerate relatively high levels of ammonia, giving it an advantage over other mushrooms trying to make a home in nitrogen-rich manure-based composts. Once a compost has hosted a few flushes of shaggy manes, it can then be used for other mushroom species that don't tolerate high ammonia levels, such as the almond portabella or blewit. Once the cap of shaggy mane elongates, the lower edge of the cap separates from the stem, signaling spore production and a meltdown phase. (The mushroom autodigests in a matter of hours after emerging from the soil.) Caps can be smooth and white, but most have brownish fibrils or tufts. The spore print is black.

Common Strains and Ideal Fruiting Conditions

Local strains you can harvest yourself are best, since shaggy mane is triggered to fruit by interactions with microbial populations in the soil, and you'll want to try to replicate those soil and microbe conditions. Shaggy manes tend to fruit during periods of cooler nights, when temperatures drop below 60°F (16°C), and foggy mornings, which helps them prolong sporulation. Sporulation destroys this mushroom, usually before lunchtime, so you'll have an advantage if you can find and develop a late- or nonsporulating strain.

Wild Spawn Expansion Techniques

A shaggy mane can quickly melt down before you can get it to your worktable, so don't be disappointed when you find one that you've just harvested curled up and decaying, like it's wearing a black-rimmed hat.

Go ahead and use it. The best choice for expansion with such a moody little shaggy is a spore solution slurry: Wash the mature mushroom in a bucket of clean water to release the spores, and add the solution generously to a growing medium rich in organic debris, such as composted manures, decomposing wheat bales, or your home compost pile.

Laboratory Isolation and Spawn Culture

Shaggy manes clone easily from interior tissues of the upper and lower stem. The caps are thin and offer less-than-desirable real estate to pull tissue from. Shaggies also do not enjoy antibiotic media, so once the culture is purified, switch back to regular agar when producing grain spawn.

Preferred Fruiting Substrates

Shaggy mane prefers composted mixtures of manure and cereal straws with cotton hulls or any other agricultural by-products, such as bean hulls, cornstalks, cotton waste, beet pulp, and sugarcane bagasse.

Casing Soil

A microbial casing soil is required. When the entire surface of the substrate, or just below if you scratch around in a few spots to check, is a continuous white mat, dress the bed with a microbial casing soil, which stimulates primordia formation (see chapter 20).

Outdoor Cultivation Notes

Media preparation. Locate beds at an interface of shade and sunlight; shaggy manes don't mind popping up in the very sun that makes them melt away so quickly. We plant shaggy mane spawn directly under rotting wheat bales near the edges of our greenhouses, where they receive the bulk of the rainwater runoff.

Fill raised beds with the growing medium 8 to 10 inches deep, poke narrow holes in a diamond pattern every 6 to 8 inches across the surface, and fill the holes with spawn. Cover the spawned holes with wet compost to keep the spawn from drying out and to keep insects from attacking it immediately. Cover the substrate with perforated plastic or with plywood that has had holes drilled through it to preserve moisture while still allowing for modest gas exchange. When colonization is complete, dress with a microbial casing soil.

Preferred spawn type. Supplemented sawdust or grain spawn.

Inoculation to fruiting time. From six to twelve months, depending on temperature and inoculation rates. These mushrooms typically fruit in the fall during periods of cool weather.

Expected yields. Extremely low. Cultivate this one for yourself, not for commercial cultivation, and count yourself lucky if you can add them to a few dinners. Inspect your beds daily in the morning or late at night with a flashlight. I like to use a blacklight, which makes the mushrooms glow bright white—especially if I am hunting for mushrooms in cow fields!

Indoor Cultivation Notes

Indoor cultivation of shaggy manes is not recommended, since they yield poorly and decay rapidly. These mushrooms are better suited for outdoor cultivation.

Harvesting

Shaggy manes like to congregate in groups but rarely fuse, making them easy to pull from the soil. Trim the substrate-laden base away and consume or store them immediately.

Storage

Fresh shaggy mane mushrooms will keep for just a few days at 38 to 42°F (3–6°C). They'll store a little longer if you submerge them in water in the refrigerator. Dried caps rehydrate well, so drying is a good option. You can also can the caps. I cook and then freeze a few small portions for the occasional shaggy treat in the off-season.

CAUTION

Note that some people have an allergic reaction to a few species of *Coprinus* mushrooms, notably the alcohol inky (*Coprinus atramentarius*), which generally has a smoother and more linear silky cap than *C. comatus* and tends to fruit in clusters.

Marketing

The shaggy mane mushroom is almost entirely water, but when the water evaporates, the remaining biomass is surprisingly delicious when sautéed with eggs or potatoes. Given the extremely short shelf life after harvest, shaggy manes are almost impossible to cultivate and sell commercially. Enjoy them with your friends.

Nutritional Value and Medicinal Uses

Coprinus mushrooms have a very high moisture content, averaging around 92 to 93 percent water, and have protein values, in the range of 25 to 29 percent of dry weight. Research suggests that *Coprinus comatus* can help regulate blood pressure and sugar levels. It has antibacterial agents (extracted from both the mycelia and the fruitbodies) and anti-inflammatory compounds, and it also has shown to suppress many cancer cell lines.

Uses in Mycoremediation

Coprinus species are efficient hyperaccumulators of mercury, cadmium, and arsenic. Enzymes from this mushroom, curiously, have been used as a dye remediation additive, and some studies have used them in detergents for washing clothes. Living mycelium from this fungus also attacks nematodes.

The Genus *Fistulina*

Common Species

F. hepatica (beefsteak)

Difficulty Level

Outdoor cultivation—2
Indoor cultivation—4

General Description and Ecology

Fistulina hepatica is a fleshy polypore that, when cut, resembles fresh, marbled red meat and exudes a reddish juice—thus the name beefsteak. Although it tastes nothing like meat—rather, it has a lemony, sour flavor—you can marinate this mushroom and cook it like your favorite tenderloin or slice and dehydrate it to make jerky.

Mycelium and Spawn

The mycelium is clear at first and then becomes white and linear, exuding a dark red metabolite on overcolonized spawn and fruiting substrates. The smell is slightly astringent to sour. Commercial spawn is usually available in the form of cereal grains, supplemented sawdusts, and wooden dowels for outdoor cultivation.

Fruitbody Development

This is an annual fan-shaped fleshy polypore, meaning the fruitbodies grow and completely decay in one season, with an off-center, lateral stem. It usually fruits from oak stumps and trunks. The stem can be absent if the mushroom is fruiting off aboveground wood or elongated if it's arising from buried wood or an underground cavity. It fruits primarily in singles or sometimes in groups gently fused near the base of the stem attachment, and only rarely in overlapping clusters. The underside of the fruitbody is white and covered with minute pores when the mushroom is fresh. It exudes a reddish juice when cut, squeezed, or injured. The spore print is pinkish to dark salmon in color.

Common Strains and Ideal Fruiting Conditions

This mushroom is rarely available from spawn suppliers since most beginners prefer more common varieties such as shiitakes and oysters. Some commercial strains are in circulation, but they're cultivated almost entirely on an experimental basis in Europe and probably will continue that way until a larger market develops. The fruiting conditions vary based on the strain. The particular *Fistulina* strain I experimented with, growing it on sawdust blocks, was extremely photosensitive, producing hundreds of baby beefsteaks all over the sides and surface of the sawdust within just two weeks of inoculation. This type of reaction suggests that this particular strain should be

The beefsteak mushroom (*Fistulina hepatica*) fruiting on partially buried hardwood rounds.

grown not in clear bags or containers but in something opaque that allows only a few small areas of the fruiting blocks to get any light. Otherwise the entire flush of mushrooms will overpin and abort. Burying the sawdust blocks outdoors in a shallow layer of wood chips is an easy way to control this tendency.

Wild Spawn Expansion Techniques

Only young, fresh mushrooms are able to be expanded. You can use a young, firm fruitbody to make a cardboard starter culture by clipping the mushroom at its base, near the attachment point, wrapping the stem bases in wet cardboard, and rolling it up tightly. The best method I have found is to submerge the mushroom upside down in water, leaving the attachment point sticking up above the water, seal it in a plastic bag, wait for mycelium to reform on the stem, and then cut away this growth for a first transfer onto either cardboard or antibiotic agar. (For more discussion of this technique, see "Special Techniques for Culturing Finicky Mushrooms" in chapter 18.)

Laboratory Isolation and Spawn Culture

Wild specimens are difficult to clone, even on laboratory agar, given their soft and juicy interiors. Attachment point is the best strategy. I like to harvest bits of wood from the tree or stump itself, using a knife to scrape them into a clean plastic bag, for transferring to agar later. Once it's on agar, the mycelium is faint; growth is irregular, not symmetrical, and may begin to produce dark red metabolites before it reaches the petri edges. I prefer to let this fungus cover a third of the plate then convert it to liquid culture, which I transfer to grain with supplemented sawdust for fruiting. I would strongly advise incubating this fungus in the dark: It is highly photosensitive, and linear vegetative growth of the mycelium is inhibited by light.

Preferred Fruiting Substrates

Beefsteak mushrooms prefer hardwood logs and stumps outdoors and supplemented sawdust indoors. Oak trees seem to be preferred, but matching up the wood type with your particular strain is important to guarantee compatibility and fruiting.

Outdoor Cultivation Notes

Media preparation. Fresh logs or stumps will do. Make sure logs or rounds are cut no more than one month before inoculation to ensure the mushroom has a chance to outcompete any wild fungal species. Much like lion's mane (*Hericium erinaceus*), the beefsteak prefers to emerge from a small cavity, developing its primordia naturally inside shallow holes, such as those made by woodpeckers. Once colonization nears completion, use a paddle-type boring bit to drill a few holes around the wood, about 2 inches deep. The delicate primordia will slowly form in these holes over a few days, gaining some momentum before emerging into the light.

Preferred spawn type. Sawdust, plug, or cardboard spawn.

Inoculation to fruiting time. From six to ten months, depending on temperature and inoculation rates. Clusters typically form at the interface of the ground and the wood, on the shadier side.

Expected yields. Low. This mushroom has yet to be cultivated on a large commercial scale due to its relatively poor return on investment. It is more appropriate for smaller farms or home cultivators who wish to add something unusual or exciting to their local markets or dinner plates.

Indoor Cultivation Notes

Media preparation. Sterilized, supplemented hardwood sawdust, such as oak, is the ideal growing medium for *Fistulina*.

Preferred spawn type. Grain or supplemented sawdust spawn.

Inoculation to fruiting time. From two to three weeks. The spawn run is rapid, initiating fruitbodies within two weeks. Larger primordia bear the hallmark pink juice oozing from the baby beefsteaks when they're grown under high carbon dioxide and elevated humidity levels.

Expected yields. Fair, but improving with substrate modification and techniques to limit overpinning.

Harvesting

These mushrooms form singly on logs and sometimes with two overlapping fruitbodies from bag openings. Indoor cultivators should grow these mushrooms in black bags or other opaque containers that minimize light exposure, and they should also limit the holes they make to allow gas exchange and fruitbody emergence to just a few, compared to the number given other mushrooms such as oysters. Beefsteaks begin as tight, reddish balls that flatten and become hoof-like, fanning out until they become soft and darken. Pull the entire mushroom out from its wood cavity or off its substrate, and trim off the attachment point before storing.

Storage

Fresh beefsteak mushrooms will keep for one to two weeks at 38 to 42°F (3–6°C). The caps dry and rehydrate well; consider slicing them into thin strips, marinating them, and drying them to make vegetarian jerky.

Marketing

This mushroom is not currently being cultivated on a large scale, and may never be a big hit at the market, but mushroom lovers and enthusiasts often want to try something they've never had before, and this one often fits the bill. The meaty appearance and texture make it attractive to vegetarians, especially if you can hand out some marinated, sautéed samples for tasting. You can also soak these mushrooms in lemon juice or brine them ceviche-style to make a refreshing appetizer. Its slight lemony flavor makes this one a summer favorite.

Nutritional Value and Medicinal Uses

Beefsteak mushrooms are loaded with ascorbic acid (vitamin C), which gives them a unique lemony flavor. Extracts from this fungus have shown high activity in suppressing *Staphylococcus aureus* in lab studies.

Uses in Mycoremediation

This mushroom is commonly found in logged sites, even in full-sun environments, making it a good candidate for landscape remediation in logged areas where most other fungi will not thrive. Given its antibacterial properties, I suggest experimenting with this mushroom for mycofiltration of water runoff from livestock operations.

The Genus *Flammulina*

Common Species

F. populicola, F. velutipes (enoki, velvet foot)

Difficulty Level

Outdoor cultivation—1
Indoor cultivation—2

General Description and Ecology

The beloved enoki mushroom gets its common name velvet foot from the wild form, which exhibits a dark, slightly hairy stem base at the attachment point. Like the wood ear, the enoki has little flavor but a wonderful crunch when fresh and slightly cooked. Overcooking these delicate mushrooms causes them to just about disappear into a hot pan, so be gentle. Enokis are primarily winter-fruiting species, fruiting at around the same time as blewits (*Clitocybe nuda*), when falling temperatures trigger fruiting.

Mycelium and Spawn

The mycelium is white and cottony, exuding a light yellow metabolite on overcolonized spawn and fruiting substrates. Its smell can be somewhat unpleasant compared to other sweeter-smelling mycelium, like *Pleurotus*, for example. Commercial spawn is usually available in the form of cereal grains, supplemented sawdusts, and wooden dowels for outdoor cultivation.

Fruitbody Development

Wild specimens look very different from indoor cultivated specimens. Wild *Flammulina velutipes* mushrooms have a tough, dark orange to yellow cap and short, fuzzy black stems, whereas a clone of the same mushroom grown indoors under low light and high carbon dioxide levels can look like a long pasta noodle with a small albino cap on top. Enoki are gilled mushrooms, with the cap centered on the stem and attached gills that are yellowish to cream-colored, or sometimes darker orange in cold weather. The spore print is always white and the stem base (in wild specimens) has a distinctive gradual darkening, almost to black at the very base, where some strains thicken to a dark fuzzy attachment with fibrils. Be sure to compare this stem base to that of the poisonous deadly galerina (*Galerina autumnalis*), another winter mushroom that is common in many temperate regions of the world. Deadly galerina lacks the distinctive darker and fuzzier stem base at the point of attachment and also has a rusty brown sporeprint, compared to white for enokis.

Common Strains and Ideal Fruiting Conditions

Enoki mushrooms are grown in large numbers in Korea, Japan, and China but are enjoying growing popularity in other regions around the world. Commercial strains abound. Wild strains typically fruit at a cool 45 to 55°F (7–13°C), although there are strains and similar species within this genus that fruit at warmer temperatures. Enokis could easily share an indoor growing space with shiitake sawdust blocks, since both do well with high carbon dioxide and low light levels (these conditions promote stem elongation in enokis).

Wild Spawn Expansion Techniques

In the wild, enoki mushrooms typically grow from wounded bark peeling away from a stump or fallen tree. If you find an enoki growing in this manner, you can start a cardboard culture from the stem base, and if you are careful in removing or bending the bark back you can usually find an additional inch or two of fresh myceliated mass, which would be perfect for starting your own cultures.

Wild-harvested enoki mushrooms (*Flammulina velutipes*), showing the darker, fuzzy base that is a significant identification feature separating them from deadly galerinas (*Galerina autumnalis*). These winter mushrooms can survive freezing temperatures and continue to grow after thawing.

Laboratory Isolation and Spawn Culture

I have cloned both wild specimens and store-bought enokis directly from the attachment point. Wild-collected mushrooms are very rubbery and elastic, making cloning more difficult, so for them I prefer spore germinations.

Preferred Fruiting Substrates

Enokis prefer hardwood logs and stumps outdoors, and supplemented sawdust indoors.

Outdoor Cultivation Notes

Media preparation. Fresh logs or stumps will do. Make sure logs or rounds are fresh (cut no more than one month before inoculation) to ensure the mushroom has a chance to outcompete any wild fungal species. This is a good candidate for the log raft method: Inoculate the logs aboveground and let them colonize for six months, and then bury them in a thin layer of fresh hardwood chips (see chapter 5 for more details). Cultivate enokis on partially buried large-diameter rounds to produce clusters, which will typically fruit at the interface between the ground and the wood, on the shadier side. Enokis can form by the thousands when grown on these partially buried rounds, but the beds don't typically have the lasting power of colonized logs grown mostly aboveground, such as in the stacked log cabin or raft methods.

When enoki mushrooms are cultivated on supplemented sawdust indoors under elevated carbon dioxide and lower-than-natural light levels, the stems elongate and the pigments never develop. Cultivating fibrous mushrooms in this manner makes the entire mushroom edible, compared to wild specimens, which have very chewy stems.

Preferred spawn type. Sawdust, plug, or cardboard spawn.

Inoculation to fruiting time. From six to ten months depending on temperature and inoculation rates. The fleshy mushrooms typically are ready to harvest within five to eight days after emerging from the bark, depending on temperature and humidity. Because enokis fruit during colder months, if freezing weather bears down, maturation may take several days longer, with the mushrooms hiding out under the snow, waiting for warmer weather to finish maturing. And yes, I have picked these mushrooms frozen solid, allowed them to thaw slowly, and found them still able to produce a spore print.

Expected yields. Low. Fruiting enokis outdoors is not a high-yielding proposition, since the mushrooms limit themselves to a narrow fruiting temperature. You can expect two flushes several weeks apart during cool or cold weather.

Indoor Cultivation Notes

Media preparation. Use supplemented hardwood sawdust, sterilized in an autoclave for two hours. Once cooled and inoculated, the substrate needs to be incubated and most strains need to be given a cold shock to initiate fruiting.

Preferred spawn type. Grain or supplemented sawdust spawn.

Inoculation to fruiting time. From three to four weeks. This mushroom prefers a dense growing medium, such as finely ground sawdust mixed with wheat or rice bran. Containers used for indoor production need to have high walls to allow the carbon dioxide to pool, which creates the signature stem elongation developed by growers. Lowering the light in the fruiting room to a minimum also reduces fruitbody pigment; if you prefer a more natural-looking mushroom, increase the light and use a lower-lipped container, or remove the sides of the bags when mushrooms begin to form.

Expected yields. Fair. Expect two flushes several weeks apart.

Harvesting

These mushrooms form small clusters in the wild and countless masses indoors in containers and bags. Pull the entire cluster of mushrooms out of the substrate and trim the substrate off the bases before storing.

Storage

Fresh enoki mushrooms will keep for up to two weeks at 38 to 42°F (3–6°C). When dried, the caps rehydrate a bit chewy—not ideal. So if they are not sold fresh, consider sautéeing the mushrooms and freezing them in water. Many commercial producers shrink-wrap these mushrooms to protect them, but doing so may increase the risk of food pathogens.

Marketing

These mushrooms are best grown indoors to achieve that classic "bundle of noodles" appearance. Wrap them in bundles and arrange them in a shallow tray of cold water, with the stems slightly submerged, much like selling herbs. The mushrooms are fragile and dry quickly, so you can also keep them in a clear container with a little ice so your market folks can see the quality and, if you have it, a fruiting container with some enokis growing. You will need to describe how to use these mushrooms; suggest your best and easiest salad recipe to ensure immediate success. These mushrooms are commonly eaten fresh, sprinkled over salads or floated in clear Asian-style soup broth.

Nutritional Value and Medicinal Uses

Enoki mushrooms, though fragile and flimsy, have a protein content ranging from 20 to 30 percent (dry weight), depending on the strain and growing conditions, making it one of the most protein-rich of all cultivated mushrooms. Studies have found that regular consumption of these mushrooms can help boost and modulate immune system function, regulate blood pressure and sugar levels, lower cholesterol, and provide cardiovascular and digestive support. The mushrooms contain antiviral and antibacterial properties as well as anti-inflammatory compounds, and they also have been shown to suppress many cancer cell lines.

Uses in Mycoremediation

Enoki is a well-known stump recycler, giving it a possible advantage if used to remediate a logging site, perhaps along with compatible species such as beefsteak polypore and wood ear mushrooms, which also can grow in these otherwise fungus-inhospitable locations.

The Genera *Fomes,*
Fomitopsis, and *Laricifomes*

Common Species

Fomes fomentarius (iceman polypore, amadou)
Fomitopsis officinalis, L. officinalis (agarikon)

Difficulty Level

Outdoor cultivation—3
Indoor cultivation—4

General Description and Ecology

The easily identifiable iceman polypore is one of the many species of mushrooms found in possession of "Otzi," a nomadic traveler who lived around 3300 BCE and was discovered frozen in the Swiss and Austrian Alps in 1991. The discovery of this mushroom in Otzi's possession has profound historical implications regarding the use of fungi for food, fuel, and medicine. Although the mushroom is not considered edible as a food (it can be ingested for medicinal properties), it can be used for making products such as medicinals, textiles and fabric, and paper.

Sometimes *Laricifomes* and *Fomitopsis* are lumped together in the same genus; it just depends on which mycologist you are talking to. Recently, *Laricifomes officinalis* has joined *Fomitopsis* based on molecular data. This is happening more and more every day as published papers attempt to identify the differences among groups of mushrooms and how they are related and may have evolved. Since these genera are very uncommon in their growth requirements and cultivation techniques, I have included them in this book to offer some insight into cultivating slow-growing conk fungi, many of which are rare or endangered.

Mycelium and Spawn

The mycelium of *Fomes* is white, cottony, delicate, and slow growing, eventually oxidizing to dark, chocolate brown as it ages. The mycelium of *Fomitopsis* is also white but somewhat powdery, reminiscent of chicken of the woods mycelium. The smell of both types is sweet and pleasant. Commercial spawn is rare; however, when it's available, it is typically available in the form of wooden dowels for outdoor cultivation.

Fruitbody Development

This is a pored mushroom that lacks a stem and attaches to the sides of downed or standing trees or logs, and rarely on stumps. The fruitbody is grayish and banded, extremely tough to hard in texture, and hoof-shaped to beehive-like in appearance. These fruitbodies persists year-round and sometimes grow in layers, with new layers added annually; a few known wild specimens have been growing for over fifty years. The spore prints of these genera vary from white to creamy yellow.

Common Strains and Ideal Fruiting Conditions

Strains are mostly obtained from wild specimens. New *Fomes* fruitbodies emerge in late spring throughout temperate climates at higher elevations. These mushrooms are typically perennial conks that add growth layers every season, persist for years, and can endure "stop and go" growing conditions such as rain and drought. *Fomitopsis* is no different, as many polypores need to experience a high-humidity initiation cycle, but require very little watering to coax them along into maturity.

Wild Spawn Expansion Techniques

Harvested conks can be submerged upside down in water, weighted down to keep them from floating, with their attachment point sticking up above the water, and capped with wet cardboard. Mycelium leaps to the cardboard in one week, giving you viable samples to plant into downed trees and stumps by wafering (inserting small pieces of the cardboard into wounds created with a machete or hatchet; see chapter 12).

Laboratory Isolation and Spawn Culture

The tissue of *Fomes* species is extremely tough, with a hard outer rind, so tissue samples are difficult to excise. I have found that popping one off from the tree, harvesting the attachment point where there is a bit of the wood exposed, and surface-sterilizing, or washing the bits in diluted alcohol or 3 percent hydrogen peroxide, then draining them can yield a few bits of colonized tissue capable of leaping off to agar media. Specimens of the rare *Fomitopsis officinalis*, or agarikon, should not be removed, since they are extremely endangered. Instead, use small boring tools to collect plugs of tissue, which can be dissected and cultured, leaving the beautiful conks in place.

Preferred Fruiting Substrates

Fomes conks occupy a wide range of hardwood trees in North America. *Fomitopsis* species are more particular in their habitat, hiding out in old-growth forests, particularly those of Douglas fir and larch. Outdoor snags and downed trees are preferred for this core collection of conks.

Outdoor Cultivation Notes

Media preparation. These species prefer to fruit from standing dead trees (snags) or large-diameter downed trees. For agarikon (*Fomitopsis officinalis*), after a year of colonization, use a paddle-type boring bit to drill a few holes around the wood, about 2 inches deep. The delicate primordia will slowly form in these holes over a few days, gaining some momentum before emerging into the light.

Preferred spawn type. Sawdust, plug, or cardboard spawn.

Inoculation to fruiting time. From one to three years depending on temperature and inoculation rates.

Expected yields. Very low. These conks are not high yielding and are extremely slow growing, making them increasingly rare in the wild.

Indoor Cultivation Notes

Indoor cultivation is not recommended.

Harvesting

Fomes fruitbodies are ready for harvest when the conks begin to slow growth (not that you can tell), and when the pore or tube layers on the underside of the cap or conk lose their distinctive white appearance, or once the pores look like they are not so fresh. Gently tap the conks upward with a soft rubber mallet; they will pop loose from trees and logs more easily than if you try to pry them off.

Though these conks are extremely slow growing in culture, grain extracts may be the best method of harvesting the mycelium for making medicinal extracts. I cultivate the fungi on sterilized grain, like making tempeh culture (see chapter 13), until the mycelium has gelled the grain together, and I either cook the grain like a "space cake" (see chapter 13) or powder it; the powder can be taken as a supplement or made into an extract (again, see chapter 13).

Storage

Fresh *Fomes* and *Fomitopsis* mushrooms will keep for one month at 38 to 42°F (3–6°C); however, molds can infect the tube layer under higher humidity. The dried conks can be used to make powders, tinctures, and even fire starter. Since they do not contain much water, dehydration is my preferred method to preserve these conks to prevent molds from developing on the pores.

Marketing

These mushrooms are slightly bitter and very fibrous; they're not suitable for direct consumption unless they are completely dried and powdered. People may not know what to do with them unless you educate

Fomes fomentarius (iceman polypore) fruits primarily on dying birch trees at higher elevations. Creating wild patches of *Fomes* and its conk-like relatives, such as agarikon (*Laricifomes officinalis*), will help keep these species perpetuating in the wild, since they are extremely slow growing and becoming well sought after as medicinal species.

them, so you might consider preparing the extracts or powders and selling them in a ready-to-use form. Cite scientific journals for the health benefits of these small, bracket-like polypores so the public can gain a respect for these wonderful fungi and include them in their diet on occasion.

Nutritional Value and Medicinal Uses

These mushrooms are wonderfully rich in compounds similar to those of turkey tail (*Trametes versicolor*), including polysaccharide-K, a protein-bound polysaccharide commonly used in Chinese medicine for treating cancer patients during chemotherapy. Studies have found that these mushrooms can help boost and modulate immune system function, regulate blood pressure and sugar levels, lower cholesterol, and provide cardiovascular and digestive support. They contain antiviral and antibacterial properties as well as anti-inflammatory compounds, and they also have been shown to suppress many cancer cell lines.

Uses in Mycoremediation

The antibiotic properties of this fungus makes it a great candidate for mycofiltration of bacteria in contaminated soil and water. Colonized biomasses of myceliated wood chips can be used in trenches downslope of livestock operations to absorb and provide filtration of surface runoff.

The Genus *Ganoderma*

Common Species

G. applanatum (artist conk)
G. curtisii, G. lucidum, G. oregonense, G. resinaceum
(reishi, ling chi)
G. tsugae (hemlock reishi)

Difficulty Level

Outdoor cultivation—1
Indoor cultivation—2

General Description and Ecology

Reishis, commonly referred to as the "mushrooms of immortality," are a beautiful group of lacquered polypore mushrooms that have been highly revered for their medicinal properties for thousands of years. In traditional Chinese medicine, these mushrooms are seen as a "cure-all" and used for their beneficial effect on many regulating functions of the human body. In particular, they have been shown to have a positive effect for those who suffer from sugar regulation complications such as diabetes, liver ailments, immune dysfunction or impairment, and more.

These mushrooms are easy to fruit indoors and outdoors anywhere in the world. They are versatile in their ability to grow on many types of hardwoods, regardless of strain—the only exception being *Ganoderma tsugae*, the hemlock reishi, which has a preference for the wood of its namesake tree. Reishi mycelium is aggressive and capable of outcompeting most wild fungi and bacteria, due to its antibiotic and antifungal properties. Even growers without the luxury of a laboratory can expand a bag of supplemented sawdust spawn to wood chips and achieve near-commercial-grade fruiting yields. Add in its widely documented medicinal benefits and this mushroom should be among your top ten for cultivation. It's a

great fungus to start with, helping you to gain more experience with the life cycle of fungi before you take on more difficult species. And if you're interested in medicinal qualities, why pay an arm and a leg for extracts and mushrooms grown under mysterious conditions when you can grow them yourself? Take special care of these mushrooms and they will take care of you as you strive for immortality.

Mycelium and Spawn

The mycelium is white and linear. It will hug an agar surface, lying almost completely flat, and with age produces a signature yellow metabolite that darkens to orange. The smell of the mycelium can be sour, much like that of a *Ganoderma* fruitbody, and somewhat unpleasant to most. Commercial spawn is typically available in the form of cereal grains, supplemented sawdusts, and wooden dowels for outdoor cultivation. If you leave it at room temperature, spawn can transform into solid bricks that are almost impossible to break up. Some growers have actually made bricks from the mycelium and used them to construct living structures to demonstrate the strength of this fungus.

Fruitbody Development

This is a stemmed mushroom that begins as an antler with a whitish tip, flattening to a cap positioned off-center, with white pores underneath. The growing edge is typically white or off-yellow when the mushroom is actively growing and then dulls and darkens at full maturity. The spore print is cinnamon to chocolate brown. The spores usually end up covering the entire cap surface, due to convecting air currents sweeping the spores up and around the cap edge.

Ganoderma tsugae, the hemlock reishi, fruiting at 3,000 feet near Asheville, North Carolina. The dying hemlocks have made growing substrate available for this mushroom in great quantities.

Common Strains and Ideal Fruiting Conditions

Many different strains are available from spawn suppliers and culture collections worldwide. Before purchasing one, be sure to inquire as to the preferred wood type, as species can vary in their ability or likelihood to fruit on various trees. If you are seeking specific medicinal qualities, note that they may be strain-specific.

Wild Spawn Expansion Techniques

Cloning works well; take tissues from the soft interior or from the stem attachment and transfer them onto sterilized agar. For cardboard culture, surface-sterilize a freshly harvested fruitbody with a peroxide solution to get rid of bacteria, add it to a plastic bag sandwiched between a few small pieces of wet cardboard, and place it under refrigeration or in the coolest location possible. The mycelium will leap off and begin to engulf or consume the bits of cardboard, which can then be transferred to additional cardboard or cleaned up in the lab by sectoring on agar. (This method works well for many species of polypores that engulf their surroundings.) Likewise, small toothpicks or dowels that have been thoroughly soaked in distilled water can be inserted just into the growing edge; with days the mycelium will have colonized the wood, and you can removed them for transfer to agar or cardboard.

Laboratory Isolation and Spawn Culture

Culturing tissue from the point of attachment on the stem works best for me in cloning mushrooms from this entire genus. Spore germinations are possible, but they're less reliable and typically covered with a contaminating yeast complex. This culture needs to be transferred from agar before it reaches the edges of

Golden reishi mushrooms (*Ganoderma curtisii*) grown on supplemented sawdust in a closed bag create beautiful antlers that elongate in the carbon dioxide–rich environment.

the plate; otherwise it becomes difficult to cut cleanly, usually peeling up as you drag the scalpel.

Preferred Fruiting Substrates

Reishis prefer hardwood logs, stumps, and sawdust, except for *Gamoderma tsugae*, which typically prefers hemlock logs, stumps, and downed trees.

Outdoor Cultivation Notes

Media preparation. This mushroom will not fruit aboveground. Use buried hardwood logs and stumps. Make sure logs and stumps are fresh (cut no more than one month before inoculation) to ensure the mushroom has a chance to outcompete any wild fungal species. Trenched log cultivation works well: Inoculate a series of small-diameter logs, let them colonize for six to eight months, then completely bury them, packed tightly together to make a raft (see

chapter 5 for more detail). The fungi produce antlers typically at the interface between the ground and the wood, on the shadier sides.

Preferred spawn type. Sawdust, plug, or cardboard spawn.

Inoculation to fruiting time. From ten to twelve months, depending on temperature and inoculation rates. Reishis typically fruit in the spring and keep producing all summer, persisting into the fall.

Expected yields. Good. Reishi mushrooms are dense and quite prolific fruiters.

Indoor Cultivation Notes

Media preparation. Use supplemented sawdust, sterilized in an autoclave for two hours.

Preferred spawn type. Grain or supplemented sawdust spawn.

Inoculation to fruiting time. Sawdust fruits almost immediately after colonization is complete, typically in as little as two to three weeks. The antlers that form inside a bag will keep growing for weeks as they try to find a way out, until the bag is cut open to allow the mushrooms to flatten and sporulate.

Expected yields. Good.

Harvesting

The mushrooms are slow to grow and can be harvested at just about any stage of development, depending on their appearance and intended use. The antlers are believed to contain more of the beneficial medicinal properties than mushrooms allowed to fruit and flatten naturally, but those claims have not been substantiated. Mature specimens that are harvested while they still have their growing edge, just before they release spores, can be dried and preserve well, and they retain the energy they would have spent in spore production.

Storage

Fresh reishi mushrooms will keep for one month at 38 to 42°F (3–6°C); under prolonged refrigeration molds can infect the tube layer. The dried antlers and fruitbodies can be used to make powders. Reishi are low-water-content fungi, so the best method is to dry and powder them or make medicinal extracts while they are fresh.

Marketing

These mushrooms may look a little odd to people who have never seen them before, and it may take a little explanation to win over first-time users. I like to bring a fruiting kit of the beautiful antlers and describe how to make the tinctures. Bring copies of published studies that support their medicinal efficacy, and you make your case on the spot. I have found that at organic growers' conferences I sell very few medicinal species, whereas at herbal conferences or natural healing seminars our reishi stock usually sells out.

Nutritional Value and Medicinal Properties

Ganoderma species have a long history of use in Chinese medicine and have one of the highest beta-glucan levels of any medicinal fungus. Although not directly edible, these mushrooms can be powdered and consumed, but hot water and alcohol extracts are preferred. Studies have found that reishi mushrooms can help boost and modulate immune system function, regulate blood pressure, lower cholesterol, and provide cardiovascular and digestive support. They contain antiviral and antibacterial properties as well as anti-inflammatory compounds, and they also have been shown to suppress many cancer cell lines. Probably the best-known mechanism that reishis are capable of regulating is blood sugar, which makes this mushroom one of the best natural treatments for diabetes patients.

Uses in Mycoremediation

Reishi mycelium has an extremely high tensile strength and an affinity for inhibiting and lysing bacterial cells, making it a good candidate for mycofiltration of water. Its ability to bind a substrate with great tenacity can help keep a biofilter from disintegrating. In fact, myceliated substrate can form an almost rubber-like material that can be dried and cut into any shape you like, so that it can fit tightly into a manufactured filter module.

The Genus *Grifola*

Common Species

G. frondosa (maitake, hen of the woods)

Difficulty Level

Outdoor cultivation—2

Indoor cultivation—4

General Description and Ecology

Maitake is a polypore with flat, overlapping petals that curl back toward their stem like large fractals. Although this mushroom can be cultivated indoors, many mushroom hunters crave the wild species, searching for them each fall and winter. Wild maitakes typically weigh 12 to 15 pounds apiece, and they have been reportedly found at upward of 100 pounds apiece. Their location is often kept a guarded secret. The mushrooms can fruit for several years, until they have depleted their host tree. I have only found maitakes fruiting from buried wood, most prominently on the ground around the bases of large dying oaks, although I once found them fruiting under a large sycamore in downtown Liberty, South Carolina. In one of our own secret spots my wife Olga and I once noticed a few babies coming up. We covered the entire base,

Maitake mushrooms (*Grifola frondosa*) fruiting from large oak stumps that were inoculated with spawn two years prior. Stumps can take a long time to fruit but are well worth the wait, fruiting for many years after. These fruiting bodies ended up weighing 13 pounds each, with smaller clusters scattered around the perimeter of the base.

including the baby mushrooms, with sticks and leaves to hide them from "predators" (other mushroom hunters) while they grew. When we returned a few days later the clusters had doubled and had also lost much of their pigment where the cap surfaces were not exposed to light. We had grown some "albino maitakes" by mistake—the same way white asparagus is grown. They still tasted delicious, but I started to wonder about the nutritional and medicinal properties of sun-exposed mushrooms versus those that are grown indoors with no natural light. Some companies even cultivate medicinal species of fungi on grain in the dark, powdering them before they are ever stimulated by light. This light exposure has a significant effect on the quality of the nutritional and medicinal properties of the fruitbodies.

Mycelium and Spawn

The mycelium is white and cottony and exudes a light yellow metabolite (which curiously fluoresces bright yellow under a black light) on overcolonized spawn and fruiting substrates. The mycelium is extremely slow to grow on agar and during the colonization period on its growing media. Its smell is sweet when fresh and fishy when old. Commercial spawn is usually available in the form of supplemented sawdusts and wooden dowels for outdoor cultivation.

Fruitbody Development

This is a multistemmed polypore mushroom, arising from a single mass, or knot, that differentiates into petals that eventually overlap to form tight bundles of curling rosettes. The mushroom arises from buried wood and develops the knot right at the soil surface, below the leaves and surrounding debris, filling its living quarters with carbon dioxide, causing the elongated portions to explore up and through the leaf matter. Once it's completely aboveground the stems and petals expand and develop their signature grayish-brown pigments, triggered by filtered light exposure. The underside is completely white, with minute holes when fresh, and the fronds are crispy or brittle when bent. The spore print is white but does not produce until very late in maturity, when the mushroom is almost too old and tough to eat. Maitakes can be confused with Berkeley's polypore (*Bondarzewia berkeleyi*), black-staining polypore (*Meripilus sumstinei*), and cauliflowers (*Sparassis* spp.), among others.

Common Strains and Ideal Fruiting Conditions

Wild specimens of maitake have a distinct preference for the type of wood on which they originally grew, so if you are purchasing spawn, inquire as to the wood type, and try to at least match your strain to the tree's genus—any type of oak, for instance. Not all strains of wild-collected maitakes fruit well indoors on sawdust blends. I have cloned tissue from clusters I bought at the grocery store with average luck, but these cultures have been expanded to the point that they are miles away from their originating parent culture and need to be retrained on agar or a fruiting substrate before generating any spawn. Commercial strains are available from most tissue culture libraries, but again, you need to know a particular strain's substrate preference. Some culture vendors advertise (often expensive) package deals with their strains, agreeing to divulge their "secret substrate formula" for each strain when you buy it. I prefer to take the cheaper and more challenging route—to discover the mushroom myself and fine-tune my own secret formula to match it. That's how people find new strains and develop better fruiting formulas.

Fruiting temperatures are variable, but most strains prefer a cooler 55 to 65°F (13–18°C) growing environment to initiate primordia formation. Outdoor strains typically prefer lower initiation temperatures than indoor strains and only fruit once each year; commercial strains that were bred to fruit indoors consistently initiate at the upper edge of this window and complete maturity at 60 to 70°F (16–21°C). Indoor fruitings can produce two flushes a few weeks apart.

Wild Spawn Expansion Techniques

This hardwood-loving species is finicky, but worth the ups and downs of learning exactly what it needs to flourish. The best method to perpetuate it outside of a laboratory is in cardboard. Wrap thin slices of stem bases in wet cardboard and allow the mycelium to colonize for a few weeks. Once the mycelium has colonized a very small portion of the cardboard, remove the stem bits; otherwise they will rot and bacteria will eat up the cultures. Roll the cardboard back up, adding more layers as necessary to increase the biomass of the mycelium as you notice it growing.

Laboratory Isolation and Spawn Culture

I prefer to clone maitake mushrooms from older specimens, not fresh ones. It took me years to figure out that all the tissue from those beautiful, crunchy, prime maitake mushrooms did not want to clone on agar. Very frustrating. Finally I started to experiment with older, inedible specimens, and voilà. The tissue of older fruitbodies is less dense and fluffier, making it easy to find a few pieces of cottony tissue able to be cloned. First place the fruitbodies in the refrigerator in a paper bag for two weeks to help remove some of the excess water from the tissue, and then take a tissue sample and transfer it to agar. You'll get a good clone almost every time. Adding sawdust from the species of tree the maitake specimen grew on to the agar is highly recommended. You can gradually train maitake strains to adapt to other types of wood in the lab, but this can take years. Cultivators with lab skills can pursue this option using monospore cultures from spore dilutions to germinate fresh prodigy on enriched agar based on the new hardwood species. Monokaryons will explore the agar in search of mates capable of inhabiting and breaking down the wood type. But the ability to adapt to a new wood source may not necessarily equate to prolific fruiting, so if you do this, isolate multiple sectors and colonies for parallel experimentation.

Preferred Fruiting Substrates

Maitake prefers buried hardwood logs and stumps outdoors and supplemented hardwood sawdust indoors. This mushroom prefers oak trees in North America but presumably can be grown indoors on a wider range of substrates if trained to do so.

Outdoor Cultivation Notes

Media preparation. This mushroom will not fruit aboveground. Use buried hardwood logs, rounds cut from logs, and stumps. Make sure logs and stumps are fresh (cut no more than one month before inoculation) to ensure the mushroom has a chance to outcompete any wild fungal species. Inoculate the wood, let it colonize for six to eight months, and then bury it. The fungi typically produce clusters of fruitbodies at the interface between the ground and the wood, on the shadier side. (See chapter 5 for more details.)

Preferred spawn type. Sawdust, plug, or cardboard spawn.

Inoculation to fruiting time. From one to two years, depending on temperature and inoculation rates.

Expected yields. Fair. Maitakes take longer to fruit than other stump-loving mushrooms such as oyster and reishi, but the fruitbodies are extremely large, and they can be dried and used all year. Expect one flush per year outdoors.

Indoor Cultivation Notes

Media preparation. Use supplemented sawdust, sterilized in an autoclave for two hours.

Preferred spawn type. Grain or supplemented sawdust spawn.

Inoculation to fruiting time. From thirty to fifty days. Maitakes can be relatively difficult to fruit indoors because gas exchange is critical if the mushroom is to transition from knotting to petal formation. Some commercial facilities move the fruiting blocks to sequential rooms with specific gas and humidity levels to streamline and automate the process. It's possible to orchestrate the steps of the maitake fruiting cycle in a single room by taking advantage of the difference in weight of the gases, such as carbon dioxide and oxygen, moving fruiting containers higher or lower within an enclosure to put them at the height where the gas ratio they require at a particular stage predominates.

Fruiting blocks should be incubated in bags to allow carbon dioxide to pool at the upper surface of the blocks and promote primordia formation there. After colonization is complete, cut the top corners of the bags (or small slits on the sides, level with the top of the block) to allow modest gas exchange while preserving a high-humidity environment. This small shift in the gas exchange rate is enough to trigger the block to begin "knotting," a precursor to fruiting. Of course, you have to appreciate that you are trying to mimic a situation that, in a natural setting, can take place several inches underground. This knotting can take several days to form into anything recognizable but will soon begin to look like an irregularly shaped dark gray mass. Midway during the knotting phase you will need to begin to gradually increase oxygen exposure, to simulate the underground mushroom arriving at the soil surface. This causes the structure to elongate and flatten or curl into what you typically see with a small, mature maitake.

Expected yields. Fair. Two flushes three to four weeks apart. Most growers plan on only one fruiting; the second flush is unreliable, and given valuable space in a fruiting room, most growers dump the bags or cultures after the initial fruiting. I would suggest dumping this biomass in an outdoor trench and covering it with a casing soil to produce maitakes later in the same or the following year, not letting this valuable biomass go to waste.

Harvesting

Pull the entire attachment of mushrooms out of the substrate and trim the base. Brush away any debris or leaves from the fronds, and remove leaves that may have landed and become trapped between the layers of fronds. These mushrooms, like many polypores, will engulf objects such as sticks and twigs that they cannot push out of the way, wrapping their tissue around the objects, so inspect the mushrooms carefully. Indoor-grown fruitings are extremely clean and only need the excess sawdust cut free from the solid base.

Storage

Fresh maitakes will keep for one month at 38 to 42°F (3–6°C). Maitake is one of the best mushrooms to dry; rehydration restores its crunchy texture almost entirely.

Marketing

This is a crunchy, earthy mushroom that is excellent in stir-fries, casseroles, and egg dishes. It is one of the best edible mushrooms you can find in the wild, with a strong and unique flavor all of its own. Maitakes are more presentable and marketable in small clusters, so I recommend adjusting the size of your fruiting containers, or exposed surface area, when you are cultivating indoors to achieve a desirable size. Promote this mushroom as an edible medicine, touting the health benefits along with its exceptional flavors and textures. Cooking this mushroom and distributing small samples at open-air markets can cast a hypnotic, odiferous trance over the crowds, drawing them to your display of edible goods!

Nutritional Value and Medicinal Uses

Maitakes have a protein content ranging from 27 to 30 percent (dry weight), depending on the strain and growing conditions. They also contain the highest vitamin D levels of any known mushroom, with over 1,250 IU per 100 grams. Studies have found that the mushrooms can help boost and modulate immune system function, regulate blood pressure and sugar levels, lower cholesterol, and provide cardiovascular and digestive support. They contain antiviral and antibacterial properties as well as anti-inflammatory compounds, and they also have been shown to suppress many cancer cell lines.

Uses in Mycoremediation

Much like reishi mycelium, except much slower growing, maitake mycelium has an extremely high tensile strength, making it a good candidate for mycofiltration of water. Its ability to bind a substrate with great tenacity can help keep a biofilter from disintegrating. This mushroom is also great for recycling stumps in forested habitats in temperate climates.

The Genus *Hericium*

Common Species

H. americanum, H. erinaceus (lion's mane, pom-pom)
H. coralloides, H. ramosum (bear's head)

Difficulty Level

Outdoor cultivation—2
Indoor cultivation—2

General Description and Ecology

The *Hericium* species—more commonly called by their market names, lion's mane or pom-poms—are delicious edible mushrooms with health and medicinal benefits. They are primarily found on the eastern coastal plains of the United States, typically fruiting out of wounds and crevices on hardwood trees and

A large, heavily branched bear's head mushroom (*Hericium ramosum*). The related lion's mane (*Hericium erinaceus*) is a dense, more compact species that is ideal for indoor cultivation and ships well.

Wild specimens average a few pounds each, although I have found a few in my lifetime that weighed over 15 pounds. On one occasion I spotted a large pom-pom growing in a vacant lot. Rooted deep inside a massive water oak, the entire ball was about 2 feet in diameter and stuck up inside the rotted-out center of the tree, so I had little leverage for pulling it out. Instead I used a long knife to slice the ball as close to the interior of the tree cavity as I could. Even with leaving a large chunk behind, the mushroom still weighed over 15 pounds, and I traded it to a local chef in exchange for a fine seven-course meal! Two weeks later I happened to be driving through the same town and checked out the tree as I passed the lot. Another lion's mane had regenerated from the tissue I'd left behind, and it had grown into another 15-pound specimen!

snags. Their flavor has been described as like that of crab lobster. To enhance the lobster-like experience, pull the mushroom apart by hand into shreds and then sauté it in oil or butter, with a small amount of garlic.

Mycelium and Spawn

The mycelium is white and very irregular, forming small primordia directly on agar even before it reaches the edges of the plate. It is initially difficult to see on agar due to its extremely fine appearance. I prefer to use a colored agar (I use blue dye with white mycelium) to make the faint mycelium more visible. The mycelium exudes a deep yellow metabolite on overcolonized spawn and fruiting substrates. Its smell can be sweet to slightly sour. Commercial spawn is typically available in the form of cereal grains, supplemented sawdusts, and wooden dowels for outdoor cultivation.

Fruitbody Development

The fruitbody attaches laterally and looks like a velvety ball that eventually differentiates into elongating and cascading teeth that are soft and white when fresh, and yellow when mature or overmature. The mushroom prefers to grow off the sides of logs and trees (and bagged substrate) rather than from above, allowing the growth to hang downward, as it develops its spore-producing spines. The spore print is white.

Common Strains and Ideal Fruiting Conditions

Wild strains can be cloned from older specimens just as they mature and begin to dry out. I have had no luck isolating fresh tissue from young fruitbodies, which are generally dense and watery. Commercial strains are available from many suppliers, and most fruit indoors on supplemented sawdust media sterilized in bags. This mushroom prefers cooler temperatures, most commonly initiating at temperatures from 55 to 65°F (13–18°C).

Wild Spawn Expansion Techniques

Hericium mycelium is extremely weak and fragile, making it difficult to propagate in the wild. My best recommendation is to slice dried specimens into thin sheets and layer them between wet sheets of cardboard. When mycelium has leaped off onto the cardboard, you can propagate it by wafering (inserting colonized cardboard into wounds made with a machete or hachet); see chapter 12 for more details.

Laboratory Isolation and Spawn Culture

This mushroom clones easily from mature specimens. Let them air-dry for a day or two to reduce their water content before taking a tissue sample; otherwise bacteria will dominate the transfers, even on antibiotic agar.

Preferred Fruiting Substrates

Use hardwood logs, primarily oak, maple, and cherry. Strains may vary in their wood preferences; it is essential to match a strain with its preferred wood type to get it to fruit.

Outdoor Cultivation Notes

Media preparation. Fresh logs or stumps will do. Make sure logs or rounds are fresh (cut no more than one month before inoculation) to ensure the mushroom has a chance to outcompete any wild fungal species. This mushroom prefers to emerge from a small cavity, developing its primordia naturally inside shallow holes, such as those made by woodpeckers. Once colonization nears completion, use a paddle-type boring bit to drill a few holes around the wood, about 2 inches deep. The delicate primordia will slowly form in these holes over a few days, gaining some momentum before emerging into the light.

Preferred spawn type. Sawdust, plug, or cardboard spawn.

Inoculation to fruiting time. From six to ten months, depending on temperature and inoculation rates.

Expected yields. Low. *Hericium* mushrooms are not high yielding. Partially burying the inoculated logs or rounds can improve yields. Expect one flush per year. Trim a mushroom when it has half developed, or is about to produce spines, using a sharp knife, and it will regenerate to produce another nice pom-pom in just a week or so.

Indoor Cultivation Notes

Media preparation. Use supplemented hardwood sawdust, sterilized in an autoclave for two hours, spawn, and incubate for two weeks. Many wild strains will fruit within two to three weeks of incubation after a brief cold shock under refrigeration for one day.

Preferred spawn type. Grain or supplemented sawdust spawn.

Inoculation to fruiting time. From three to four weeks. Commercial strains usually fruit within three weeks. High levels of carbon dioxide and humidity are necessary for primordia to form. Once the primordia have formed, they need to be lightly misted; they can rot, so allow the mushrooms to dry slightly between watering. If you're growing the mushrooms in bags, for best results deflate the bags when colonization is complete, typically two weeks, by poking a small hole high in the upper air space and pushing out all the air. Then wrap the plastic flat and tight around the substrate, tape it down tight, and poke two fruiting holes on one side of the bag, facing your aisles or growing area so you can monitor them.

Expected yields. Fair. Commercial operations typically average 1 pound of mushrooms for every 5 pounds of supplemented sawdust substrate.

Harvesting

These mushrooms form singly, rarely producing more than one fruitbody per log or bag. If more than one mushroom is initiated, it typically dries and drops off. Harvest the mushrooms when the spines are approximately ½ inch. As they approach full elongation the shelf life decreases and the spines may yellow, which is not visually attractive to consumers. Harvest by twisting or cutting the mushroom clean, placing the cut side of the mushroom down in your harvest tray to protect the spines. Wrap it individually or deliver several in tight bundles side by side.

Storage

A fresh *Hericium* mushroom will keep for two to three weeks at 38 to 42°F (3–6°C). *Hericum* species do not rehydrate well, becoming chewy when reconstituted. If you must dry them, consider powdering them. You could also cook and then freeze these mushrooms, or can or brine them.

Marketing

With its wonderful seafood flavor, this mushroom sells itself. It can look strange to a first-time consumer, but it is often a favorite for adventurous beginners who want to try something that no one would think was edible just by looking at it.

Nutritional Value and Medicinal Uses

Cultivated *Hericium* have a protein content ranging from 28 to 30 percent (dry weight), depending on the strain and growing conditions, making this one of the most protein-rich of all cultivated mushrooms. Studies have found that this genus is highly antioxidative, which can help in cases of nerve cell damage. It has been shown to reduce symptoms in Alzheimer's and Parkinson's patients.

Uses in Mycoremediation

This genus has shown activity against and inhibition of several bacterial species, most prominently *Staphylococcus*. Mycelium from this mushroom has been used for the treatment of paper bleaching and dye waste.

The Genus *Hypholoma*

Common Species
H. capnoides (conifer brick top)
H. sublateritium (brick top, chestnut)

Difficulty Level
Outdoor cultivation—1
Indoor cultivation—2

General Description and Ecology
Hypholoma sublateritium, commonly sold in Europe and the United States under the name brick top, is relatively easy to grow on hardwood tree species. It can be grown using the log raft method similar to nameko (*Pholiota nameko*) or in wood chip beds using the same parameters for king stropharia

The brick top mushroom (*Hypholoma sublateritium*) fruiting from hardwood chips cased with leaves. Brick tops often share a raft bed with nameko mushrooms (*Pholiota nameko*). Namekos have much slimier cap cuticles when young, which makes them easy to differentiate from the dry-capped brick tops.

(*Stropharia rugoso-annulata*). My only warning for anyone growing this mushroom is that it is a little more difficult than some of the other cultivated mushrooms to distinguish from wild natives such as sulfur tufts (*Hypholoma fasciculare*), which unlike brick tops have greenish-yellow gills and are poisonous. Brick tops have a distinct reddish cap and smoky gray gills, and if you look at these mushrooms side by side you will clearly see the differences. I recommend becoming very confident and experienced with identification of your brick top fruitings before you consume or sell them.

Mycelium and Spawn

The mycelium is white and linear and exudes a light yellow metabolite on overcolonized spawn and fruiting substrates. Its smell can be extremely sweet. Commercial spawn is usually available in the form of wooden dowels or supplemented sawdusts for outdoor cultivation and is rarely available on cereal grains for bulk substrates, making this one a good selection for lab expansion if commercial cultivation indoors is desired.

Fruitbody Development

Brick top is a stemmed mushroom, with the cap centered on the stem. The attached gills are whitish to dull yellow at first and darken to smoky gray at maturity. The cap is orange-red with a lighter yellow margin. The spore print is smoky gray to blackish.

Common Strains and Ideal Fruiting Conditions

Since this mushroom requires some skill to identify, many spawn producers do not carry it on account of liability. If you can find a commercial strain, be sure to inquire about its preferred wood habitat. Wild specimens can be cultured or produced easily from spore germinations; just be sure you are isolating the brick top and not a sulfur tuft! I often find brick tops popping up in my king stropharia beds, like weeds, in which case I escort them directly to the kitchen. This entire group of mushrooms prefers buried wood but can fruit from stacked rounds.

Wild Spawn Expansion Techniques

A hardwood-loving species, this mushroom fruits from stumps and logs but does not reproduce easily from the harvested stem bases. Instead, pull back the bark and harvest some of the elongated stem base or colonized wood near the attachment point. Mycelium from colonized wood will leap off to wet cardboard quickly, allowing you to reinsert the mycelium back into your own logs and stumps in just a few weeks.

Laboratory Isolation and Spawn Culture

Hypholoma species are very easy to get into culture by any method of tissue cloning or spore germination. The caps have enough tissue above their gills that you can easily tease out enough tissue for plating. Adding trace amounts of sawdust from the host tree to the agar medium greatly benefits the culture of these species, especially *H. capnoides*, which is a conifer-degrading mushroom.

Preferred Fruiting Substrates

Most *Hypholoma* mushrooms will fruit from most deciduous hardwood trees, the exception being *H. capnoides*, which prefers conifer wood.

Outdoor Cultivation Notes

Media preparation. This mushroom fruits best from buried logs and stumps. Make sure they are fresh (cut no more than one month before inoculation) to ensure the mushroom has a chance to outcompete any wild fungal species. Trenched log and raft methods work well (see chapter 5). Fruitbodies form typically at the interface between the ground and the wood, on the shadier sides.

Preferred spawn type. Sawdust, plug, or cardboard spawn.

Inoculation to fruiting time. From six to ten months, depending on temperature and inoculation rates. Mushrooms of this genus fruit primarily in cooler months.

Expected yields. Good. *Hypholoma* mushrooms are very prolific fruiters.

Indoor Cultivation Notes

Media preparation. Use supplemented hardwood sawdust, sterilized in an autoclave for two hours.

Preferred spawn type. Grain or supplemented sawdust spawn.

Inoculation to fruiting time. From six to eight weeks. These mushrooms love to fruit down the sides of bags, pushing up against the plastic, then eventually tunneling upward to flatten out their caps. Be sure to cut the bag tops off fairly low (just above the substrate surface); otherwise these mushrooms can become stemmy, and the caps are far more desirable than the fibrous stems. Fogging the fruiting room can help maintain moisture on exposed fruiting blocks to initiate primordia. A cold shock is required to fruit this mushroom consistently indoors, meaning that you will need to expose rested and hydrated containers to a temperature below 60°F (16°C) at intervals that coincide with flushing records.

Expected yields. Fair. Expect two flushes several weeks apart.

Harvesting

These mushrooms form singly or in small clusters. Pull the entire attachment of mushrooms out of the substrate and trim the excess waste away before storing.

Storage

Hypholoma mushrooms will keep fresh for two to three weeks at 38 to 42°F (3–6°C). Dried caps rehydrate well, and the stems can be powdered into flour for use in breads, pastas, soups, and sauces.

Marketing

Hypholoma mushrooms, the most recognizable being the brick top, are similar in texture to black poplar mushrooms (*Agrocybe aegerita*), with a mild nutty flavor and crunchy bite. They are delicious sautéed and can be added to stir-fries and warm salads. Sell them wrapped in bundles, since they tend to fuse at the base into clusters.

Nutritional Value and Medicinal Uses

Hypholoma mushrooms are fibrous and have a protein content ranging from 20 to 30 percent (dry weight), depending on the strain and growing conditions, putting them among the most protein-rich of all cultivated mushrooms. Studies have found that *Hypholoma* species contain antiviral, antibacterial, and anti-inflammatory compounds, and they also have been shown to suppress many cancer cell lines.

Uses in Mycoremediation

Given their cooler fruiting temperature window and ability to inhibit bacteria and degrade chemicals at lower temperatures, this entire genus should be explored for mycoremediation projects in cooler climates.

The Genus *Hypsizygus*

Common Species

H. tessulatus (shimeji)
H. ulmarius (elm oyster)

Difficulty Level

Outdoor cultivation—1
Indoor cultivation—1

General Description and Ecology

Elm oysters and shimeji are relatives, both grouped in the genus *Hypsizygus*. However, they look and taste completely different. Both are relatively easy to grow on hardwood species and sterilized sawdust blends and prefer to fruit under cooler temperatures, normally during spring and fall. Elm oysters can become very large, particularly if the primordia or the fruiting sites are limited to just a few. I have harvested caps that were more than 2 feet across from columns in my greenhouse; when only one cluster emerges, it is quite impressive! Shimeji mushrooms are best cultivated as smaller fruits and in clusters, which shows off their marbled caps.

Mycelium and Spawn

The mycelium is white and linear and exudes a light yellow metabolite on overcolonized spawn and fruiting substrates. Its smell can be sweet to slightly anise-like. Commercial spawn is usually available in the form of cereal grains, supplemented sawdusts, and wooden dowels for outdoor cultivation.

Fruitbody Development

These are stemmed mushroom, with the cap centered on the stem, with attached (shimeji) or descending (elm oyster) gills that are white. The spore print is white for both.

Common Strains and Ideal Fruiting Conditions

Elm oyster strains are common and available from most spawn suppliers, although you'll rarely find this species in the wild, unlike most of the other mushrooms profiled in this book. Shimeji strains are available in a few different cap colors, which include white, brown, and watermarked (tortoiseshell). Fruiting temperatures vary but are usually confined to a window of 55 to 70°F (13–21°C). Strains grown indoors generally initiate at the lower edge of this window; increase temperatures to 60 to 70°F (16–21°C) to bring the mushrooms to maturity before harvesting. Outdoor fruitings generally produce two flushes about a month apart or when moisture becomes available.

Wild Spawn Expansion Techniques

As hardwood-loving species, these mushrooms are well suited for natural cardboard expansion methods for preliminary isolation. Much like oyster mushrooms (*Pleurotus* spp.), the stem bases readily produce mycelium when wrapped in wet cardboard, and within a few weeks sheets of mycelium can be transplanted back into logs or onto pasteurized growing media.

Laboratory Isolation and Spawn Culture

This genus of mushrooms is extremely easy to culture with tissue cloning and spore germinations. I prefer to crack the caps open and peel them apart from the cap down to the stem base, looking for the best cottony, dry tissue to clone from. I rarely perform spore germinations for isolating this genus unless the genetic variability would be useful, such as when I am trying to culture a strain that will be able to break down a specific food source, for example in a specific mycoremediation project.

Elm oysters (*Hypsizygus ulmarius*) are typically much larger than their *Pleurotus* look-alikes. These massive clusters are common on stumps in the wild.

Preferred Fruiting Substrates

In the wild these mushrooms prefer hardwood trees, logs, and stumps. They will also fruit on a straw-based substrate so long as it is finely shredded, since they prefer a high-density fruiting substrate.

Outdoor Cultivation Notes

Media preparation. Grow these species on logs, stumps, or wood chips. Make sure the wood is fresh (cut no more than one month before inoculation) to ensure the mushroom has a chance to outcompete any wild fungal species. Elm oysters can also grow on straw-based substrate, such as might be used in and around garden plants or for mulching pathways.

Preferred spawn type. Sawdust, plug, or cardboard spawn.

Inoculation to fruiting time. From six to ten months depending on temperature and inoculation rates.

Shimeji mushroom (*Hypsizygus tessulatus*) strains vary in their color and cap cuticle ornamentation. These brown shimeji, cultivated on supplemented sawdust, show off their unique watermarked, almost "cracked glass" caps.

Expected yields. Good on stumps and rounds; poor in gardens. Nevertheless, *Hypsizygus* mushrooms are excellent additions to garden; as brown rot mushrooms, they bond with the nitrogen-fixing bacteria that collaborate with mycorrhizae around vegetables and fruit trees.

Indoor Cultivation Notes

Media preparation. Use pasteurized agricultural waste, supplemented sawdust, or wood chips.

Preferred spawn type. Grain or supplemented sawdust spawn.

Inoculation to fruiting time. From three to four weeks.

Expected yields. Good. These mushrooms, when cultivated on straw or supplemented sawdust, are heavy yielding. Highly recommended for commercial production.

Harvesting

These mushrooms can easily be twisted away from their growing substrate and trimmed before storing. Shimeji mushrooms are best harvested as tight clusters, while elm oysters should be allowed to flatten and expand laterally before harvesting.

Storage

These mushrooms will keep fresh for two to three weeks at 38 to 42°F (3–6°C). Older elm oyster mushrooms are better off dried and powdered or sautéed and frozen for long-term storage.

Marketing

Elm oysters resemble white oyster mushrooms but are larger and have a more anise-like flavor. Shimeji mushrooms are crunchier and form in tighter clusters. Both can be sautéed in stir-fries and added to sauces and soups. The two mushrooms, although paired in the same genus, are quite different in appearance, taste, and texture, and you should make these distinctions clear to consumers to help them appreciate the differences.

Nutritional Value and Medicinal Uses

Hypsizygus mushrooms can vary in their protein content, ranging from 4 to 8 percent (dry weight), depending on the strain and growing conditions. The most prominent research centered on this genus is curently investigating its antiviral and antibacterial properties, along with the inhibition rates of many cancer cell lines.

Uses in Mycoremediation

These brown rot mushrooms are efficient wood chip and stump recyclers that are highly compatible with mycorrhizal fungi, making them great companion species on slopes and in highly eroded or disturbed habitats for rebuilding soil.

The Genus *Laetiporus*

Common Species

L. conifericola, L. gilbertsonii, L. persicinus, L. sulphureus (chicken of the woods)

Difficulty Level

Outdoor cultivation—3
Indoor cultivation—5

General Description and Ecology

The many chicken of the woods mushrooms species are distinguished by three characteristics: cap color, pore color, and host tree preference. The mushrooms can be brilliant orange, yellow, and sometimes peach or pinkish in color. Chickens fruiting on conifer and other fragrant wood types may be inedible, since

The many species of chicken of the woods can be distinguished by the color of their upper cap surface and their pores. This strain we found growing on a buried oak log at Isaqueena Falls, South Carolina.

A "tamed" chicken of the woods harvested from a stack of hardwood logs sandwiched with supplemented sawdust spawn. Vince Bambino and Deb Bates cultivated this beautiful specimen from one of our wild cloned ecotypes from the Southeast, which we named "Chicken Phil," after the collector Phil Simms, who sent us the original fruitbody from his farm.

these species tend to absorb the qualities of their host, including resinous compounds, so I prefer hardwood chicken of the woods. They all have a reputation for being good edibles, whether they are foraged or cultivated, so long as they are picked young, when they are silky and tender. The mushrooms do not produce indoors (yet), possibly because of the volum and density of biomass they require. I have one strain collected from Palmetto Bluff, South Carolina, possibly a new species, that creates primordia on agar, which could be promising for indoor cultivation. For now, the easiest method of cultivation remains inoculation of stumps or large wood rounds, situated outdoors in woodland gardens.

Mycelium and Spawn

The mycelium is orange and powdery. It smells like rotten eggs or sulfur. Commercial spawn is usually available in the form of cereal grains, supplemented sawdusts, and wooden dowels for outdoor cultivation.

Fruitbody Development

These mushrooms form either overlapping layers or large soft brackets on wood aboveground, sometimes in large rosettes fruiting from buried wood. The color of the cap and pore surface varies from species to species. Most species are salmon to orange-colored above, with either white, orange, or yellow pore surfaces beneath. The spore print of all species is white.

Common Strains and Ideal Fruiting Conditions

This is a hardwood-loving genus. Strains are best isolated from wild fruitings, so that you can match them up with their preferred wood type. If you are purchasing spawn, be sure to ask your supplier what wood type the strain was isolated from, since each strain prefers its own unique wood chemistry and density and lacks the ability to grow on anything other than its original wood type. This mushroom is not cultivated indoors, and is best suited for outdoor stump or round cultivation.

Wild Spawn Expansion Techniques

I have had no luck with cardboard expansion, although the fruitbodies are easy to clone from slightly drier specimens.

Laboratory Isolation and Spawn Culture

As with maitake, I struggled for years to clone chicken of the woods fungi when they were fresh and in their prime, with no success, only to find that taking tissue samples from older, drier specimens did the trick. I now prefer to clone specimens of this genus once they are almost completely dried. My best results have come from taking a chunk of the base, near the point of attachment, and storing it in a paper bag in the refrigerator for a few days. Then I crack open the flesh, looking for a dry, somewhat powdery texture and no signs of moisture wicking when I press it. Taking multiple samples of the fluffiest tissue with the brightest color typically produces the best isolation results.

Outdoor Cultivation Notes

Media preparation. This mushroom prefers larger-diameter stumps or cut rounds, spawned with layers of supplemented sawdust spawn in between. Make sure the stumps or rounds are fresh (cut no more than one month before inoculation) to ensure the mushroom has a chance to outcompete any wild fungal species. Ground contact greatly enhances the movement of moisture to the developing mushrooms.

Preferred spawn type. Sawdust, plug, or cardboard spawn.

Inoculation to fruiting time. From one to two years, depending on temperature and inoculation rates.

Expected yields. Low. Chicken of the woods is not a high-yielding or predictable fruiter. These mushrooms can become quite large, forming colonies of overlapping clusters. I cultivate these mushrooms mainly as ornamentals and enjoy the occasional fruitings for dinner. Expect one flush per year.

Indoor Cultivation Notes

Indoor cultivation is not recommended, as I have never been able to produce a single fruiting body under sterile conditions, or on supplemented sawdust blocks. This fungus may require microbial symbionts to complete its reproductive cycle and produce a fruiting body. However, grain and sawdust formulas can be used to make antibiotic extracts without even fruiting the mushroom, by stimulating the blocks with a microbial solution. The mycelium responds by producing metabolites that can be harvested and used for medicinal or research purposes.

Harvesting

Harvest these mushrooms with a sharp knife, trimming the outer, silky tips back toward the attachment point. The closer you get to the wood, the tougher and more fibrous it becomes, with a texture more like cardboard than chicken! You can also bend the tips; they will break at the "magic spot" that separates the edible flesh from compost.

Storage

Fresh chicken of the woods mushrooms will keep for two to three weeks at 38 to 42°F (3–6°C). These mushrooms have a longer shelf life than other fleshier mushrooms such as shiitakes and oysters, and cooking them and freezing works quite well. Do not dry this mushroom unless you plan to make powder; it does not reconstitute well.

Marketing

This mushroom should be marketed just as is, with the common name appearing as the headline. Let customers know that the mushroom is best prepared by parboiling the sliced fruitbodies and then cooking the "meat" as if it were chicken. This tends to be a big hit with vegetarians, so you can promote it as a meat substitute.

Nutritional Value and Medicinal Uses

Chicken of the woods mushrooms are fibrous and have protein contents ranging from 12 to 14 percent (dry weight). Studies have found that the mushrooms can help regulate blood pressure and sugar levels and have high levels of anti-inflammatory compounds. This genus has also shown incredibly high activity against *Staphylococcus*, making it a good candidate for further research for those suffering from systemic staph infections.

Uses in Mycoremediation

The species of this genus are well known for their enzymatic activity in degrading a wide range of chemical dyes, making them good candidates for mycofiltration units in textile operations and also for solid-state decolorization of clothing scraps.

The Genus *Lentinula*

Common Species

L. edodes (shiitake)

Difficulty Level

Outdoor cultivation—1
Indoor cultivation—3

General Description and Ecology

Shiitakes are probably the most recognized specialty mushroom in the world. In the late 1970s, a shiitake cultivation revolution shook up an industry of white button domination that had lasted for over a hundred years. Although shiitake mushrooms have been cultivated in Asia for possibly thousands of years, it wasn't until recently that they began to see worldwide popularity, driven in part by their nutritional and health properties, which can make *Agaricus* species seem inferior in comparison. The ease with which shiitakes can be cultivated—especially as opposed to the complex composting process adopted by larger *Agaricus* farms—has made them a darling of small commercial and hobby farmers. Shiitakes are one of the mushrooms I routinely recommend for beginners.

The easiest method of cultivating shiitakes is on hardwood logs outdoors. They can be grown just about anywhere in any climate as long as you use strains that are appropriate for the fruiting temperature windows and weather fluctuations of your locale. The indoor method is more intensive, requiring relatively fresh sawdust (or other substrates) to be sterilized in autoclaves or steam sterilization units, a process that is typically used by larger commercial growers, but can be financially out of reach for beginners.

Mycelium and Spawn

The mycelium is white and linear, eventually becoming bumpy in areas and turning slightly brown. The

Young shiitakes fruiting. Notice the iconic tufts, or scales, that edge the cap margins. These tufts will eventually flatten and almost disappear at maturity or when it rains.

These shiitake mushrooms are just ready to harvest. It is best to pick them just before the cap edges turn upward as they flatten out to produce spores. The curled-in, denser caps of younger shiitakes are preferable to those of mature specimens that have wavy margins, which dehydrate quickly, attract flies, and have a shorter shelf life.

A supplemented sawdust block inoculated with shiitake mycelium eventually darkens to form its own "bark" or outer rind, which eventually differentiates into blistering regions, where shiitake primordia form and quickly swell into mature fruitbodies, ready for harvest in just two to three days.

spawn gradually browns to a dark, chocolate-colored crust when it's exposed to air or after prolonged storage. The odor is distinct, identical to the smell of fresh shiitakes when the spawn is fresh and viable. Small shiitakes commonly fruit on spawn; they can be removed prior to using the spawn. Commercial spawn is usually available in the form of cereal grains, supplemented sawdusts, and wooden dowels for outdoor cultivation.

Fruitbody Development

The shiitake is a stemmed mushroom, with the cap centered on the stem and attached cream-colored gills. The cap is honey brown to dark chocolate in color, nearly black in colder weather, and sometimes covered with distinct whitish fibrils or tufts arranged in zones. The spore print is creamy white.

Common Strains and Ideal Fruiting Conditions

There are many different strains of shiitake mushrooms. Most are uniform in flavor. They are typically described according to the ideal temperature at which they initiate fruiting, as well as according to their cap ornamentation; some, for example, feature unique tufts of fine fibrils arranged in concentric patterns around the cap margin. Spawn producers are constantly changing the names and cross-breeding varieties to create new strains that perform optimally under specific environmental conditions or that have

The same spot on the sawdust block of shiitake four days later, showing how quickly primordia can mature into fruitbodies.

a particular affinity for a growing substrate. Growers interested in year-round production outdoors on logs should consider the temperature fluctuations in their climate when deciding on which strains to choose from, matching strains to evolving weather conditions to create a calendar of cultivation.

I prefer to categorize strains by their temperature fruiting windows, as described below, which is how you can shop for and ask for strains suitable for your area and wood types.

- **Warm strains.** The fruiting window for warm strains generally rises above 75°F (24°C), making them suitable for tropical climates, or for summer months in a temperate climate (with cold and wide-range strains to fill the gaps for the other months of the year). We have had some of our most prolific fruiting the past few years from warm strains in August when temperatures reached 105°F (41°C) during an extreme drought. With supplemental watering, the shaded fruiting beds were much cooler, allowing for the evaporation to cool the surfaces of the logs, drawing out the mushrooms at a rapid rate.

- **Cold strains.** Cold strains can be grown year-round at colder latitudes or at higher elevations, or during the cooler months of the year in temperate climates. Fruiting occurs between 35 and 50°F (2–10°C), although a cold shock below 50°F and lasting a few days can initiate production at any time of year. As with all mushrooms, cold-fruiting strains form darker caps when they're grown at colder temperatures, making them appear dark brown to nearly black. This may be where the common name black forest mushroom comes from.

- **Wide-range strains.** While the warm and cold shiitake strains occupy the ends of the fruiting spectrum, the wide-range strains offer more flexibility, fruiting generally in temperature ranges of 55–80°F (13–27°C). In temperate climates, wide-range strains can fill gaps in the spring and fall when the other strains are resting. This group represents the most common strains of shiitake worldwide. The majority of wide-range strains are being hybridized and trained to specific kinds of wood species and agricultural substrates.

- **Indoor strains.** Indoor (or bulk substrate) shiitake strains have been bred for specific kinds of decomposing wood supplemented with agricultural wastes to create "synthetic" logs or bricks. Genetically these strains have been cultured for their enzymatic efficiency in utilizing the available nutrients and their ability to form a thick outer rind during the browning process, a trait that is sometimes weak or absent from log-grown varieties. Their temperature window typically reflects the consistent and tightly controlled conditions of indoor operations. The most typical indoor strains tend to fruit best between 60 and 70°F (15–21°C), just below room temperature, and flush about every twenty-one days once fruiting has started.

Wild Spawn Expansion Techniques

The stem bases of shiitake mushrooms do not clone as well on cardboard as, say, oyster mushrooms. The dense fibrous stems do not readily produce mycelium, making it more difficult to culture without a laboratory, and shiitake mycelium is much slower growing and weaker than most mushroom cultures in the lab. One method to use is the disk stack method (outlined in chapter 12), where once you have a colonized piece of wood, you can stack new ones on its end and allow the mycelium to run into the next one, then flip them over to colonize the other end. In just a few weeks, the process can be repeated by again adding more rounds and shuffling them into the stacks, using wet cardboard in between to help maintain moisture.

Laboratory Isolation and Spawn Culture

Shiitakes can be cultured by either tissue cloning or spore germinations.

Preferred Fruiting Substrates

Shiitake mushrooms strains are most fond of hardwood trees and sawdusts. They enjoy a substrate that is higher in density to fruit prolifically, so supplemented sawdusts should be packed tightly and particle size kept to a minimum. Spore dilutions on supplanted agar formulas using substrates other than hardwoods can be trained to acclimate to a new growing substrate, as many indoor strains are very specific to the exact wood type, and shiitake mycelium can lose interest in its food source after just a few transfers, meaning that a cultivator cannot overexpand spawn on the same growing media more than five times without risking strain senescing, or loss of vigor.

Outdoor Cultivation Notes

Media preparation. Shiitakes are typically grown on logs or rounds. Be sure the wood is the same type preferred by the strain. Make sure logs or rounds are fresh (cut no more than one month before inoculation) to ensure the mushroom has a chance to outcompete any wild fungal species.

Preferred spawn type. Sawdust, plug, or cardboard spawn.

Inoculation to fruiting time. From six to ten months, depending on temperature and inoculation rates. If the logs or rounds dry considerably during colonization, soak them by submerging them in water overnight. White mycelium will be partially visible on the ends of the logs when colonization nears completion. The mycelium will then gradually brown, further indicating that it is shiitake mycelium and not a competitor fungus. I encourage growers to scratch and smell this mycelium, which has the same odor as fresh-cut shiitake mushrooms, as another confirmation.

Once the logs or rounds are colonized, I recommend submerging them in water for twelve to twenty-four hours (not any longer), which will help initiate fruiting. Once the logs have have been soaked, try to maintain a daily misting schedule to keep the shiitake primordia moist so they do not dry out. If they are drying out, you may need to set up a timer to mist several times a day for just a few minutes at a time, especially during the daylilght hours when evaporation is high. Shiitakes are fleshy to fibrous and form more slowly than other cultivated gilled mushrooms. They are typically mature and ready to harvest within five to eight days, depending on temperature and humidity. Allow the logs to dry between flushes, a span of about eight to ten weeks. Then soak them again completely to stimulate the next flush, and keep the logs misted over the next week to keep the primordia from drying out until they approach maturity. Once the primordia have formed, reduce watering to prevent bacterial blotch and rot.

Expected yields. Very good. Shiitake mushrooms are high yielding compared to other log-grown mushrooms. Expect two to three flushes, yielding 1 to 2 pounds per log that is about 3 feet long and 6 to 8 inches in diameter each year.

Indoor Cultivation Notes

Media preparation. Use supplemented sawdust sterilized in an autoclave for two hours in heat-tolerant filter bags. Once inoculated, the blocks are pressed tightly to form compressed blocks and allowed to colonize and brown naturally, when they are removed from the bags and transferred into a fruiting room with a fog-like environment.

Preferred spawn type. Grain or supplemented sawdust spawn.

Inoculation to fruiting time. From seven to ten weeks. At about the seven-week mark, begin daily inspection of the bags. Gently rub them to feel for slightly bumpy formations, which signal the onset of the browning phase. Browning is a phase of fruiting unique to shiitakes in which the outer surface of the myceliated mass forms a brownish rind, from which primordia form. Browning is triggered by increased exposure to oxygen under high humidity and the secondary metabolite, which is a reddish-brown fluid. It takes about two weeks for the browning cycle to complete, which best occurs under fog-like conditions or simply by leaving the shiitakes in the bag. Some cultivators brown their shiitake blocks inside their bags, which lowers the risk of contamination by green molds during this phase. Others, like me, remove the blocks entirely from the bags to provide a uniform contact with greater oxygen levels. This encourages an even browning around the entire surface of the block. Either way has its advantages and disadvantages, and both result in labor saved and spent. What's important is that you experiment with your specific conditions to determine which method is best given your situation. (See the sidebar for the specifics on each method.)

Once the blocks have browned and appear corky, or are spongy when squeezed, they will be readying themselves for production within a few days. Ideally, you should transfer the cured blocks to a fruiting room with slightly lower humidity (70 to 80 percent) and allow them to dry out slightly between waterings to prevent bacteria and mold contamination. Although I prefer to fog when the mushrooms are pinning,

THE BROWNING PHASE

In order to stimulate the browning phase inside a cultivation bag, you must provide increased ventilation by poking more holes in the bags. Since carbon dioxide is heavier than oxygen, it makes sense to poke the holes just above the surface of the substrate to create a convection current: Fresh air will enter through the micron patch at the top of the bag and flow out of the holes you make at the block surface, essentially bathing the top of the block with the fresh air. Another in-bag trick relies on the use of a custom-made roller with small needles to perforate the surface of the bag around the mass of the colonized block, allowing for hundreds of uniformly placed ventilation points. Of course, having good ventilation in the room itself is critical. The more shiitakes you're cultivating in a room, the greater will be the buildup of carbon dioxide in the room as you ventilate the bags.

Browning outside the bag requires additional humidity to prevent the blocks from drying out. Misters and fogging units that can produce a uniform fog-like environment are perfect for this phase. Workers should wear gloves whenever they are handling the exposed blocks, since this is their most vulnerable state.

Whether in the bag or out, place the browning blocks on shelving to cure, either individually or stacked into walls of fruiting blocks.

Maintain the curing blocks under high humidity—about 80 to 90 percent—at all times. And keep them at a slightly lower temperature—about 10°F (5.5°C) lower—than the optimum fruiting temperatures. These conditions allow for a slow and steady release of gases and metabolites. These gases (like traffic on a highway) can only move so fast through the mycelium and substrate to escape, so keeping the temperature down will slow traffic at the cellular level to prevent blistering and the destruction of the block surface.

hand watering is ideal once the mushrooms begin to mature, especially in fruiting rooms where you have multiple batches at various stages of the fruiting cycle. Hand watering allows you to target the batches that need the watering and leave the rest to dry out a bit between fruitings.

Expected yields. Good.

Harvesting

Shiitakes form singly or in small clusters. Pull the entire mushroom, including its attachment point, out of the substrate and trim it before storing.

Storage

Fresh shiitakes will keep for two to three weeks at 38 to 42°F (3–6°C). Shiitakes dry well but tend to reconstitute a bit chewy; powdering the dried mushrooms is a better option. Sautéing and freezing also works well, helping to preserve the natural rich color of the caps.

Marketing

Shiitakes enjoy a sort of superstardom and have been cultivated commercially worldwide for the past twenty-five years. Log-grown shiitakes command a higher price than those grown on sawdust because the cap-to-stem ratio is higher, and that equates to more usable product. Many growers bring colonized logs to markets for those who want to try to fruit a few pounds themselves.

Nutritional Value and Medicinal Uses

One of the most well-known medicinal foods, shiitake mushrooms have protein contents ranging from 16 to 22 percent (dry weight), depending on the strain and growing conditions, setting them among the most protein-rich of all cultivated mushrooms. They contain nearly all of the essential amino acids, making this as close to a "complete protein" as meat. Studies have also found that the mushrooms can help boost and modulate immune system function, regulate blood pressure and sugar levels, lower cholesterol, and provide cardiovascular and digestive support. They contain antiviral and antibacterial properties as well as anti-inflammatory compounds, and they also have been shown to suppress many cancer cell lines.

Uses in Mycoremediation

Given the antibiotic properties of this genus, I would recommend using shiitakes for creating biological filters for water contaminated with human pathogens and parasites.

The Genera *Macrocybe* and *Calocybe*

Common Species
C. indica (giant milky)
M. titans (giant macrocybe)

Difficulty Level
Outdoor cultivation—2
Indoor cultivation—3

General Description and Ecology
One of the largest gilled mushrooms in the world, the giant macrocybe is usually (though infrequently) found in subtropical to tropical regions of the Caribbean, Central America, Mexico, and occasionally the Gulf States. The architecture of this robust fungus is impressive, with its thick, leathery trunk of a stem and its massive cap, which can be as wide as 4 feet across. It can exist aboveground for as long as two months!

Similar species in the genus *Macrocybe*, such as *Calocybe indica*, cultivated primarily in India and Indonesia, are both edible and cultivatable, but giant macrocybe remains a mystery. We are still uncertain of its edibility and cultivation parameters—and its unique offerings. As an example, a giant macrocybe fruitbody sent to me from the coast of South Carolina had a powdery fungus—similar in appearance to blue cheese—colonizing most of its stem base. Perhaps this fungus was evidence of an interspecies symbiotic relationship? The fruitbody was also insect-free, with no evidence of the larval infestations typical of most gilled mushrooms. This lack of insect pressure, coupled with the giant macrocybe's incredible durability once it has fruited, may signal that this species is unique in ways that are worth trying to understand.

Mycelium and Spawn
The mycelium is white, linear, and clean and does not exude metabolites unless there is contamination in the growing medium. Its smell is like cherries.

Fruitbody Development
This is a stemmed mushroom, with its cap centered on the stem and attached gills. The spore print is white. The stem is rough and covered with granules or a powdery layer of speckled molds. It is extremely tough on the outside, with a thick rind, presumably to hold the tremendous weight of the large cap above, but is soft in the center. The cap is wavy along the margin when mature.

Common Strains and Ideal Fruiting Conditions
There are no known widely circulated or described strains; most are clones from locally collected specimens. These are tropical giants, and probably best suited for those climates or simulated conditions. These heat-loving mushroom prefer to fruit at temperatures above 80°F (27°C).

Macrocybe titans is one of the largest gilled mushrooms in the world. This impressive cluster was found fruiting in a landscaped bed in Bluffton, South Carolina. The giant stems were covered with a thick, powdery layer of microbes and other fungi, possibly suggesting a complex of symbionts needed for fruiting. Tropical fruitings of this species, such as in Mexico or Costa Rica, can reach diameters over 2 to 3 feet across when mature. *Photograph by Ross Pascall.*

Wild Spawn Expansion Techniques

Macrocybe titans is a robust mushroom full of cottony tissue suitable for cloning. I have not attempted cardboard expansion since it is a terrestrial mushroom, not a typical wood inhabiter; this one is probably best cultured in the lab.

Laboratory Isolation and Spawn Culture

This mushroom is easily cloned from the soft, cottony interior tissue of the stem. Score the thick outer rind around the perimeter and then snap it in half to reveal the sterile interior tissue. The extreme upper and lower portions of the stem have cloned best in my experience; the middle can be fibrous, hollow, and more differentiated than the solid portions at the extremities.

I had been storing cultures of this mushroom on dehydrated agar media, but recently I have been using a liquid culture method, where I blend those agar cultures with sterile water, transfer them to storage tubes, and keep them at room temperature as oxygen-deprived slurries, since the liquid mycelium runs out of oxygen quickly and goes into a "hibernation state." The results so far have been surprisingly excellent, with slow but great recovery from the liquid cultures compared to dehydrated agar culturs or grain cultures stored in sterile water under refrigeration. This may provide valuable culture and storage information that can be applied to other tropical mushrooms!

Preferred Fruiting Substrates

Giant macrocybes prefer composted mixtures of grasses, cereal straws, wood mulch, spent coffee grounds, and any other agricultural by-products common in your area, such as bean hulls, cornstalks, cotton waste, beet pulp, and sugarcane bagasse. These warm-weather fruiters enjoy a complex growing medium over a basic one, so be sure to give them a nice variety.

Casing Soil

A microbial casing soil is required. When the entire surface of the substrate is a continuous white mat, dress the bed with the microbial casing soil, which stimulates primordia formation (see chapter 20).

Outdoor and Indoor Cultivation Notes

Media preparation. Shred agricultural by-products such as grasses, cereal straws, and hulls into a fluffy consistency and pasteurize in a hot water bath for one to two hours. Drain and cool, mixing in spawn. Fill beds or containers with the growing medium 8 to 10 inches deep. Once the substrate is colonized, dress the tops of the beds or containers with 2 to 3 inches of hydrated, microbial casing soil to initiate fruiting, and water generously.

Preferred spawn type. Grain or supplemented sawdust spawn; grain spawn will fruit much sooner and produce larger yields.

Inoculation to fruiting time. From three to five weeks, depending on temperature and the type of spawn you use.

Expected yields. Average. These mushrooms are quite large and fruit either singly or in fused, smaller clusters. Expect two rapid flushes several weeks apart.

Harvesting

These mushrooms are extremely large and can be harvested by grasping the massive stalk and twisting it free from the growing substrate. Trim the base to remove any substrate adhering to it before storing.

Storage

This mushroom is amazingly resistant to decay and rot, lasting four to six weeks fresh at 38 to 42°F (3–6°C). The interior of the cap and stem is as soft as marshmallow, whereas the cap cuticle and stem rind are extremely tough.

Marketing

This mushroom is not yet grown or marketed on a large scale, but it deserves attention from growers in tropical regions worldwide.

Nutritional Value and Medicinal Uses

This complex of tropical giants is not well known, and therefore no data is currently available for nutritional properties. The mycelium, however, is highly antibiotic; at our lab facility we've found that it exhibits activity against many strains of *E. coli* and *Staphylococcus*.

Uses in Mycoremediation

With its antibiotic action, I would suggest using the mycelium of this fungus for mycofiltration modules. Also, the natural insecticidal nature of this mushroom group suggests that extracts of the mycelium could be used as natural foliar sprays to reduce pest pressure in fruiting rooms, though this use has not yet been tested.

The Genera *Macrolepiota* and *Lepiota*

Common Species

L. americana (reddening lepiota)
M. procera (parasol)

Difficulty Level

Outdoor cultivation—2
Indoor cultivation—4

General Description and Ecology

The parasol is one of the world's most recognizable mushrooms and a prized edible in Europe and Russia. The tallest and most regal of the gilled mushrooms, the parasol is widespread in temperate climates at lower elevations. A temperate to tropical species, the parasol is part of a larger group of *Lepiota*, common

A lineup of beautiful parasols collected from fruiting beds full of decomposed wood chips.

A towering parasol mushroom stands out in a grassy area or clearing in the woods. Parasols prefer interface environments, such as composting grass near a tree drip line, or the edge between a meadow and the woods.

throughout many regions of the world, including an edible reddening species *Lepiota americana*, implying that the similarities in cultivation of the parasol can be used to explore the cultivation parameters for many more species to be used for food or medicinal compounds, as this family of mushrooms has shown potent antibiotic properties in lab studies. It is an interface species, meaning that it prefers to grow in areas where different environments meet, such as at the edges of grassy fields. This mushroom is enjoyed not only by humans but also by a number of ant species worldwide. Leaf-cutter and other ant species have been cultivating related mushrooms within the *Macrolepiota* genus for millions of years.

With its superior mycelial speed, the parasol mushroom is well suited to take advantage of the early phases of composting biomass, when it is still green and just beginning to lower its antifungal defenses.

Parasol mycelium races through this premature substrate, well ahead of competitors, using its "infantry cells" to colonize as much territory as possible.

Mycelium and Spawn

The mycelium is white and linear and almost never exudes a metabolite in culture. The grain spawn of my particular strain also exhibits consistent sclerotia-like mass formations at about four weeks of incubation. Once the spawn is transferred to supplemented sawdust, the number of sclerotia decreases but the size of each one magnifies, usually forming a single massive node that appears to put a great deal of pressure on the growing bag or block—a phenomenon that can occur even under cold storage. It is unclear why this phenomenon consistently exists in strains from the southeastern United States and not in those from European stock.

Fruitbody Development

This is a stemmed mushroom, with the cap centered on the stem. It has an extremely large stem-to-cap ratio, meaning it is very tall. The cap is light brown and darker in the middle, with a slightly raised umbo or nipple. Young, tight fruitbodies are smooth on top, with streaks of brown fibrils and minute scales that make them appear somewhat powdery. Later they become more or less scaly, depending on the strain. The stem is also streaked and fibrous; there is always a double-edged annulus, or ring, near the upper third of the stem that you can easily slide up and down. The spore print is creamy white. The poisonous green-spored parasol (*Chlorophyllum molybdites*) can look similar, but its cap is mostly white, it has larger scales on the cap, and it has a distinctive green spore print that makes it hard to confuse with the true parasol.

Common Strains and Ideal Fruiting Conditions

Parasols are easy to collect and clone from wild stock. Although the soft tissue in the cap is only a thin layer compared to other gilled mushrooms, it is typically dry and transfers easily to agar media. There are many commercial strains that have demonstrated successful outdoor fruitings, but if you're looking to buy one, ask about its original fruiting substrate and environmental conditions to make sure they match up with your own conditions and resources. The reddening *Lepiota* (*Lepiota americana*) can also be found and cloned, although proper identification is needed to make sure you are not culturing and collecting poisonous *Lepiota* specimens. I like to visit NAMA-sponsored mushroom hunts (namyco.org) and remove specimens from the identification tables at the end of the weekend for culturing, double-checking with the mycologist on duty to verify in person.

Parasols are truly "wild" mushrooms, enjoying the freedom to wander and colonize substrates outdoors. The primordia prefer tall, grassy interface environments on the edges of lawns and woodland gardens, where debris is plentiful and the excessive water from drip lines of trees rains down upon them. Once the primordia form, the mushrooms are very resilient in drier conditions, and can still mature into a large, developed fruitbody even after several days without rain or watering.

Wild Spawn Expansion Techniques

You can use stem bases to inoculate cardboard or spent coffee grounds for expansion techniques. Parasols are dependent on interaction with microbial communities for fruiting; you'll also want to harvest the stem bases to prepare a microbial slurry for use in a casing soil for your cultivation.

Laboratory Isolation and Spawn Culture

You can easily clone tissues from parasols. Take tissue samples from the interior, preferably from the base of the stem or from the cap.

Preferred Fruiting Substrates

Parasols prefer composted mixtures of grasses, cereal straws, wood mulch, spent coffee grounds, and any other agricultural by-products common in your area, such as bean hulls, cornstalks, cotton waste, beet pulp, and sugarcane bagasse. These temperate fruiters prefer a complex growing medium over a basic one, so be sure to give them a nice variety.

Casing Soil

A microbial casing soil is required. When the entire surface of the substrate, or just below if you scratch around in a few spots to check, is a continuous white mat, dress the bed with a microbial casing soil, which stimulates primordia formation (see chapter 20).

Outdoor Cultivation Notes

Media preparation. Shred any agricultural by-products you intend to use in the growing medium, such as grasses, cereal straws, and hulls, into a fluffy consistency. Soak the growing medium mixture in water overnight, then drain. Prepare a cultivation bed by digging a series of holes or trenches in the garden or near a tree line. Fill beds with the growing medium 8 to 10 inches deep, poke narrow holes in a diamond pattern every 6 to 8 inches across the surface, and fill the holes with spawn. Cover the spawned holes with wet compost to keep the spawn from drying out and to keep insects from attacking it immediately. Cover the substrate with perforated plastic or with plywood that has had holes drilled through it to preserve moisture while still allowing for modest gas exchange. When colonization is complete, dress the beds with 2 to 3 inches of a microbial casing soil. Cover the casing with an annual grass seed to provide a short grassy covering and microclimate where the mushrooms can form.

Alternatively, you could use sawdust or grass seed spawn and take advantage of the work of ants (and other terrestrial insects) to tend your mushrooms. Scatter the sawdust or grass seed spawn in the fall or spring. Ants will grab particles of the spawn and transfer them underground, where they will thread through the subterranean environment. You can expect fruitings in the following late summer or fall. You may find it useful to broadcast the sawdust or grass seed spawn over large areas of freshly cut lawn or in piles of grass clippings using a fertilizer spreader.

Preferred spawn type. Grain or supplemented sawdust spawn.

Inoculation to fruiting time. From four to eight months.

Expected yields. Average to poor. This mushroom is not known to produce commercial quantities and is best suited to home gardens.

Indoor Cultivation Notes

Media preparation. Shred agricultural by-products such as grasses, cereal straws, and hulls into a fluffy consistency and pasteurize in hot water bath for one to two hours. Drain and cool. Fill beds or containers with the growing medium 8 to 10 inches deep, poke narrow holes in a diamond pattern every 6 to 8 inches across the surface, and fill the holes with spawn. Cover the spawned holes with wet compost to keep the spawn from drying out and to keep insects from attacking it immediately. Cover the substrate with perforated plastic or with plywood that has had holes drilled through it to preserve moisture while still allowing for modest gas exchange. Once the substrate is colonized, dress the containers with 2 to 3 inches of microbial casing soil to initiate fruiting, and water generously. Maintain trays or containers in a high-humidity or fog-like environment until primordia form, then reduce humidity and watering as mushrooms enlarge and mature.

Preferred spawn type. Grain or supplemented sawdust spawn; grain spawn will fruit much sooner and produce larger yields.

Inoculation to fruiting time. From six to eight weeks, depending on temperature and the type of spawn you use.

Expected yields. Average. These mushrooms are quite large and fruit either singly or in fused, smaller clusters. Expect two rapid flushes several weeks apart.

Harvesting

Parasol mushrooms are loosely attached to their substrates, and have a bulbous base, which can be gripped with two fingers and simply twisted to snap the mycelial cords attaching it to the substrate. Trim excess compost and store. Save stem bases for future microbial casing slurries to stimulate fruiting.

Storage

Fresh parasols will keep for two to three weeks at 38 to 42°F (3–6°C). They can be dried as well; the caps rehydrate well, while the stems should be powdered.

Marketing

The caps are the most desirable part of this mushroom—they look great in packaging and also dry well. The stem separates cleanly from the cap, partly due to its free gills, making it quick and easy to harvest and clean the caps.

Nutritional Value and Medicinal Uses

Parasols are fibrous mushrooms, with the caps being the best edible component, with protein contents ranging from 12 to 22 percent (dry weight). Studies have shown that the mycelium and fruitbody contain antiviral, antibacterial, and anti-inflammatory compounds, and they also have been shown to suppress many cancer cell lines.

Uses in Mycoremediation

These mushrooms are terrestrial and have an affinity for mulch and compost piles, which means they can be used for filtering contaminated compost sediment, such as leachate from livestock waste, using their antibacterial properties to reduce coliform counts.

The Genus *Pholiota*

Common Species

P. nameko (nameko)

Difficulty Level

Outdoor cultivation—1
Indoor cultivation—2

General Description and Ecology

Although namekos are commonly cultivated in Japan, they have yet to go mainstream in the rest of the world—perhaps because of their slimy caps. The sliminess factor can make the nameko a hard sell, which is a pity because it is delicious, and you can easily cook off the extra moisture on the cap surface.

One spring we inoculated a log raft with a new strain of nameko we had purchased from another spawn supplier. Earlier we had fruited that strain indoors on supplemented, sterilized sawdust blocks but had harvested only a few clusters per block. So our expectations for outdoor log raft cultivation were fairly low. The fall after we inoculated the logs, following some cooler weather, Olga and I walked out to the log raft site and were stunned to find over a thousand namekos fruiting from this very small area. We could not even see the surface of the logs between these crowded mushrooms. The raft fruited heavily for two years and then quickly diminished as it was reduced to soft, decomposed humus.

Mycelium and Spawn

The mycelium is white and linear and exudes a light yellow metabolite on overcolonized spawn and fruiting substrates. Its smell can be sweet to slightly anise-like. Commercial spawn is usually available in the form of cereal grains, supplemented sawdusts, and wooden dowels for outdoor cultivation.

Fruitbody Development

Nameko is a stemmed mushroom, with the cap centered on the stem. Its descending gills are smoky gray at first and then darken at maturity. The gills are never yellow or greenish. The spore print is charcoal gray to black.

Common Strains and Ideal Fruiting Conditions

Fruiting temperatures are confined to a narrow window of 55 to 65°F (13–18°C). Strains grown indoors generally initiate at the lower edge of this window. The mushrooms will mature as temperatures increase to 60 to 70°F (16–21°C). Outdoor fruitings generally produce two flushes a few weeks apart.

Wild Spawn Expansion Techniques

Namekos are a hardwood-loving species and another mushroom that merits some experimentation with natural cardboard expansion methods for preliminary isolation. Since this mushroom fruits from stumps and logs but does not reproduce easily from the harvested stem bases, pull back the bark and find some of the elongated stem base or colonized wood near the attachment point to culture.

Laboratory Isolation and Spawn Culture

Namekos and most *Pholiota* species are easy to clone from the interior cap tissue. Supplement your agar with screened samples of wood from the host tree to help the mycelium recognize and colonize the growing susbtrate.

Preferred Fruiting Substrates

Namekos prefer hardwood logs and stumps outdoors and supplemented sawdust indoors.

Nameko mushrooms pushing up through leaf cover, which helps maintain humidity during primordia formation. These mushrooms are triggered to fruit by a drop in temperature, typically at the time when deciduous leaves begin to fall to the ground to provide a beneficial ground cover.

Outdoor Cultivation Notes

Media preparation. For outdoor cultivation no heat treatment is necessary, since this technique requires inoculating fresh logs or stumps. This mushroom fruits better from buried hardwood logs. Make sure the logs are fresh (cut no more than one month before inoculation) to ensure the mushroom has a chance to outcompete any wild fungal species. The trenched log method works well; see chapter 5 for details.

Preferred spawn type. Sawdust, plug, or cardboard spawn.

Inoculation to fruiting time. From ten to twelve months, depending on temperature and inoculation rates. This is a cold-season mushroom, so bear in mind that after colonization is complete, fruiting will occur only when the appropriate temperature window occurs.

Expected yields. Very good. Namekos are dense and quite prolific fruiters, covering the faces of buried logs or the bottom edges of rounds and stumps. Expect two flushes several weeks apart at the beginning of your cold season, lasting two to three years.

Indoor Cultivation Notes

Media preparation. Use supplemented hardwood sawdust, sterilized in an autoclave for two hours.

Preferred spawn type. Grain or supplemented sawdust spawn.

A carpet of nameko fruiting all at once from the fused mycelia of adjacent buried logs. Cultivating namekos on buried logs is highly productive, but the logs rot quickly and typically last for only two years.

Inoculation to fruiting time. From four to six weeks.

Expected yields. Fair. Expect two flushes several weeks apart. This mushroom may need a cold shock, mimicking outdoor fruiting temperatures, to trigger fruiting. Fogging the indoor environment greatly improves primordia formation with this species.

Harvesting

These mushrooms form dense clusters, arising from a slimy mass, looking as if they are covered with clear jelly. The mushrooms are mature when the cap separates from the stem and begins to expand. The longer the cap is left to mature, the less of the slime layer persists, although the risk of rot increases. Harvest namekos with a pair of scissors and a plate or pan to catch the mushrooms as you cut, since they are too slippery to grab and pull free from the substrate.

Storage

Fresh namekos will keep for two to three weeks at 38 to 42°F (3–6°C). These mushrooms do not reconstitute well when dried; powder them instead. You can also sauté and freeze them or can them in a brine.

Marketing

This is a soft, meaty mushroom with a texture similar to that of oysters. It has a sweet, nutty flavor and is one of the main ingredients in traditional miso soup. If you have a good and reliably producing crop, consider approaching Asian (especially Japanese) markets and restaurants, since these mushrooms are highly regarded in Eastern culture and you may find a local buyer willing to pay a good price for fresh product.

Nutritional Value and Medicinal Uses

Cultivated nameko mushrooms are known to have a protein content ranging from 20 to 22 percent (dry weight), depending on the strain and growing conditions. The most well-known studies show that they have antiviral, antibacterial, and anti-inflammatory compounds, and they also have been shown to suppress many cancer cell lines.

Uses in Mycoremediation

With their antibacterial properties, this mushroom could be useful for mycofiltration of water contaminated with bacterial pathogens. Studies suggest that namekos are highly active against coliforms and other human pathogens found in livestock waste runoff.

The Genus *Piptoporus*

Common Species
P. betulinus (birch polypore)

Difficulty Level
Outdoor cultivation—3
Indoor cultivation—5

General Description and Ecology
Piptoporus betulinus is a corky polypore that fruits primarily from birch trees in temperate climates. The primordia start out as small white balls that eventually flatten and differentiate into hoof-shaped polypores with a tightly inrolled margin. The entire mushroom is white, although it sometimes has dingy streaks or fibrils covering the upper cap surface, and it occasionally hosts algae or moss near the attachment point, depending on the humidity and climate of the growing region. Like the iceman polypore (*Fomes fomentarius*), it was found in the pouch of Otzi, the prehistoric nomad found in the Alps frozen in ice.

Mycelium and Spawn
The mycelium is clear at first, becoming white and linear as it develops. Its smell is slightly astringent to sour. Commercial spawn is usually available in the form of wooden dowels for outdoor cultivation.

Fruitbody Development
This is a hoof-shaped corky polypore with an off-center, lateral stem usually fruiting from oak stumps and trunks. Sometimes the mushroom lacks a stem and fruits directly from aboveground wood. The fruitbody may elongate if it's arising from buried wood or an underground cavity. It fruits primarily singly, although it sometimes fuses in groups near the base of the stem attachment or, rarely, in overlapping clusters. The underside of the cap is white when it's fresh, with minute pores covering its entire surface. The spore print is white, but is rarely attainable since this is a slow-growing mushroom and the pores are tightly bound, making spore release improbable on young specimens.

Common Strains and Ideal Fruiting Conditions
Spawn is rarely available for this species, meaning wild collection is the best option for cultivators. Finding and using local ecotypes will prove to be the best cultivation strategy.

Wild Spawn Expansion Techniques
You can clone this mushroom by slicing back the bark near the point of attachment for a wild

The birch polypore (*Piptoporus betulinus*), a medicinal mushroom exhibiting a wide range of beta-glucans and naturally occurring antibiotics, fruiting outdoors on a log half buried in the soil via the "totem method."

fruitbody and inserting small pieces of wet cardboard. Water occasionally, or cover the spot with plastic. Within several weeks the mycelium will have leaped off onto the cardboard, which can then be used for expansion.

Laboratory Isolation and Spawn Culture

This mushroom clones easily from slightly older, dried specimens, even months after being harvested. The best method I have found is to submerge the mushroom upside down in water, leaving the attachment point sticking up above the water, seal it in a plastic bag, wait for mycelium to re-form on the stem, and then cut away this growth for a first transfer onto either cardboard or antibiotic agar. (For more discussion of this technique, see "Special Techniques for Culturing Finicky Mushrooms" in chapter 18.)

Preferred Fruiting Substrates

Birch polypores prefer birch logs and stumps outdoors.

Outdoor Cultivation Notes

Media preparation. These species prefer to fruit from standing trees or large-diameter downed trees. If you're bringing in logs for cultivation, you can lay them on their sides or stand them up, one end buried for stability, totem-style. Make sure the logs are fresh (cut no more than one month before inoculation) to give the mushroom a chance to outcompete wild fungal species. After a year of colonization, use a paddle-type boring bit to make a series of shallow cavities in the wood, 8 to 10 inches apart, mimicking large beetle galleries or woodpecker holes. The primordia will slowly form in these holes, gaining some momentum before emerging into the light.

Preferred spawn type. Sawdust, plug, or cardboard spawn.

Inoculation to fruiting time. From one to three years, depending on temperature and inoculation rates. The mushrooms are corky and form slowly. They typically mature and are ready to harvest within thirty to fifty days from the time when the primordia first emerge.

Expected yields. Very low. These polypores are extremely slow growing and are not high yielding, making them increasingly rare in the wild.

Indoor Cultivation Notes

Indoor cultivation is not recommended. I have never had any luck fruiting this species indoors. I do sometimes inoculate grain and then, after colonization, prepare extracts from the myceliated grain.

Harvesting

Birch polypores snap away easily, and cleanly, from their host tree; just pull the cap upward.

Storage

Fresh birch polypores will keep for one month at 38 to 42°F (3–6°C); then they start to develop moldy pores. For long-term storage, slice and dry them, and store the dried slices in vacuum-sealed plastic or airtight containers.

Marketing

This is a strictly medicinal mushroom and should be sold as such, although the potential for using the powder or extract in food products such as beer (see chapter 14) makes it particularly exciting.

Nutritional Value and Medicinal Uses

Since this mushroom is not directly edible, no nutritional information is currently available. This mushroom is best known for its antibiotic and antiviral properties. Studies have found that birch polypore can help boost and modulate immune system function and also suppress many cancer cell lines.

Uses in Mycoremediation

Due to its antibiotic properties, I would suggest that the mycelium of this mushroom, just like that of *Fomes fomentarius*, would be highly valuable for use in mycofiltration modules for water contaminated with fecal coliforms.

The Genus *Pleurotus*

Common Species

P. citrinopileatus, P. cornucopiae (golden oyster)
P. djamor, P. eöus, P. flabellatus (pink oyster)
P. eryngii (king oyster)
P. floridanus (Florida oyster)
P. ostreatus (tree oyster)
P. columbinus (blue oyster; possibly a variety of *P. ostreatus*)
P. pulmonarius (phoenix oyster)
P. tuber-regium (king tuber)

Difficulty Level

Outdoor cultivation—1
Indoor cultivation—1

General Description and Ecology

With their rapid rate of growth and adaptability to a wide range of growing substrates, oyster mushrooms can be grown just about anywhere. The strains themselves are common worldwide and can be trained to tolerate unique or harsh climate conditions and a variety of growing substrates. Given their adaptability and ubiquitousness, oyster mushrooms are perfect for beginning cultivators. My recommendation is to attempt to isolate a local strain to train on the substrates that you have available. In climates where temperatures fluctuate, consider alternating strains throughout the year to develop a production cycle that allows continuous fruiting. Oyster mushrooms are some of the fastest-fruiting fungi on the planet, making them perfect for educational projects and disaster relief. To make harvesting easier, choose strains that cluster, rather than those that form singly, since harvesting accounts for a lot of the labor involved in a larger operation.

A commercial strain of warm blue oyster (*Pleurotus columbinus*) fruiting on bags filled with pasteurized wheat straw and cottonseed hulls.

A white strain of tree oyster mushrooms (*Pleurotus ostreatus*) fruiting from an opening in a plastic cage. Note the long stems, which formed when the bag was pulled up high to preserve humidity, causing carbon dioxide to pool and the stems to stretch.

The fast-growing pink oyster mushroom (*Pleurotus djamor*) fruiting from pasteurized peanut hulls just seven days after inoculation.

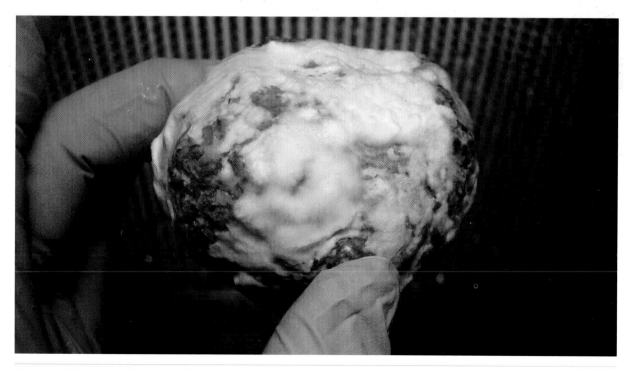

King tuber (*Pleurotus tuber-regium*) sclerotia, which are highly prized for their high levels of protein and active medicinal compounds, including beta-glucans.

The brown oyster, a strain of *Pleurotus ostreatus*, shows off its robust, meaty fruitbodies and puffy, tender, edible stems.

Mycelium and Spawn

The mycelium is white and linear and exudes a light yellow metabolite on overcolonized spawn and fruiting substrates. Its smell can be sweet to slightly anise-like. Commercial spawn is typically available in the form of cereal grains, supplemented sawdusts, and wooden dowels for outdoor cultivation.

Fruitbody Development

The flattened cap is typically off-center on the stem; it has a trumpet-like appearance with descending gills. Many commercial strains exhibit minute cross veining or webbing between the gills. The center of the developing cap is usually depressed, ideal for holding moisture between waterings, and flattens at maturity to release a large amount of spores. Depending on the strain, the fruitbodies may grow singly, in groups of singles, or in fused clusters. The spore print for most oysters is white, with the exception of the pink oyster (pinkish-white spore print) and the phoenix oyster (light lilac spore print).

Common Strains and Ideal Fruiting Conditions

Fruiting conditions will vary depending on the strain you choose. There are many different kinds available. Some of the more common ones include the following.

Massive golden oyster mushrooms (*Pleurotus citrinopileatus*) clustering at the interface between two large nursery pots filled with pasteurized cotton waste.

- *P. citrinopileatus* (golden oyster). This oyster looks like a bouquet of flowers, with branching clusters of brilliant yellow caps on white stems. The fragrance of the fresh mushroom is sweet, resembling that of watermelon or other fruit, while the cooked fruitbodies taste nutty and meaty, like cashews. This mushroom is extremely brittle, and handling it at any stage of development can shatter the cap and stem, so take care when you are growing or transporting it.
- *P. columbinus* (warm blue oyster). This is a high-yielding, clustering strain. Its pins and young primordia are steel blue, but the color quickly fades to steel gray as the mushrooms near maturity. Light plays a critical role in pigment development; too little can prevent pigments from forming, while too much can bleach the mushrooms nearly white. If you want to retain the blue color, adjust your lighting and harvest the clusters a bit earlier than you would other strains. This strain has a long shelf life and does not crack or bruise easily, making it a great choice for shipping beyond local markets. It also enjoys a much wider fruiting temperature window than most of the oyster strains you can buy, fruiting at anywhere from 55 to 90°F (13–32°C). However, temperatures between 70 and 80°F (21–27°C) will encourage rapid flushing cycles (only two to three weeks apart).

- *P. columbinus* (cold blue oyster). Another blue oyster and sister to the warm blue, this is a strain that picks up on the lower temperatures that tend to make the warm blue go dormant. The cold blue also retains more of the blue pigments during maturity, possibly due to the cooler temperatures. The primordial mounds that first emerge are nearly black! Use the same caution with lighting as with the warm blue strain to retain as much blue color as possible. These mushrooms are strikingly beautiful, especially when you display them alongside golden and pink oysters. Fruiting is triggered by a drop in temperature down to 45 to 65°F (7–18°C), so the best cultivation plan is to colonize this strain in a warmer environment and transfer it into a cooler, humidified room for fruiting. Raising the temperature slightly after fruiting will reduce the time until the next flush.

- *P. djamor, P. flabellatus* (pink oyster). This is a stunning oyster with cap colors ranging from light pink to deep red, with a more intense gill color below. For most of the fruiting phase the cap is flat, but it will curl at extreme maturity. The fruitbodies

King tuber sclerotia fruiting in moist gravel after being planted two weeks earlier.

form in dense, overlapping clusters. This meaty mushroom loves extreme heat, with a fruiting temperature window of 80 to 100°F (27–38°C). It has an unusually fast rate of growth; fruiting can occur in as few as seven to ten days after inoclation, with rapid flushes every ten days in optimal warm growing conditions. During colonization, this mushroom does not form as dense a mycelial mat as related *Pleurotus* species. Sometimes this fungus fruits so quickly that the substrate looks lighter in color than when it was just pasteurized. The pink oyster has a mild seafood flavor and a more meat-like texture than other oysters. And it is also well known for outcompeting molds in culture—I have repeatedly seen the mycelium of this species race toward and consume mold species!

- *P. floridanus* (Florida oyster). As the name suggests, this oyster is commonly found at the southern tip of Florida, but its distribution is global in the subtropical bands. Fruiting generally occurs at a relatively high temperature, in the range of 75 to 90°F (24–32°C).

- *P. ostreatus* spp. (white tree oyster). This pure white oyster strain is geared for a relatively cool fruiting temperature window of 55 to 65°F (13–18°C).

- *P. ostreatus* spp. (brown tree oyster). Some of my favorite strains of *P. ostreatus* are the brown oysters. This strain represents most native oyster collections worldwide, commonly found in the same regions as white-capped strains. Brown oysters typically form singly, rather than in clusters, which makes harvesting more laborious, and they are more photosensitive than other strains, so you'll need darker cropping containers in order to prevent too many primordia from forming. The best strain I have found is one with an edible stem that is puffy and tender all the way down to its attachment point. The stems of most oyster strains are generally tough, but you can cut the stem of this brown oyster into thin rounds and prepare them like fresh scallops, which may be a unique feature of this particular mushroom I cloned from my area, whereas most oyster mushrooms have a dense and fibrous stem too tough to eat. Brown oysters can

Phoenix oyster mushrooms (*Pleurotus pulmonarius*) fruiting from logs that are half buried in the ground can mature quickly and become quite large in the summer heat.

fruit at anywhere between 45 and 80°F (7–27°C), depending on the strain, but the cooler strains are more meaty and tend to have less dense (more edible) stems and a longer shelf life.

- ***P. pulmonarius* (phoenix—indoor commercial white).** Phoenix strains are a broad collection from primarily European stock that fruit between 75 and 85°F (24–29°C). The yields are quite high when they're grown indoors on pasteurized substrates, especially supplemented sawdusts or cereal grain straws.

- ***P. pulmonarius* (phoenix—wild pine).** Some strains being offered by spawn suppliers are claimed to grow well on a few conifer species, namely Douglas fir and Scotch pine. However, I have not had any luck fruiting them in any substantial quantity compared to strains particular to hardwood trees. Despite the low yield, phoenix oysters isolated from conifers may make good candidates for recycling conifer stumps and reclaiming habitat in conifer-dense logging regions.

Wild Spawn Expansion Techniques

Mushrooms in this genus are among the easiest to clone and generate spawn from without the use of lab equipment. The easiest method is to wrap the stem bases from harvested mushrooms in wet cardboard and place them in plastic bags to preserve humidity. The stem bases revert to a mycelial state, quickly fuzzing over and colonizing the cardboard. Once the cardboard has been colonized, remove the fleshy stem base to prevent molds and bacteria from developing.

Laboratory Isolation and Spawn Culture

This entire genus clones easily from slightly dried fruit-bodies. Take tissue samples from the interior of the cap or stem. Spore germinations are also extremely reliable.

Preferred Fruiting Substrates

Pleurotus species generally prefer hardwood logs and stumps outdoors and pasterurized agricultural waste indoors.

Outdoor Cultivation Notes

Media preparation. These mushrooms benefit from ground contact, so use a submerged log technique, such as trenched logs or a log raft (see chapter 5). Make sure the logs are fresh (cut no more than one month before inoculation) to ensure the mushroom has a chance to outcompete any wild fungal species. After six to eight months of colonization, use a paddle-type boring bit to make a series of shallow cavities in the wood, 8 to 10 inches apart, mimicking large beetle galleries or woodpecker holes. The primordia will slowly form in these holes, gaining some momentum before emerging into the light. An oyster mushroom log patch can fruit for two to three years.

Oysters can also be grown on wood chips alone, using supplemented sawdust spawn, planted in between rows of vegetables in the garden, or along woodland pathways. You could also use colonized cardboard as spawn, propagating by means of the log disk or wafering methods (see chapter 12).

Although oysters are among the easiest mushrooms to cultivate, outdoor fruitings of this genus are extremely prone to bug infestations, such as fungus gnats and mushroom beetles, which can greatly reduce the quality and shelf life of the mushrooms. Use fine-mesh netting or row cover fabric to shield developing mushrooms while they are maturing; the coverings will also provide shade and moderate gas exchange.

Preferred spawn type. Sawdust, plug, or cardboard spawn.

Inoculation to fruiting time. From six to eight months, depending on temperature and inoculation rates.

Expected yields. High. Expect two to three flushes several weeks apart during the temperature fruiting window specific to the strain, lasting for several years.

Indoor Cultivation Notes

Media preparation. Shred any agricultural by-products, such as grasses, cereal straws, and hulls, into a fluffy consistency. Pasteurize the substrate in a hot water bath for one to two hours. Drain, cool, and inoculate.

Preferred spawn type. Grain or supplemented sawdust spawn; grain spawn will fruit much sooner and produce larger yields.

Inoculation to fruiting time. From three to five weeks, depending on temperature and the type of spawn you use.

Expected yields. Very high. Expect four or five rapid flushes several weeks apart.

Harvesting

Depending on the strain, oysters will grow singly or in fused, small clusters. Harvest by twisting them to remove the entire base from the growing medium, and trim them to remove any residual substrate. Handle them carefully to avoid damaging the caps. Arranging these mushrooms uniformly in boxes, facing the same direction, allows for tighter and more protective storage. Golden oysters are extremely brittle and should be handled with great care from harvest to delivery; they're ideal for local markets, since out-of-town producers cannot ship them without considerable damage. Pink oysters ship very well, and I prefer to stack them in boxes with their gills facing up to show off the brilliant pink to red color below, which is much brighter than the cap surface.

Storage

Fresh oyster mushrooms will keep for one to two weeks at 38 to 42°F (3–6°C). When dried, the caps and stems do not rehydrate well; they should be powdered instead. Sautéing and freezing or canning these mushrooms in brine also works well for long-term storage.

Marketing

Oyster mushrooms are extremely variable in flavor, so you will need to explain to consumers the individual characteristics of the strains you're offering. Most oysters have a sweet, slightly anise-like flavor and are high in protein compared to other mushrooms.

Nutritional Value and Medicinal Uses

Oyster mushrooms vary in their protein content, ranging from 15 to 30 percent (dry weight) depending on the strain and growing conditions. Studies have found that oyster mushrooms have a remarkable ability to shrink tumors in mice, decrease obesity, and regulate blood sugar, suggesting them as a possible nutritional adjunct therapy for diabetes patients.

Uses in Mycoremediation

Oyster mushrooms are well known for their ability to degrade chemicals, such as hydrocarbons and pesticides. We have used wild strains of *Pleurotus ostreatus* to degrade atrazine in both soil and water. This mushroom also has potential for use in mycofiltration modules against coliform bacteria.

The Genus *Sparassis*

Common Species

S. crispa, S. herbstii, S. spathulata (cauliflower)

Difficulty Level

Outdoor cultivation—3

Indoor cultivation—4

General Description and Ecology

Cauliflower mushrooms can be found in most temperate forests worldwide. Collecting these mushrooms is exciting, since even individually they can be quite large, and sometimes they form numerous clusters around the base of a dying tree or stump. They have a pungent, spicy odor.

Many *Sparassis* species partner with particular tree species, primarily conifers. Since conifers resist decay and rot, they can host cauliflower mushrooms for years. Even after the trees crack and fall, the stumps continue to produce mushrooms, which arise from bits of buried root fragments. Most tree experts consider cauliflower mushrooms to be weak tree parasites as well as an "indicator species" for healthy forests; in

A cauliflower mushroom fruiting from a pine round buried in the soil, showing the distinctive "egg noodle" folds endemic to this type of fungus.

other words, where cauliflower mushrooms grow, forests are generally healthy, and conversely, the absence of cauliflowers can signal poor ecosystem health.

Mycelium and Spawn

The mycelium is white and cottony and exudes a light yellow metabolite on overcolonized spawn and fruiting substrates. It grows slowly on agar. Its smell can be spicy and cinnamon-like. Commercial spawn is usually available in the form of supplemented sawdust (which should include a minute amount of pine) or wooden dowels for outdoor cultivation.

Fruitbody Development

The fruitbody is a large, rounded mass of tightly curled or flattened blades; it resembles a large brain coral (or a mass of wet egg noodles piled up at the base of a conifer). The colors are typically off-white, yellow, or banded combinations of the two. The fruitbody browns at the tips with age or drying. The spore print color is white and can be obtained by laying a piece of aluminum foil loosely fitted and shaped over the wavy surface overnight.

Common Strains and Ideal Fruiting Conditions

Cauliflower mushrooms are usually specific to tree types and not very adaptable, so collecting local strains that have evolved with locally available wood types, soil microbes, and other specific conditions should prove to be a rewarding strategy with good results. You can find commercial strains that have been isolated from eastern and western hemlocks and a variety of other coniferous trees around the world. Commercial hardwood strains are mostly isolates from Europe. You can experiment with spore germinations and sectoring on agar supplemented with hardwood sawdust to isolate strains that will thrive on hardwood; see chapter 18 for details.

Wild Spawn Expansion Techniques

Place wet pieces of cardboard between the fronds of the fruitbody, cover with plastic to make a humidity tent, and let sit for several days. The mycelium will leap off onto the cardboard, which you can then use for expansion. This method takes advantage of the natural tendency of cauliflowers (and other polypores) to engulf bits of debris, such as sticks, pine needles, and leaves—and your cardboard. You can also work with the fruitbody directly: Harvest the fruitbody, cutting away the entire upper, wavy portions, and cutting the firm, dense base into a clean cube, then cutting this large cube into smaller bits. Rinse these in hydrogen peroxide and place them in a plastic bag for a week to surface-sterilize the mycelium. You can then expand these mushroom pieces by wrapping them into sheets of wet cardboard, putting them in a plastic bag, and waiting for the mycelium to leap off.

Laboratory Isolation and Spawn Culture

When cut open, cauliflower mushrooms reveal an intricate network of thin, overlapping folds and blades, which makes it difficult to take a tissue sample. None of my attempts succeeded until I started to culture the area where the fruitbody meets the taproot, which is generally thick, with a dense, cottony center from which it is possible to remove a few samples.

Preferred Fruiting Substrates

Most strains prefer conifer logs and stumps.

Outdoor Cultivation Notes

Media preparation. Use buried or partially buried logs, stumps, or rounds. With conifer-inhabiting species, make sure the wood has aged for several months before inoculation. After inoculation, let the fungus colonize for six to eight months before burying the logs or rounds. Cover the buried wood (or the base of an inoculated stump) with a thin layer of leaves to regulate humidity and protect primordia. As with many terrestrial mushrooms that fruit from buried wood, cauliflowers may benefit from having small cavities or holes drilled into the wood to create humid microclimate pockets where a developing mass can form the mycelial taproot structure. (And see chapter 5 for more information about cultivating mushrooms on buried wood.)

Preferred spawn type. Sawdust, plug, or cardboard spawn.

Inoculation to fruiting time. From one to two years, depending on temperature and inoculation rates. Water heavily just before a scheduled seasonal fruiting, which is typically in fall in temperate climates, and then reduce the watering to a daily misting once the primordia are visible as a tight, amorphous mass around the base of the wood. If you do not see any signs of growth around the base of the stump or edges of your buried logs and assume the spawn didn't take, it's still worth snooping around after a year or two to see if there are any fruiting bodies you have missed. Don't give in to the temptation to rake back the leaves every time you feel impatient and want to take a look; it will disrupt this fragile interface environment. Instead, be patient and observe your own and wild fruitings. With time and experience, you'll learn the vagaries of cauliflower life cycles in your region and come to recognize the changes that signal an imminent fruiting.

Expected yields. Low. Conifer logs and stumps can take years to fruit. Cauliflowers are very unpredictable fruiters and may skip a year, since conifers are slow to decompose and the mycelium may need extra time to recharge itself enough to produce a fruitbody. Expect one flush per year, and possibly every other year. The yield depends on the mass of the substrate wood, along with water and temperature conditions. But the average fruitbody from wild specimens can range from 2 to 5 pounds!

Indoor Cultivation Notes

Indoor cultivation is not recommended. This mushroom prefers to fruit from buried wood, so indoor cultivation efforts are currently experimental and very low yielding.

Harvesting

Harvest these mushrooms by twisting or cutting them clean from the growing substrate. Many species or strains have taproots that you'll need to trim and discard. Inspect the folded tissues carefully and pull out any debris, leaves, or twigs.

Storage

Fresh cauliflower mushrooms will keep for two to three weeks at 38 to 42°F (3–6°C). For drying, cut the fruitbody into slices; they reconstitute well. The stems and taproots can be peeled (to remove dirt) and dried for powdering.

Marketing

These mushrooms have a strong, earthy odor and can have a spicy flavor and crunchy texture. One of the best recipes I give to people who have never tried a fresh cauliflower is to sauté crumbled bits of this mushroom and scramble them with eggs. It makes great quiche and casseroles and maintains its crunchy texture when baked.

Nutritional Value and Medicinal Uses

Wild cauliflower mushrooms have protein contents ranging from 8 to 12 percent (dry weight), depending on the strain and growing conditions. Studies have shown that *Sparassis* species help regulate blood pressure and sugar levels, contain antibacterial properties (even against methicillin-resistant staph, or MRSA), have high levels of anti-inflammatory compounds, and suppress many cancer cell lines.

Uses in Mycoremediation

Cauliflower mushrooms are perfect candidates for decomposing conifer stumps at logged sites. They also have the ability to withstand sunlight better than fleshier gilled mushrooms, making them useful for habitat renewal and soil building projects.

The Genus *Stropharia*

Common Species

S. rugoso-annulata (king stropharia, garden giant, wine cap)

Difficulty Level

Outdoor cultivation—1
Indoor cultivation—2

General Description and Ecology

People around the world enjoy cultivating, collecting, and eating these mushrooms since they are not only easy to grow but also easy to identify. If you are a beginner, king stropharia is one of the top three (along with oyster and shiitake) to get started with.

Mycelium and Spawn

The mycelium is white and linear and exudes a light yellow metabolite on overcolonized spawn and fruiting substrates. Its smell can be very sweet.

Baby king stropharia primordia pinning near larger developing fruitbodies on the surface of a cased, supplemented sawdust block indoors.

Commercial spawn is usually available in the form of supplemented sawdust; grain spawn does exist but is hard to find.

Fruitbody Development

The cap is centered on the stem and is tan to dark burgundy in color, depending on the strain and its exposure to light. The smaller buttons are darker and more firm, and as they grow and mature the pigments spread and soon the cap lightens to a very light reddish brown. Older mushrooms are often filled with fly or beetle larvae; however, you can still collect the stem bases to use for expansions. The spore print is dark purple-black and easily obtainable on paper or aluminum foil.

Common Strains and Ideal Fruiting Conditions

Wild strains are easily collected and cloned or propagated through spore germinations. The wild strains are very aggressive saprophytes, unique to their fruiting substrates, which can vary from beauty bark to sapwood chips, composting grasses, and composts. This mushroom is resilient to drier conditions and needs only minimal misting, even in the primordial state, to mature effectively. Saturating the outdoor beds or indoor growing media is critical to the timing of the primordia formation, and more abundant fruitings can be achieved if you maintain accurate observations about these mushrooms and their schedules. Reduce watering as these mushrooms mature, since larger specimens rot quickly, where the cap tends to liquefy if kept wet, making the mushrooms inedible in a hurry. I prefer to keep the buttons heavily misted and then stop watering when they appear to be about to open their caps. I have found that landscaped areas around parks, colleges, and businesses in both urban

King stropharia mushrooms (*Stropharia rugoso-annulata*) fruiting from recycled nursery pots filled with pasteurized wheat straw and cased with potting soil.

and rural environments can support large communities of king stropharia.

Wild Spawn Expansion Techniques

Along with blewits (*Clitocybe nuda*), this is one of the easiest wild mushrooms to isolate on cardboard. Wrap the stem bases—full of thick mycelium and bits of wood chips and soil that harbor beneficial microbes—in wet cardboard and place them in plastic ziplock bags to preserve humidity. I prefer to place these in the fridge if available, since it can reduce bacterial contamination and allow the mycelium to still grow just fast enough to leap onto the cardboard. The stem bases will revert to a mycelial state, quickly fuzzing over and colonizing the cardboard. Once the cardboard has been colonized, remove the fleshy stem base to prevent molds and bacteria from developing. (You can place this bit of stem in a blender with distilled water to make a microbial slurry that you can later apply to your planted beds. Store the slurry in a jar in the refrigerator for a few months, until colonization is complete in your planted beds.)

Laboratory Isolation and Spawn Culture

This genus clones easily from cap tissue; stem tissue is stringy and less likely to produce clean, viable mycelium on agar. Spore germinations are also very reliable.

Once mushrooms begin to pin, they can mature very quickly under optimum conditions. These mushrooms were pinning just two days earlier.

DOWN TO THE LAST THREAD

My king stropharia strain is one of the best wood chip recyclers in my collection. I collected it from a single, wild fruiting specimen I found at a botanical garden at Clemson University. It was the first stropharia that I ever collected in the wild, so I decided to show off my prize to a local mushroom club after one of my lectures. While I was lecturing, I passed around a basket with my single stropharia so that people could take a look at it. Imagine my surprise when the basket came back to me and there was nothing in it! Stunned and upset that someone would take my mushroom without asking, I returned home, looked again through my empty basket, and found a few small pieces of the wood chips that the stropharia had been clinging to when I found it in the mulch at the botanical garden. A few remnants of mycelium clung on the chips. My moment of desperation turned into an opportunity to expand my skills at isolating a mushroom culture from almost nothing to work with.

Having never tried this technique before, I placed the chips in a small ziplock bag, covered the chips with hydrogen peroxide, and placed the ziplock in the refrigerator. A few days, later I removed the chips, rolled them up tightly in small pieces of wet cardboard, and put them back into a plastic bag and into the lab fridge—still certain that my culture was lost forever. About two months later, long after I had given up, I noticed the bag again. Imagine my excitement when I unrolled the cardboard and discovered a small fan of mycelium on the cardboard!

I took my dental tool pick and lifted a few strands of mycelium off the cardboard surface, transferring them gently to a few fresh agar plates. After a few days, it was clear that my prized culture was not going to be lost. A few weeks later, to validate that I had indeed copied the culture and not some random fungus, I transferred it to grain spawn and then finally to pasteurized wheat straw, which I cased with an inch of unpasteurized, moist potting soil. Just under thirty days later, my nursery pots were filled with king stropharia. It had come a long way from those few remnant chips in my basket, a poignant revenge against the thief who came to my class that night. (You know who you are!)

Now this same mushroom culture fruits in the thousands all over the United States and has completely naturalized the gardens around my home—all from a few small fragments. Let this be a lesson for your own cultivation challenges, which you are sure to face: Be persistent and patient; life wants to survive and often can if you help it along. Okay?

Preferred Fruiting Substrates

King stropharia prefers hardwood chips and shredded cereal straw, or any other agricultural by-products common in your area, such as bean hulls, cornstalks, cotton waste, beet pulp, and sugarcane bagasse.

Casing Soil

A microbial casing soil is required. When the entire surface of the substrate, or just below if you scratch around in a few spots to check, is a continuous white mat, dress the bed with a microbial casing soil, which stimulates primordia mushroom formation (see chapter 20).

Outdoor Cultivation Notes

Media preparation. Use hardwood chips layered with cardboard to a depth of at least 6 to 8 inches. As a general rule, the deeper the mulch, the larger the mushrooms will be. Make sure the chips are fresh (cut no more than one month before inoculation) to ensure the mushroom has a chance to outcompete any wild fungal species. Mixing the chips with straw can help the bed retain moisture. You can also use other agricultural by-products in the beds; I suggest using whatever is locally available and has a small particle size, such as dried hulls from rice, soy, or peanuts. Cover the beds with mesh or additional cardboard to prevent wildlife such as turkeys, squirrels, and other scratching or rooting animals from destroying the beds; they are attracted to the sweet smell of the mycelium and the masses of earthworms. When colonization is complete, dress the beds with 2 to 3 inches of a microbial casing soil.

Preferred spawn type. Supplemented sawdust spawn.

Inoculation to fruiting time. From three to six months, depending on temperature and inoculation rates.

Expected yields. Good. King stropharia are known to produce massive flushes in outdoor wood chip beds. Expect approximately 20 pounds of fresh mushrooms per cubic yard of wood chips, and more if your wood chips are mixed with agricultural by-products. Expect several flushes per year outdoors, typically in the late spring and again in the fall.

Indoor Cultivation Notes

Media preparation. Shred agricultural by-products such as grasses, cereal straws, and hulls into a fluffy consistency and pasteurize in a hot water bath for one to two hours. Drain, cool, and inoculate, stuffing the substrate into trays, buckets, or bags. Once colonization is complete, dress the substrate with 2 to 3 inches of moist casing soil to initiate fruiting, and water generously.

Preferred spawn type. Grain or supplemented sawdust spawn; grain spawn will fruit much sooner and produce larger yields.

Inoculation to fruiting time. From three to five weeks, depending on temperature and the type of spawn you use.

Expected yields. Average. These mushrooms are quite large and fruit either singly or in fused, smaller clusters. Expect two rapid flushes several weeks apart.

Harvesting

Harvest a king stropharia by twisting it to remove the entire base from the growing medium, and then trim the base to remove any residual growing substrate. Harvest the mushrooms when they are in their mid-button stage, just as they are beginning to open, when the cap separates from the stem. The buttons should be firm at this stage, with few insects and little bug damage.

Storage

This mushroom does not have a great shelf life, so you should sell or eat it as soon as possible after harvesting. Open caps will keep for up to one week at 38 to 42°F (3–6°C), while buttons can last for up to two weeks. Refrigerate them until you are ready to eat or deliver them. The dried caps rehydrate well; dried stems should be powdered.

Marketing

This mushroom tastes like potatoes cooked in a light wine marinade; the stringy stem is reminiscent of steamed asparagus. Since this is a seasonal mushroom and does not store well, you need to advertise well in advance. I recommend contacting potential buyers, such as restaurants, when your stropharia approach complete colonization to let them know the fungi are on their way, so you can set up sales before they start to deteriorate. These mushrooms can be grown to different sizes depending on the depth and volume of the substrate. Growing them indoors in containers can offer predictable yields and size, which may help if you're marketing them for a particular purpose, such as stuffing or grilling.

Nutritional Value and Medicinal Uses

King stropharia have a protein content ranging from 20 to 24 percent (dry weight), depending on the strain and growing conditions. Studies suggest that this mushroom may contain trace amounts of anticoagulant compounds.

Uses in Mycoremediation

This mushroom can be used for creating simple filtration modules or living biofilters near livestock and pond water runoff to reduce coliform entry into watersheds. King stropharia have also shown activity against molecular disassembly of hydrocarbons, where the mycelium can be used for soil remediation by capping an area with wood chips or growing substrate and watering the colonized mycelium, thereby sweating the enzymes into the contaminated soil. Another method is to use a countertop filtration or harvesting unit to collect the enzyme solution to use as a soil treatment or wash for potentially cleaning fuel tanks.

The Genus *Trametes*

Common Species
T. versicolor (turkey tail)

Difficulty Level
Outdoor cultivation—1
Indoor cultivation—2

General Description and Ecology
Turkey tails can be found on just about every continent, and the various strains can fruit on a variety of woods. In my own collection I have cloned mushrooms that I've found growing on sweetgum trees, eastern cedar trees, pressure-treated wood, untreated pallet wood, fruit trees, and old, dying hemlocks. The distinctive color bands on the upper surface and white pores underneath make them easily identifiable. The color banding can vary even within a strain, which may mean that it is directly related to the micro- and macronutrients absorbed and made available to the fungi by their host trees. Although turkey tails are not considered great edible mushrooms due to their fibrous and paper-like texture, the fruitbodies have a wide range of medicinal properties. As with most medicinal mushrooms, the exact makeup of their active principles will depend on the strain and the substrate on which they grew.

Mycelium and Spawn
The mycelium is white and linear and exudes a light yellow metabolite on overcolonized spawn and fruiting substrates. As is the case for most polypores, the mycelium of this mushroom forms a thick mat unless it's stored under refrigeration, and it can be difficult to separate colonized dowels or grains, or even to cut your agar plates. The smell tends to be slightly acidic. Commercial spawn is usually available in the form of cereal grains, supplemented sawdust, and wooden dowels for outdoor cultivation.

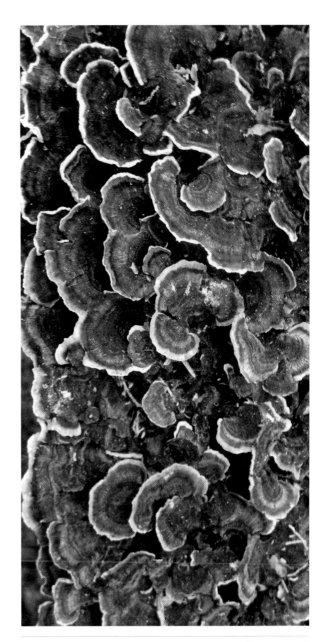

A fused, overlapping cluster of turkey tail mushrooms (*Trametes versicolor*), showing their alternating bands of colorful, velvety growth. These mushrooms can completely cover fallen logs and stumps.

Fruitbody Development

Turkey tails form flat and leathery brackets in tight, overlapping or fusing clusters. They can completely cover logs and stumps. Fresh specimens have an alpine to creamy white pore surface with minute, round pores (you'll need a hand lens) on the underside of the brackets. The spore print color is white. Lookalikes, such as *Lenzites betulina*, may be smooth on their underside and have tooth-like or even maze-like gills. Turkey tails can have a variety of distinct color bands—you may see many different bands in a single day of foraging. Be aware that this aggressive wood decomposer can become the bane of your shiitake cultivation operation. If you overwater outdoor shiitake logs during colonization, hordes of wild turkey tail spores may take over, so it's a good idea to keep your shiitake logs on the drier side.

Common Strains and Ideal Fruiting Conditions

Wild strains are so abundant that spawn producers rarely carry turkey tail cultures. However, if you purchase spawn, I would inquire as to which wood type this species was native to to best duplicate the growing parameters. Although, that being said, this mushroom is aggressive and can grow or acclimate to just about any hardwood, and I have found several strains in my area growing on fresh-cut cedar, making excellent strains for possibly developing conifer stump degrading spawn for reforestation of clear-cut areas.

Wild Spawn Expansion Techniques

Wrap fresh turkey tails up tightly in wet cardboard, place the bundle in a plastic bag, and let sit for two to three weeks under refrigeration if available. Then check on the fungi. If the mycelium has leaped off onto the cardboard, pull out and discard the fruitbodies, since they will eventually rot inside the wet cardboard. You can now use the myceliated cardboard for expansion.

Another possibility for expanding a wild strain is to harvest the myceliated bark fragments from near the base or attachment point of turkey tail brackets and insert those fragments into slices or holes you've made in new wood, as in the wafering method (see chapter 5).

Laboratory Isolation and Spawn Culture

This is a thin, leathery little mushroom. It is somewhat rubbery when it's fresh and becomes more brittle when dried, which makes it difficult for beginners to take a good tissue sample for cloning. The method I've had most success with is to pull the caps apart slowly, revealing a thin but fluffy and fibrous layer of tissue that clones easily. Placing small fragments of this tissue on sterilized agar is almost a guarantee of successful expansion.

Preferred Fruiting Substrates

Turkey tails mostly prefer hardwood trees and stumps outdoors and supplemented sawdust indoors.

Outdoor Cultivation Notes

Media preparation. This mushroom will grow on just about any type of hardwood log or stump. Using local strains and local wood (of the same type the strains were found growing on) will improve results and yields. Make sure logs or rounds are fresh (cut no more than one month before inoculation) to ensure the mushroom has a chance to outcompete any wild fungal species. After inoculation, rounds or logs can simply be placed on the ground to make soil contact, where the wood soaks in ground moisture, which is what turkey tails love. These mushrooms are resilient to drying out, so burying the wood is a waste of effort and not needed.

Preferred spawn type. Sawdust, plug, or cardboard spawn.

Inoculation to fruiting time. From eight to twelve months, depending on temperature and inoculation rates. These mushrooms are slow to mature after the appearance of the primordia; fruitbodies can take anywhere from several weeks to months to mature, depending on temperatures. These mushrooms can persist side by side with the previous year's fruitbodies on the same trees. The older fruit are not as perky and fresh looking, are usually riddled with beetle holes along the underside, and lack the pure white pore surface that fresh mushrooms have.

Expected yields. Good. Turkey tail mushrooms are aggressive; in fact, they're considered "weed" fungi that can outcompete your other cultivated mushrooms. These brackets typically persist all year long, eventually succumbing to beetle damage and turning into dust. Expect one flush per year outdoors.

Indoor Cultivation Notes

Media preparation. Use sterilized or super-pasteurized supplemented sawdust. The supplements of choice are wheat or rice bran, at 5 percent (dry weight) of the sawdust mixture. Bags are the container of choice. After three weeks of colonization bags can be fruited indoors or out. To initiate fruiting, make just a couple of long (3- to 4-inch) slices in the bag, either on opposite sides or as an X on the side of the bag where you wish the mushrooms to form (such as if you were stacking the bags and creating aisles of fruiting walls, which makes harvesting easy.)

Preferred spawn type. Grain or supplemented sawdust spawn.

Inoculation to fruiting time. From twenty-five to thirty-five days. Expect two flushes several weeks apart.

Expected yields. Low. Up to 0.5 pound for every 5 pounds of supplemented sawdust substrate.

Harvesting

Harvest turkey tails by peeling the clusters away from their host substrate, trimming away excess bark or sawdust.

Storage

Fresh turkey tails will keep for over a month at 38 to 42°F (3–6°C); just be sure to check for molds developing on the underside, or pore surface, before using them. Although not directly edible in their fresh form, turkey tails can be pureed with water to make a delicious soup base. Dried brackets can be powdered into flour for adding to breads, soups, pastas, and sauces.

Marketing

This mushroom is primarily sold as a medicinal supplement, highly valued in the cancer treatment field for its ability to improve white cell counts following chemotherapy. These mushrooms are best powdered and made into pills or extracts to make them more palatable and digestible.

Nutritional Value and Medicinal Uses

Turkey tails have a protein content estimated at 9 to 12 percent (dry weight). Studies have shown that the mushrooms can help boost and modulate immune system function, regulate blood pressure and sugar levels, lower cholesterol, and provide cardiovascular and digestive support. They contain antiviral and antibacterial properties as well as anti-inflammatory compounds, and they also have been shown to suppress many cancer cell lines. This is one of the panacea mushrooms, similar to reishi, which when used in combination with other polypores creates a powerful holistic therapy for many common and serious aliments.

Uses in Mycoremediation

Trametes mushrooms show great promise for the mycoremediation of biological and chemical pollutants in water and soil.

The Genus *Volvariella*

Common Species

V. bombycina (tree volvariella)
V. volvacea (paddy straw, horse mushroom, rice straw mushroom)

Difficulty Level

Outdoor cultivation—2
Indoor cultivation—3

General Description and Ecology

Paddy straws are among my favorite summer mushrooms—they're a great sight when temperatures are soaring, especially since most mushrooms struggle in intense heat. Not this one. It thrives at temperatures from 80° and 100°F (27–38°C), and it won't fruit if temperatures—even nighttime temperatures—drop below 80° (27°C). The culture itself can be damaged or killed if temperatures drop below 50°F (10° C) for any great length of time, making for an interesting spawn storage dilemma, given that most spawn storage requires refrigeration or freezing.

I tried everything I could think of to fruit this mushroom for over ten years, to the point that it became an obsession. I tried strains from many different collections. I followed the instructions from several popular cultivation guides. But none of my outdoor beds ever fruited—that is, until one summer.

My friend Garrett Steadman and I had been trying to grow paddy straw on a variety of substrates—cotton hulls mixed with straw, straw alone, cotton alone, sawdusts, beet pulp. We thought that if we could just get one mushroom to fruit, we would have a reference point, an idea of the food source or the process that could work, and with that little bit of information we could figure out what the next step would be. But nothing fruited. The summer crept on and we were running out of heat.

After giving up on one batch of buckets of inoculated substrate, which had been baking in our greenhouse, Garret and I tossed them into our massive compost pile, depressed that our efforts were for nothing. Two weeks later, while I was emptying out my oyster mushroom fruiting room, I noticed something unusual, something I had only seen in books: paddy straws, fruiting from the compost.

As I examined and harvested my new, prized paddy straws, I dug around, looking for clues. I slowly excavated the substrates that appeared to be attached to the paddy straws, tracing them back to a large mass of partially composted, spent oyster mushroom media. Intrigued, I took some fresh paddy straw spawn and dumped it around another pile of spent oyster media. The new piles fruited prolifically in just under two weeks. Wow. Turns out that was easy after all.

Mycelium and Spawn

The mycelium is clear to yellowish brown, like dirty fiberglass, and on agar can resemble molds like *Rhizopus*, without the black pin heads. As it ages pinkish-red zones of spores form directly on the mycelium. The smell is to me is unpleasant, like grease or soap.

Fruitbody Development

The cap is centered on the stem and emerges from a large volva, or egg-like structure. The cap is streaked with fine, silky fibrils that are darker than the background whitish-gray color. The gills free and the spore print pink.

Common Strains and Ideal Fruiting Conditions

My best luck has been with strains that I have purchased from Penn State University that originated in Puerto Rico in the early 1980s. It took me several years

Paddy straw mushrooms (*Volvariella volvacea*) showing the beautiful sac-like volvas and pinkish gills they develop as they mature.

to figure out how to maintain these cultures, since they die if it exposed to temperatures below 50°F (10° C) for a considerable amount of time. I have found that simply allowing grain spawn to dehydrate on its own at room temperature works quite well; covering colonized grain with sterilized water or mineral oil under refrigeration can prolong storage time for up to a year. Harvesting mature specimens and taking spore prints is also recommended so you have a backup in case the stored mycelium does not survive.

Wild Spawn Expansion Techniques

This mushroom has some of the fastest-growing mycelium on the planet, capable of growing about 2 inches per day, which, adjusted for size, is like a human running 22 miles per hour nonstop for twenty-four hours, covering a distance of 528 miles. It manages to colonize substrates in just days and fruits edible mushrooms in as little as four to seven days after inoculation, speedily converting complex organic material into edible protein. To expand a wild strain, just transplant a clump of the colonized substrate and stem bases into fresh compost.

Laboratory Isolation and Spawn Culture

This mushroom clones easily from slightly dried fruitbodies. Spore germinations seem to me to be the easiest, since spores germinate and mycelium spreads so rapidly that contamination cannot keep up, making a second transfer very easy to purify. Once you notice a clear, wispy growth and pinkish spore colonies on the mycelial surface, you know you have isolated a culture correctly.

Preferred Fruiting Substrates

This mushroom prefers partially decomposed or fermented substrates, from cereal straws (oat, rye, wheat, and rice) and prairie grasses to dried cacti and aquatic

Cultivator Garrett Steadman grows paddy straw mushrooms on alternating layers of partially composted wheat straw and sun-pasteurized spent oyster mushroom compost, using supplemented sawdust spawn.

plants such as hyacinth and water lettuce. All of these can be air-dried and composted or pasteurized. Spent oyster mushroom media also support heavy fruitings. Fermented or composted straw and cotton waste is the best formula for commercial production indoors.

Casing Soil

Indoor commercial cultivators have found that casing the substrate after three to four days of colonization with a microbial slurry greatly enhances yield. This technique is rarely used outdoors, but I would encourage anyone to try it. See chapter 20 for more details.

Outdoor Cultivation Notes

Media preparation. Pasteurize the growing medium mixture. Pile the medium into raised beds, inoculate with spawn, and cover with plastic to preserve moisture. Be sure to shade the beds if you are cultivating in a sunny location.

Preferred spawn type. Grain or supplemented sawdust spawn.

Inoculation to fruiting time. From one to two weeks, depending on temperature and inoculation rates. Mushrooms tend to form where the rows of the different fruiting substrates meet, such as all along the border of shredded wheat straw and cotton hulls, in a perfect line.

Expected yields. Low. This mushroom is not high yielding but makes up for it in flavor and protein content.

Indoor Cultivation Notes

Media preparation. Shred agricultural by-products such as grasses, cereal straws, and cotton hulls into a fluffy consistency and soak in cold water for twelve hours to pretreat, or ferment the media. The fermented (and smelly) media can then be pasteurized in a hot water bath for one to two hours or steam-pasteurized in an enclosed unit or box with live steam for six to twelve hours. Drain, cool, and inoculate, stuffing the substrate into trays, buckets, or bags. After four days of colonization, dress the soil with 2 to 3 inches of moist casing soil to initiate fruiting, and water generously.

Preferred spawn type. Grain spawn.

Inoculation to fruiting time. From seven days to two weeks, depending on temperature.

Expected yields. Average. These robust, egg-like mushrooms fruit either singly or in fused, smaller clusters. Expect two rapid flushes a week or two apart.

Harvesting
Harvest paddy straws by twisting them to remove the entire base from the growing media, and then trim them to remove any residual growing substrate.

Storage
If picked in the closed "egg" state, paddy straws will keep for one week at 38 to 42°F (3–6°C); mushrooms with open caps will keep for just a few days. Dried caps can be a little chewy when rehydrated; you may choose to reserve them for powdering.

Marketing
Paddy straw mushrooms, though well known in Asian cultures, are still a bit new to Western markets. However, they are delicious and have a strong flavor, superior to the common white button and portabella types. The egg-like primordia can also be harvested and sold, cleverly packaged in egg containers to protect them from damage.

Nutritional Value and Medicinal Uses
Cultivated paddy straw mushrooms have protein content ranging from 26 to 30 percent (dry weight), depending on the strain and growing conditions, making them among the most protein-rich of all cultivated mushrooms. As of the time of this publication, their medicinal properties have not yet been identified.

Uses in Mycoremediation
Paddy straw mycelium is fragile and lacks many of the enzymes found in white rot fungi, making it a poor candidate for mycoremediation trials. The rapid rate of growth, however, should not be discounted, and paddy straw may be a good candidate for remediating heavy metal contamination in soil.

Acknowledgments

I would like to thank most of all my beautiful wife, Olga, for allowing me to make my dreams come true and for carrying the weight of the work at home and with the business for all these years as I returned to school, took a tremendous amount of time to write this book, and continually snuck off to the lab throughout the nights to make it all happen. I love you.

I'd also like to thank my sister Kim, who also has a passion for mushroom hunting, and has been a great companion; we shared many great memories growing up together. I cherish the time we spent in the Sinai Desert with the scorpions and venomous snakes, and I'll never forget the 1981 car bombing in Damascus near our home. I often refer to that as the worst and best day of my life. I wouldn't trade those moments for anything.

My mom and dad, who for years watched me struggle with the pursuit of happiness, not only helped financially with my education but have always been some of my best fans, supporting all of our research and watching me evolve into an active contributor who has a positive impact on the world and can promote lasting change. Thank you for not giving up on me, for supporting my decisions, and for being there through the bumpier moments to help me keep my dreams afloat.

Thank you also to the following:

Miro and Dragica Katic, for all of their amazing support and help with building up our business and sharing in the growth of our passion. I cannot remember any moment when they didn't offer financial or physical help in building, helping to pay for, or designing our greenhouses and laboratory, from where we could finally get our mycelium growing outward and into the world! None of this book could have been possible without the resources and experimental spaces they gifted us. Their contributions are priceless.

My relatives and extended family, for showing me the ways of nature, by taking me fishing and exploring the woods as a young child, which seems to have had a great impact in shaping the way I see the world. In particular, thanks to Miriam and A. B. Altman and Ginnie and Charlie Cotter.

The directors and conference organizers who recognized the need for this information and gave me the opportunity to talk about my passion, and kept inviting me back to share all the latest mushroom research. In particular, thanks to the committees of the Carolina Farm Stewardship Association, Georgia Organics, the Organic Growers School (Asheville, North Carolina), the Southern Sustainable Agriculture Working Group, the Telluride Mushroom Festival, the Tennessee Organic Growers Association, and the Virginia Biological Farming Conference.

My teachers, mentors, and professors, along with many who had no idea I secretly cherished every second I spent in their presence, feeling much like a mushroom in trying to absorb their worldly knowledge. I so appreciate your beautiful contributions, passionate work, and willingness to take the time to share your precious wisdom with me and many other students. Thank you especially to Todd Elliot, Dr. Mike Henson, Rick Huffman, Dr. Julia Kerrigan, Dr. Walker Miller, Dr. Allein Stanley, Dr. "Hap" Wheeler, and Dr. Patricia Zungoli, for fanning my fungal flame and helping me expand my reach of knowledge in the directions of bacteriology, mycology, native plant botany, and microbiology.

This book would not have been possible without the incredible dedication, support, and creativity from the Chelsea Green staff who took my raw information and helped me craft it into a distilled, timeless work that we could all be proud to share. Special thank-you to Brianne Goodspeed for all of her hard work and dedication, Pati Stone for her skill and perserverance to meet our looming deadlines and putting up with my hectic schedule (including a few prank phone calls), and the entire staff for the patience and encouragement. Making the transition from lecturer to author was very difficult, and you all helped me through every step, and misstep! I am proud to be a part of your extended family of authors and looking forward to the next publication (already in motion!). Group hug here.

And of course all of the students, customers, mycofans, and followers who continue to support our efforts, which allows us to continue our research and make our knowledge available to the world community. I am forever grateful. I look forward to taking all this information to the next level, and encourage anyone who reads this to do the same. Think like a mushroom.

Glossary

Agar. A derivative of seaweed used as a gel for cultivating bacteria, fungi, and plants.

Annulus. A ring-like appendage of tissue that typically is located on the stem of gilled mushrooms. It protects the *gills* of young, developing mushrooms. Also known as a veil.

Attached gills. In a gilled mushroom, *gills* that are directly attached to, or fused with, the stem. The opposite of *free gills*.

Autoclave. A sealed container or vessel using superheated steam under high pressure for *sterilization*.

Bacteria. Single-celled microorganisms that inhabit diverse ecosystems throughout the world and are noted for their biochemical and pathogenic activities.

Basidiomycete. Describes a group of fungi that produce their *spores* externally on basidia.

Biological efficiency. The ratio of mushrooms produced, measured as a dry weight, compared to the dry weight of the *substrate*.

Brown rot. Describes fungi that are capable of metabolizing cellulose in vegetative material, leaving behind the mostly brown woody tissues. See also *white rot*.

Casing soil. A soil applied as a final layer on top of a mushroom cultivation bed or container to stimulate fruiting. Most casing soils are mixed with supplements and are specifically designed for the mushroom species they are being used with. Some are made with a microbial slurry, prepared from a sample of the native soil that surrounded the base of the mushroom specimen from which the *spawn* used in the bed or container was *cultured*.

Cell-free. Describes metabolites produced by fungi, including antibiotics and enzymes, that are capable of functioning freely and away from the cells of their parent organism.

Chill hour. An hour in which the soil temperature at a depth of 4 to 10 inches is less than 32°F (0°C).

Certain types of mushrooms (such as morels) require a certain number of chill hours to trigger *fruiting* the following season.

Chitin. A semitransparent, durable nitrogenous polymer, chemically related to cellulose, that forms the shells of insects and the cell walls of fungi and specific algae.

Clamp connection. A short, cylindrical extension common on many *dikaryotic mycelia* of *basidiomycetes*.

Conditioning. The stage following *pasteurization* of compost or another growing *substrate* when the temperature is held between 113 and 122°F (45–50°C) to promote microflora that are *thermophilic*.

Conk. General term for larger perennial or annual polypores that are typically very hard and hoof- to fan-shaped.

Culture. A collection of living tissue comprising the basic building blocks of cells that is capable of exponential expansion.

Dikaryotic. Describes a cell that contains two distinct nuclei, typically the product of two mating *spores* fusing in *karyogamy*.

Ecotype. A *strain* of a species that is unique to the particular ecological conditions—soil, temperature, climate, and so on—of a particular site. An ecotype may possess a diverse spectrum of qualities that distinguish it from other members of its species.

Ectomycorrhizal fungi. A symbiotic relationship between a fungus and a plant's roots, typically woody plants and trees, in which the *hyphae* do not penetrate the host root cells, instead create a branched network of *mycelium*.

Expand. In mushroom cultivation, the process of transferring small quantities of spawn or *mycelium* to make large quantities.

Flush. A periodic *fruiting* of mushrooms. Flushing occurs in cycles, based on available energy reserves or environmental stimuli.

Free gills. In a gilled mushroom, *gills* that are do not touch the stem. The opposite of *attached gills.*

Fruitbody. The reproductive structure of a fungus, sometimes called a sporocarp, that is composed of densely packed *mycelium.* This is the so-called mushroom that we normally eat.

Fruiting. The phase in a fungus's development when it produces *fruitbodies.*

Gallery. A behaviorial or chemical observation test using one or more organisms to test for reactions.

Germination. For a fungus, the process by which a *spore* initiates reproduction, extending a thread of *mycelium* into the surrounding environment to find a compatible spore to join with and initiate a *fruiting* cycle.

Gills. Thin, blade-like projections typically on the underside of a mushroom's cap where *spores* are formed and ejected into the surrounding environment.

Gravitropism. Directional growth by an organism toward or away from gravitational pull.

Growing medium. Any nutritive substance, such as agar or a *fruiting substrate,* that supports the growth of *mycelium.*

Hydrophilic. Attracted to water, or able to dissolve in or absorb water.

Hydrophobic. Repelled by water, or unable to dissolve in or absorb water.

Hypha. A long, tubular, branching, thread-like structure of the *mycelium.* Plural hyphae.

Infantry cells. The foremost advancing tips of *hyphal* cells that explore a new environment, relaying information back to the body of the *mycelium* to orchestrate biological responses and processes.

Karyogamy. The fusion of two compatible *hyphae,* resulting in the combination of genetic material; typically two *monokaryons* fusing as a single *dikaryon* to create fertile *mycelium.*

Master culture. A purified isolation of a biological organism (a fungal *strain,* for example) that represents the base of all future expansions. Also called mother culture.

Master spawn. A zero-generation *spawn culture* that represents the base of all future spawn expansions. Also called mother spawn.

Monokaryon. A fungal cell that contains one distinct nucleus.

Mycelium. Vegetative part of a fungus or mushroom that comprises *spawn.*

Mycology. The study of fungi, including mushrooms, yeasts, molds, and other *sporulating* organisms.

Mycorrhiza. The symbiotic association between the *mycelium* of fungi and the roots of seed plants. Mycorrhizal mushrooms are *obligate* in that they must have a living host. The mycelium invades but does not harm the root cells of the host plant, instead increasing the roots' available surface area for mineral and water absorption, which benefits the plants directly. The fungus receives carbon in the form of sugars in return.

Obligate. Relying exclusively or specifically on a particular or specific growing environment or living host.

Pasteurization. Partial *sterilization,* primarily through the application of heat, at a temperature (typically 150–165°F/66–74°C) and for a period of exposure that destroys undesirable microbes or competitive organisms. In mushroom cultivation pasteurization is often used to prepare a growing medium for inoculation with a fungal *culture* or *spawn.*

Petri dish. A shallow, lidded, glass or plastic dish commonly used to *culture* or observe living organisms. Also called a petri plate.

Pin. A young mushroom at the beginning stage of *fruiting,* when the *mycelium* differentiates into *primordia* on the surface of the growing *substrate.*

Pinning. The process by which the *mycelium* produces *pins.*

Primary decomposer. A fungus that prefers or is capable of degrading fresh, nonliving organic debris, such as logs or straw.

Primordium. A small knot-like protrusion that is the first distinguishable structure produced by the *mycelial* mass in the process of forming a *fruitbody.* Plural primordia.

Pure culture. A *culture* consisting of a single isolated organism that is free of contaminants, pathogens, and any other organisms.

Rhizomorph. A thick, cord-like bundle of *hyphae* that comprises a main "artery" for nutrient and water transport in fungi.

Rhizosphere. The narrow and intimate zone between a plant's roots and the surrounding environment where specialized microbial interactions take place.

Saprophyte. An organism, typically a bacterium or fungus, that derives nourishment by absorbing dissolved organic material.

Sclerotium. A hard-surfaced, energy-rich "microtuber" or resting body, formed by fungi during times when conditions are unfavorable for development or *fruiting*, which may remain dormant for long periods of time and resume growth upon the return of more favorable conditions. Plural sclerotia.

Secondary decomposer. A fungus that prefers or is capable of degrading nonliving organic debris that has already been utilized or rendered by *primary decomposers* first.

Sector. To isolate or separate a *mycelial* growth or section of growth that has differentiated into a unique morphological condition. The goal of sectoring is to isolate a *culture* or *strain* in order to preserve its distinctive characteristics.

Septum. The wall or partition that separates fungal filaments into individual cells. Plural septae.

Slant. A test tube that has been filled with liquid agar and angled to create a greater surface area.

Spawn. The product or carrier of *mycelium* that is physically "plantable," such as sawdust, grain, or wooden dowels.

Spawn run. The stage at which the *mycelium* radiates outward from the *spawn*. Also known as the colonization stage.

Spent substrate. The growing medium left over from a mushroom cultivation operation. It is still biologically active, with viable *mycelium*, but is incapable of producing any great quantity of additional *flushes* of mushrooms.

Spore. A fungal reproductive unit, or packet of haploid DNA, bound inside a highly protective capsule that resists desiccation and heat and is capable of growing into a new organism when fused with another compatible spore to form *dikaryotic mycelium*.

Sporulation. The process of producing and releasing *spores*, the reproductive unit of fungi.

Sterile. Free from biological life.

Sterilization. A process by which biological organisms are completely destroyed, primarily through the application of heat or chemical reagents.

Strain. A variation of an organism that is unique to specific growing conditions or requirements. An *ecotype* is a strain unique to the ecology of a specific area.

Subculture. To transfer an isolated tissue or sector of a *culture* to a new culturing medium.

Substrate. A medium that provides a habitat and food source for a fungus.

Symbiotic. A long-term mutually beneficial association between two or more separate species.

Terrestrial. In reference to mushrooms, fungi that inhabit or prefer to fruit from a horizontal plane, such as wood chip-loving species or mushrooms that arise from underground *mycorrhizae*.

Thermogenesis. The production of heat as a result of a biological, physical, or chemical process.

Thermophilic. Describes an organism that requires or thrives in high temperatures.

Tissue culture. The process of cultivating viable, living cells of an organism in a sterile *culture*.

Veil. See *annulus*.

White rot. Describes fungi that degrade lignin in woody plant material, leaving the woody tissue with a white appearance after colonization is complete. See also *brown rot*.

Bibliography

Arora, D. *Mushrooms Demystified*. Berkeley, Calif.: Ten Speed Press, 1986.

Bessette, A. *Rainbow Beneath My Feet*. Syracuse, N.Y.: Syracuse University Press, 1999.

Beyer, D. M. "Impact of the mushroom industry on the environment." Penn State Extension, http://extension.psu.edu/plants/vegetable-fruit/mushrooms/mushroom-substrate/impact-of-the-mushroom-industry-on-the-environment. Accessed April 1, 2014.

———. "Managing microbial activity during phase II composting: A summary of the compost pasteurization process." Penn State Extension, http://extension.psu.edu/plants/vegetable-fruit/mushrooms.hmtl. Accessed April 1, 2014.

———. "Substrate preparation for white button mushrooms." Penn State Extension, http://extension.psu.edu/plants/vegetable-fruit/mushrooms/mushroom-substrate/substrate-preparation-for-white-button-mushrooms. Accessed April 1, 2014.

Bonito, G., T. Brenneman, and R. Vilgalys. "Ectomycorrhizal fungal diversity in orchards of cultivated pecan (*Carya illinoinensis*; Juglandaceae)." *Mycorrhiza* 21 (2011): 601–12.

Bonito, G., O. Isikhuemhen, and R. Vilgalys. "Identification of fungi associated with municipal compost using DNA-based techniques." *Bioresource Technology* 101 (2010): 1021–27.

Bonito, G., J. M. Trappe, S. Donovan, and R. Vilgalys. "The Asian black truffle *Tuber indicum* can form ectomycorrhizas with North American host plants and complete its life cycle in non-native soils." *Fungal Ecology* 4 (2011): 83–93.

Borromeo, E. S. "Integrated production of straw mushroom (*Volvariella volvacea*) and oyster mushroom (*Pleurotus sajor-caju*) on banana leaves." *Mushworld* (2004).

Currie, Cameron R. "The Tangled Bank of Ants and Microbes." *Microbe Magazine*, October, 2011.

Deppe, C. *The Resilient Gardener: Food Production and Self-Reliance in Uncertain Times*. White River Junction, Vt.: Chelsea Green Publishing , 2010.

Fairlie, S. *Meat: A Benign Extravagance*. White River Junction, Vt.: Chelsea Green Publishing, 2010.

Finlay, R. D. "Mycorrhizal fungi and their multifunctional roles." *Mycologist* 18 (2004): 91–96.

Gadd, G. M., ed. *Fungi in Bioremediation*. British Mycological Society Symposia. Cambridge, U.K.: Cambridge University Press, 2001.

Hamlyn, P. F. "Cultivation of edible mushrooms on cotton waste." *Mycologist* 3, no. 4 (1989): 171–73.

Harris, B. *Growing Shiitake Commercially: A Practical Manual for Production of Japanese Forest Mushrooms*. Summertown, Tenn.: Second Foundation, 1986.

Harris, B. *Growing Wild Mushrooms: A Complete Guide to Cultivating Edible and Hallucinogenic Mushrooms*. Seattle, Wash.: Homestead Book Company, 1989.

Hart, T. *Microterrors*. Hart, Calif.: Firefly Books, 2004.

Hobbs, C. *Medicinal Mushrooms*. Santa Cruz, Calif.: Botanica Press, 1995.

Kalarus, M., and R. B. Beelman. "Methods and Compositions for Improving the Nutritional Content of Mushrooms and Fungi." U.S. patent No. 8,337,921.

Katz, S. E. *The Art of Fermentation*. White River Junction, Vt.: Chelsea Green Publishing, 2012.

Katz, S. E. *The Revolution Will Not Be Microwaved*. White River Junction, Vt.: Chelsea Green Publishing, 2006.

Katz, S. E. *Wild Fermentation: The Flavor, Nutrition, and Craft of Live-Culture Foods*. White River Junction, Vt.: Chelsea Green Publishing, 2003.

Kuo, M. *Morels*. Ann Arbor: University of Michigan Press, 2005.

Kyte, L., and J. Kleyn. *Plants from Test Tubes*. Portland, Ore.: Timber Press, 1996.

Lindahl, B. D., and S. Olsson. "Fungal translocation: Creating and responding to environmental heterogeneity." *Mycologist* 18, no. 2 (2004): 79–88.

Line, D. *Brewing Beers Like Those You Buy*. London: Argus Books, 1984.

Logsdon, Gene. *Holy Shit: Managing Manure to Save Mankind*. White River Junction, Vt.: Chelsea Green Publishing, 2010.

Ludwig, M. "Special investigation: The pesticides and politics of America's eco-war." Truthout, http://www.truth-out.org/news/item/1515:special-investigation-the-pesticides-and-politics-of-americas-ecowar, Accessed April 1, 2014.

Marley, G. A. *Chanterelle Dreams, Amanita Nightmares: The Love, Lore and Mystique of Mushrooms*. White River Junction, Vt.: Chelsea Green Publishing, 2010.

Mau, J. L., and R. B. Beelman. "Role of 10-oxo-trans-8-decenoic acid in the cultivated mushroom, *Agaricus bisporus*." In *Mushroom Biology and Mushroom Products: Proceedings of the 2nd International Conference, June 9–12, University Park, Pennsylvania* (Penn State University, 1996).

Oei, P. *Mushroom Cultivation: With Special Emphasis on Appropriate Techniques for Developing Countries*. Leiden, Netherlands: Tool Publications, 2006.

Ohga, S., et al. "Effect of Electric Impulse on Sporocarp Formation of Ectomycorrhizal Fungus *Laccaria laccata* in Japanese Red Pine Plantation." Research Institute of University Forests, Kyushu University, no. 6 (2001): 37–41.

Ott, E. A., E. L. Johnson, and R. A. Nordstedt. "Composting horse manure." Gainesville: University of Florida, 2005.

Quimio, T. H., S. T. Chang, and D. J. Royse. *Technical Guidelines for Mushroom Growing in the Tropics*. Rome: Food and Agriculture Organization of the United Nations, 1990.

Rice, M. *Mushrooms for Color*. Eureka, Calif.: Mad River Press, 1999.

Rigas, F., K. Papadopoulou, V. Dritsa, and D. Doulia. "Bioremediation of a soil contaminated by lindane utilizing the fungus *Ganoderma austral* via response surface methodology." *Journal of Hazardous Materials* 140 (2007): 325–32.

Rogers, Robert. *The Fungal Pharmacy: The Complete Guide to Medicinal Mushrooms and Lichens of North America*. Berkeley, Calif.: North Atlantic Books, 2011.

Royse, D. J. "Specialty mushrooms: Cultivation on synthetic substrates in the USA and Japan." *Interdisciplinary Science Reviews* 20, no. 3 (1995): 205–14.

Safer, D., M. Brenes, S. Dunipace, and G. Schad. "Urocanic acid is a major chemoattractant for the skin-penetrating parasitic nematode *Strongyloides stercoralis*." *Proceedings of the National Academy of Sciences of the United States of America* 104, no. 5 (2007): 1627–30.

Sasek, V. *Why Mycoremediations Have Not Yet Come into Practice: The Utilization of Bioremediation to Reduce Soil Contamination: Problems and Solutions*. Dordrecht, Netherlands: Kluwer Academic Publishers, 2003.

Schwab, A. *Mushrooming Without Fear: The Beginner's Guide to Collecting Safe and Delicious Mushrooms*. New York: Skyhorse Publishing, 2006.

Shiva, V. *Soil Not Oil: Environmental Justice in an Age of Climate Crisis*. Cambridge, Mass.: South End Press, 2008.

Silverstein, S. *The Giving Tree*. New York: Harper and Row, 1964.

Stamets, P. *Growing Gourmet and Medicinal Mushrooms*. Berkeley, Calif.: Ten Speed Press, 2000.

Stamets, P., and J. S. Chilton. *The Mushroom Cultivator*. New York: Agarikon Press, 1983.

Stamets, P., and D. Yao. *Mycomedicinals: An Informational Treatise on Mushrooms*. Olympia, Wash.: Mycomedia Productions, 2002.

Stein, M. *When Disaster Strikes: A Comprehensive Guide for Emergency Planning and Crisis Survival*. White River Junction, Vt.: Chelsea Green Publishing, 2011.

Steineck, H. *Mushrooms in the Garden*. Eureka, Calif.: Mad River Press, 1981.

Suay, I., F. Arenal, F. Asensio, A. Basilio, M. A. Cabello, M. T. Diez, J. B. Garcia, et al. "Screening of basidiomycetes for antimicrobial activities." *Antonie Van Leeuwenhoek* 78, no. 2 (2000): 129–39.

Tani, A., M. Kiyota, and I. Aiga. "Trace gases generated in closed plant cultivation systems and their effects on plant growth." *Biological Sciences in Space* 9, no. 4 (1995): 314–26.

Webster, J., and R. Weber. *Introduction to Fungi.* Cambridge, U.K.: Cambridge University Press, 2007.

Weil, A. *Spontaneous Healing.* New York: Ballantine, 1995.

Wells, P. *Simply Truffles: Recipes and Stories That Capture the Essence of the Black Diamond.* New York: HarperCollins, 2011.

Zivanovic, S. "Identification of opportunities for production of ingredients based on further processed fresh mushrooms, off-grade mushrooms, bi-products, and waste materials." Report prepared for the Mushroom Council (San Jose, Calif.), September 2006.

Resources and Suppliers

Mushroom Spawn (Plugs, Sawdust, or Grain)
Field and Forest Products—www.fieldforest.net
Fungi Perfecti—www.fungi.com
Mushroom Mountain—www.mushroommountain.com
Mushroompeople—www.mushroompeople.com
Sharondale Mushroom Farm—www.sharondalefarm.com

Pure Mushroom Cultures
ATCC—http://atcc.org
Fungi Perfecti—www.fungi.com
Mushroom Mountain—http://mushroommountain.com
Penn State Mushroom Spawn Lab—http://plantpath.psu.edu/facilities
      /mushroom/cultures-spawn

Laboratory Equipment and Cultivation Supplies
Fungi Perfecti—www.fungi.com
Unicorn Bags—http://unicornbags.com

Plans for Solar Dehydrators, Ovens, and Water Heaters
Build It Solar—builditsolar.com/Projects/Cooking/cooking.htm

Greenhouse Equipment, Supplies, and Shade Cloth
FarmTek—www.farmtek.com
Jäderloon—www.jaderloon.com

Magazines and Newsletters
Fungi Magazine—http://fungimag.com
The Mushroom Growers' Newsletter—http://mushroomcompany.com

Mushroom Hunting Clubs in North America
North American Mycological Association—www.namyco.org

Index

Note: Page numbers in *italics* refer to photographs and figures; page numbers followed by *t* refer to tables.

About the Author

Olga Cotter

TRADD COTTER is a microbiologist, professional mycologist, and organic gardener, who has been tissue culturing, collecting native fungi in the Southeast, and cultivating both commercially and experimentally for more than twenty-two years. In 1996, he founded Mushroom Mountain, which he owns and operates with his wife, Olga, to explore applications for mushrooms in various industries and currently maintains over 200 species of fungi for food production, mycoremediation of environmental pollutants, and natural alternatives to chemical pesticides. His primary interest is in low-tech and no-tech cultivation strategies so that anyone can grow mushrooms on just about anything, anywhere in the world. Mushroom Mountain is currently expanding to 42,000 square feet of laboratory and research space near Greenville, South Carolina, to accommodate commercial production, as well as mycoremediation projects. Tradd, Olga, and their daughter, Heidi, live in Liberty, South Carolina.

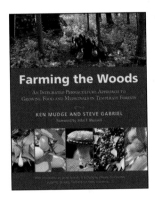

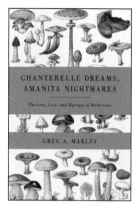
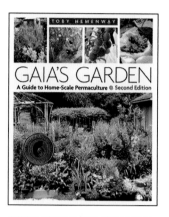